Technology for Facilitating Humanity and Combating Social Deviations:

Interdisciplinary Perspectives

Miguel Vargas Martin
University of Ontario Institute of Technology, Canada

Miguel A. Garcia-Ruiz
University of Ontario Institute of Technology, Canada & University of Colima, Mexico

Arthur Edwards
University of Colima, Mexico

Information Science REFERENCE

INFORMATION SCIENCE REFERENCE

Hershey · New York

Director of Editorial Content: Kristin Klinger
Director of Book Publications: Julia Mosemann
Acquisitions Editor: Lindsay Johnston
Development Editor: Dave DeRicco
Publishing Assistant: Casey Conapitski
Typesetter: Casey Conapitski
Production Editor: Jamie Snavely
Cover Design: Lisa Tosheff

Published in the United States of America by
Information Science Reference (an imprint of IGI Global)
701 E. Chocolate Avenue
Hershey PA 17033
Tel: 717-533-8845
Fax: 717-533-8661
E-mail: cust@igi-global.com
Web site: http://www.igi-global.com

Library of Congress Cataloging-in-Publication Data

Technology for facilitating humanity and combating social deviations : interdisciplinary perspectives / Miguel Vargas Martin, Miguel A. Garcia-Ruiz, and Arthur Edwards, editors.
p. cm.
Includes bibliographical references and index.
Summary: "This book presents socio-technical approaches to support the prevention, mitigation, and elimination of social deviations with the help of computer science and technology, providing historical backgrounds, experimental studies, and future perspectives on the use of computing tools to prevent and deal with physical, psychological and social problems that impact society as a whole"--Provided by publisher.
ISBN 978-1-60960-094-5 (hbk.) -- ISBN 978-1-60960-096-9 (ebook) 1. Information technology--Social aspects. 2. Social problems. I. Vargas Martin, Miguel, 1973- II. Garcia-Ruiz, Miguel A., 1970- III. Edwards, Arthur, 1957-
HM851.T454 2011
303.48'33--dc22
2010021052

British Cataloguing in Publication Data
A Cataloguing in Publication record for this book is available from the British Library.

Editorial Advisory Board

List of Reviewers

Table of Contents

Section 4
Moral Principles and Responsible Use of Technology

Section 5
Perspectives on Combating Cybercrime

Detailed Table of Contents

The sexual exploitation of children remains a very serious problem and is rapidly increasing globally through the use of the Internet. This chapter focuses on the child pornography and IT, and the various methods to combat this problem. It is shown that the lower cost of digital equipments, the global reach of the Internet and freely available peer-to-peer services made child pornography a very complex issue to undertake. The borderless nature of the Internet and the lack of unified criminal code among nations further escalated the complexity of child pornography.

It is estimated that there are over a million pornographic sites on the Internet, with as many as 10,000 added every week. In addition to having a major financial impact, adult pornography is strongly associated with various types of violence against women, especially sexual assault. Some studies have found that the contribution of pornography to woman abuse in dating, marriage, and during or after separation/divorce is related to male peer support, which refers to the attachments to male peers and the resources they provide that perpetuate and legitimate woman abuse. The main objective of this chapter is twofold: (1) to review the extant social scientific research on the relationship between violence against women, male peer support, and adult Internet pornography and (2) to suggest new directions in empirical work on the association between these three social problems.

The problem of providing appropriate and effective medical care to the elderly has gained importance in recent years because of the exponential growth of people older than 65 years. This research work proposes using interactive television embedded in a ubiquitous computing environment to help mitigate some of the effects of discrimination and provide health services to older adults living alone in their homes who require timely medical attention.

areas efficiently, reliably and as safely and as rapidly as possible while keeping cost to the minimu. Although demining has been given top priority, currently mine's clearing operation is dangerous, complex, time consuming, slow, labor-intensive and costly operation. The currently available technologies are not suited to achieve the objectives of humanitarian demining. In the context of humanitarian demining it is essential to have a reliable and accurate sensor and/or an integration of heterogeneous/homogeneous sensors with efficient and reliable data fusion and processing technique that can quickly discriminates mines from innocuous buried objects. In addition, it is necessary to overcome the constrain on the resources by developing innovative, cost effective and practical technology inspired by locality and real minefield needs to help in speeding up the demining process and enhance accuracy, productivity, operation and personnel safety, achieve higher quality of the service, and contribute to local economy. This chapter presents the facts and problems associated with landmines and their impact on health, economy, land and environment along with the difficulties in detecting and removing them. It highlights the main requirements for humanitarian demining action plan and list up solutions and priorities.Then, it presents the challenges facing technological development in different directions and concludes with the suitable actions to save human and environment from such complex problem facing humanity.

Most of the time, Mexicans who cannot find a possible solution for their financial problems decide to migrate to the United States looking for an improvement in their quality of life. However, people who migrate usually face a series of abuses mostly because of their illegal status, but also because they arrive into a society which is not their own. Migrants are mainly excluded from the society due to their race and religion but also because they do not speak English in most cases, do not have studies higher than elementary school, and are not trained even for the basic use of technology such as office software or e-mail. With this panorama in mind, the Colimenses sin Fronteras Web Portal becomes a tool that supports them to overcome the processes from which they are discriminated and it also provides them with information regarding the abuses they might suffer and how to notify the authorities.

This chapter presents the conceptual development, design requirements and functional descriptions of a team based multi-user, web-based and global telecooperative system as a new multidisciplinary collaborative paradigm. This system uses the distributed and collaborative intelligent environment (DCIE) developed by the author. Distance, time, and distribution are the main attributes of the telecooperative system and its applications. The developed system aims to carry out cooperative processes over geographically distributed locations that lead to make well-informed decisions, better utilization of

Section 5
Perspectives on Combating Cybercrime

In this chapter, we play the devil's advocate to those who favor strict government supervision over technology itself. Our argument is that technology is a "neutral" mean to an end, and that the use of technology to detract social deviations is dependent on public policy and social behavior. To elaborate our argument we propose the concept of "illicit appropriation," based on the Human Computer Interaction concept of appropriation. We argue that sometimes appropriation can be geared towards activities that can be considered as illicit, and in some cases criminal. We illustrate the use of illicit appropriation through a series of case studies of current events, in which we show that either a state or the individual can rely on illicit appropriation. Our final conclusion is that the use of technology to combat social deviations is not a technological problem, but a public policy issue, where a delicate balance has to be found between the enforcement of the law by technological means (approved by legislation), the user experience, the civil liberties of the individual and the checks and balances to the power of the state. This chapter is written from the expertise of the authors on Human Computer Interaction and Security Studies.

Currently, many powerful applications designed to combat social deviations are available, like the web-filtering systems, which measure the content of a website before submitting it to the user, notifying whether the content of the website presents (or not) information related to pornography, violence, racism, among others, and prompting the user to not accessing the site, or even blocking access to the website. Nevertheless, frequently the feedback of these systems is not well-designed, which may confuse users and lead to mistakes, disappointments, and misunderstandings. In order to reduce this concern, a method is provided to developers with guidance in designing usable security notifications to be incorporated in web-filtering systems. The method is structured through a library of user interface design patterns which integrates essential concepts of security and usability. We show the effectiveness of the patterns by using an illustrative example as a proof-of-concept together with a preliminary study.

Foreword

I was very honoured when the editors of this book asked me to write the Foreword to this book. And, of course, who can complain about reading engaging and substantive works, like the chapters in this book. As I approach 62 eras of age, these chapters reminded me of how far computing technologies and information systems have come since I ran thousands of data cards for the statistical analyses of my PhD thesis through a clunky (that is the technical word) mainframe computer at Syracuse University in 1980. Since that humbling, or more appropriately fumbling, time with card readers requiring ever-so-careful typing to make sure I had the right code in the right column, I moved through epochs of mainframe, then minicomputers to desktop machines. My first desktop was a Radio Shack machine that had a magnetic tape deck for a hard drive and that required me to argue long and loudly that I would need more than 16K RAM for the simulations I wanted to build and write about. I marvel at the technology and information systems now available for routine work and for research I could not imagine when I finished my PhD in the summer 1980.

Even so, then and now, I know that the machines themselves have great potential but will only do what you tell them to do. The real creativity that stuns me is the type of programming and algorithm development that has evolved, including the interesting papers in this volume. My own work began to increasingly focus on how people develop an understanding of the complexity of the real world. In particular, I study the development of pattern recognition of healthcare providers and students during episodes of clinical education and training. Within the domain of my own work, I have become very interested in the development of misconceptions among clinicians and the ethical issues related to evidence-based learning when training people who will take the lives of humans in their hands during the planning and delivery of healthcare. That is, how do faculty and clinical staff educators develop and implement training that is likely to improve clinical judgment of healthcare providers?

We have been using the term evidence-based learning, as an analog to evidence-based practice, to stimulate discussions about the ethics of education and how to understand what really works to improve learning outcomes and in particular clinical judgment in the care of patients. I do not believe that the majority of faculty and clinical staff trainers intentionally teach in ways that instil misconceptions, and healthcare students and practitioners must also take responsibility for how well they learn to care for their patients. Healthcare students and practitioners still develop misconceptions and still make mistakes that sometimes kill or badly hurt people. So I was very intrigued by this special collection of papers that examined from many different perspectives the intentional unethical uses of the Internet, computers, information technology, and information management systems. How do we think about intentionally causing harm or, as some of the authors point out, act in ways that have the potential to hurt people as well as the Earth's habitats and other inhabitants? And how do we prevent unethical uses of technologies and information systems in ways that are likely to hurt humans and the Earth around us?

Indeed, I was very intrigued by the idea of better understanding how computing, broadly defined, could support prevention, mitigation, and elimination of social deviations and that such preventions could be based, at least in part, on the help of information sciences and advanced information system and computing technologies that could provide frameworks for novel ways of dealing with social problems. I did not anticipate, however, the impact these papers would have on me. As I read the chapters, I began to reflect on my own life and a number of deep memories emerged, all painful. So, as I began to write this Foreword I wondered if I should contextualize these memories in the themes of the papers. The reflections that emerged were quite surprising to me and I finally decided to use parts of my life to try to look at the many social implications of the work described by the papers in this book. I want to be clear that I do not think of my life as special in any way, and I do believe each of us has memories that may be evoked by this book. Our memories may open up very interesting reflections on individual perceptions of and responsibilities for understanding social deviance. Thus, I hope you will indulge my memories and how I have reflected on these memories in the context of these very interesting and substantive papers. Let me be clear at the start that I enjoyed reading these papers and feel they make a significant contribution. This is a wonderful and important work, and I thank the editors and authors for their great efforts.

So let me get started then, and admit right away that I am not very imaginative, having read the chapters of this book by following the order they appeared. In this Foreword, I will talk about the specific memories that I had for each of the sections of the book, which I think will provide a coherent approach to commenting on how important I think this collection of papers will be to those of us who have strong interest in computing and information sciences as well as in the well-being of Earth, and of the many beings who live on this planet. I also want to own at the start that I am a Zen Buddhist, and so my reflections will certainly be informed by my Buddhist practices and perceptions. And when all is said and done, I want to be clear that this Foreword reflects my own views and should not be imposed on or held against the editors and authors who have provided important work on critical issues in the unethical use of computer and information sciences as well as the technologies and programming so ubiquitously pervasive in the fabric of our social structures.

In Section 1, Technology for Dealing with Crimes Against Children, I read with a growing uneasiness about technologies developed for dealing with crimes against children. My uneasiness was not because of the value of such technology, but because my oldest child was almost certainly sexually molested during her early teen years. I say "almost certainly" because my oldest child, Vanessa, was born with cerebral palsy and is mentally retarded. She was a very beautiful child, always happy, and always willing to talk to anybody. During her early teen years, there was a dramatic shift in her behaviour. She became angry and aggressive, which her physicians and psychologists described as a fairly frequent pattern in special children like Vanessa as they entered puberty. However, certain of her behaviours prompted her school to contact a social worker and that social worker then talked with Vanessa's mother and me about the possibility that Vanessa had been sexually abused.

The story was complex and convoluted, especially when told from Vanessa's perspective. As best as we could put together, it seemed that the boyfriend of the mother of one of Vanessa's friends had sexually abused Vanessa and probably also her friend. Of course, we wanted to have this alleged perpetrator arrested. As we discussed the issues with the social worker, however, we became increasingly concerned about a trial that would involve Vanessa's participation as a witness and the difficulties she might face when challenged again and again by the alleged arbitrator lawyers. Furthermore, we could not determine the extent of the sexual abuse and Vanessa's friend's mother would not discuss the allegations that her boyfriend may have abused her own daughter as well as Vanessa.

As I read the papers in Section 1 of this book, Technology for Dealing with Crimes against Children, I could not help but think of the great value afforded by investigative techniques using sophisticated information technologies and pattern recognition algorithms to better protect children from exploitation of various kinds. Indeed, as I reflected back on the time with Vanessa, I had to remember the deep and abiding fury I felt and the helplessness. The fury led to hatred for this man who allegedly sexually abused a retarded teenage girl, who happened to be my daughter. This horrifying hatred led me to consider acts of violent revenge, so clear and strong at the time, so abhorrent to me now. Of course, then I would become what I hated. The helplessness filled me as I could find no sensible legal solution and with the realization that vigilante justice was not the path I could follow. I am still ashamed of the extreme and absolute intensity of my ideation to plan ever so carefully how to make the man who allegedly had so deeply hurt my child pay for what he had done, and pay with every fibre of his being. And to this day, I remain overwhelmed by the complexity of trying to decide if that man was in fact guilty of sexually molesting my daughter Vanessa.

I really don't know if I made the right decision at the time not to take some kind of legal action against the alleged perpetrator of sexual abuse against my daughter. The uncertainties were profound and distributed broadly, including trying to figure out what Vanessa was really telling us, ascertaining the truth of what she was saying, trying to figure out how to approach and question Vanessa's friend's mother and her boyfriend, the complexity of the legal system and what recourse was available to us, and the overarching despairing sense of failure as a father. Certainly, sexual exploitation of children remains a serious problem worldwide, facilitated in part by what can be shown and shared on the Internet, as well as by the increased number of child pornography sites and the relative ease of access to these sites. The chapters in Section 1 suggest that we might be very close to having systems that could identify those who intend to commit or have committed crimes against children. And yet, a careful reading of these papers suggests to me that there are uncertainties in establishing guilt that are in part inherent in the pattern recognition capacities of the information systems but also in part due to variability among nations in terms of cooperation and lack of a unified criminal code related to crimes against children.

Stated in a slightly different way, how could we build pattern recognition systems to identify perpetrators of crimes against children and design these systems so that they were infallible or had such a low probability of being wrong that the results of these systems would provide unequivocal evidence in a court proceeding? And, as typified by Vanessa's case, what are the other levels of complexity, such as the reliability of witnesses who were victims of crimes and who were willing to bring charges against a perpetrator that an information system identified as a likely person of interest or even as exhibiting patterns of behaviour indicative of crimes against children?

Section 2, Technological Approaches for Hate Crimes, evoked memories of the racism my interracial family faced during a time when racism and segregation were much more evident in the everyday, multicultural fabric of American life. My father is of Japanese ancestry and my mother is Caucasian. I have four siblings who are half Asian, two siblings who are Caucasian, and two siblings who are African-American. As a young boy in 1954, my family lived in the Florida Panhandle because my father had returned from a deployment to Korea and had been stationed at Eglin Air Force Base near Pensacola, Florida. In those times, segregation was a routine part of Southern life in America. There were segregated bathrooms, drinking fountains, seating areas in restaurants and buses, as well as a generalized institutional racism that pervaded much of the South. We were a "colored" family and, in fact, the anti-miscegenation laws made my parents' marriage illegal in a number of states. There also were overt racist episodes, for example, when white boys beat me up a couple of times a week. These fights left me with a nose broken so badly that I had chronic throat infections until I was 21 years old and had my nose rebuilt.

Even by the early 1960s, when we lived outside of Cincinnati Ohio, there was a Ku Klux Klan rally area not too far from the town in which we lived in and that small town itself also had entrenched racist attitudes. For example the swimming pool was segregated. We had been admitted to the pool the first year we were there, because my white mother got us in before they realized how "mixed" we were. But when my African-American brother Michael was adopted into the family, we were asked to give up our pool membership. And, the majority of white girls in that town were not allowed to date "colored" boys like me or my African-American friends.

As I read the papers in Section 2, I couldn't help but become anxious at the thought those bullies and me in Florida being transported to the future where they became better organized by engaging in Cyberbullying and perhaps not only better coordinate where to meet, find me, beat me down physically, but also cyber-ferreting me out while hiding in my house and then emotionally abusing me in diverse ways through the Internet. And, those high school boys and girls in our small-town Ohio, would they have benefited from the virtual learning environment that would allow them to engage in meaningful exploration and discussion of race hate crimes? I cannot tell you how many hours I worried about my sisters and daughters, about them facing violence of various kinds. Now, they face new threats from increasing and insidious contributions from the "dark side" of the Internet. Certainly, having survived an interesting childhood and young adulthood, as a person in my 60s I feel obliged to become more active in reducing social discrimination against older adults. As in Section 1, Section 2 both opened old wounds and gave me new hope.

I thought I might get some emotional relief as I began reading the papers in Section 3 (Humanitarian Technologies) and Section 4 (Moral Principles and Responsible Use of Technology). Yet, these papers brought up memories that were contextualized at a different level for me. Whereas Sections 1 and 2 evoked memories more individualistic and personal, Sections 3 and 4 brought up memories of trying to sort through the complexities of larger social issues rather than the impacts on my personal and immediate family life. In Section 3, I was very interested in the systems that allowed a demining action plan, a system that could foster human rights defence processes for Mexican immigrants, and a multiuser distributed and collaborative environment to support emergency and relief activities. The idea of helping immigrants find support to ameliorate discriminatory practices struck a particularly resonant chord.

I had mentioned earlier that my father was in the United States Air Force and had returned from a deployment in Korea, where he served just after the end of the Korean War. I did not mention that my father and his family were threatened with "relocation" during World War II under President Franklin D. Roosevelt's Executive Order 9066. Fortunately my father, his siblings, and his parents were not put into the American concentration camps which eventually held between 117,000 and 120,000 people of Japanese ancestry, over 70% of whom were actually American citizens. Grandpa was a prominent biochemist and a faculty member at the University of Cincinnati Medical School and, as I was told by one of his colleagues, the faculty supported him remaining at the University of Cincinnati and continuing his teaching and research duties. However, some of our relatives were put into the concentration camps and I remember in particular a series of long discussions with my second cousin, who I will call SH, in which he described his relocation to the Santa Anita Racetrack Assembly Center near Los Angeles and then his transfer to the Amache concentration camp in Granada, Colorado.

I find an interesting note in American history that the vast majority of the people of Japanese ancestry who were ordered into the camps did not, for the most part, resist their relocation, and most tried to show their support of America by "doing their part" and not resisting their transfer to concentration camps. By the way, I also want to note that I use the term "concentration camps" intentionally, not because the camps in the United Sates were the horrors of the Holocaust, but because this is the term many of the internees used to describe them, including my cousin SH.

I will not go into the history of the interment of people of Japanese ancestry by America during World War II, although this is both a sad and fascinating part of our heritage as Americans. The story is much more complex than the small bits I've shared here. The real point is that as I reread the chapters of Section 3, I was so very deeply struck by how the tools described for very important humanitarian relief efforts could easily be permuted to identify individuals and subpopulations that might be the target of scrutiny or isolation from the general populace.

So, for example, immigrants who used a Web portal that actually fostered human-rights defence processes could, with today's technologies and knowledge systems, be quite easily tracked and identified. Modifications of the demining systems could certainly be used to track individuals if they had been "bugged" with a location signal-emitting device. And, the telecooperative system designed to support emergency and relief activities could have provided a more effective management of the logistics of moving 117,000 people from their homes, putting them in temporary assembly centers, and eventually moving them to concentration camps, and supplying various types of resources to those camps. Now, I've taken a somewhat extreme view in these last few paragraphs, because I want to share the shift in my perceptions as I moved through the chapters of this book. Without intentionally trying to look at each of these chapters from perspectives of the good, the bad, and the ugly, slowly but surely my reflections began to center on the complexity of the systems in regard to all of their possible outcomes.

The chapters of Section 4, therefore, provided me with a moment in which I could think more deeply about the moral principles and responsible use of technology. Specifically, I began to think about how social norms shaped patterns of technology and information systems usage. I also thought about the role that governments could play in monitoring the people in their respective countries, perhaps identifying and tracking some groups of individuals or subpopulations, and, as history would suggest, sometimes isolating or imprisoning or killing those with dissident opinions. For example, during the Vietnam War, and when the draft lottery was in place, I received a low lottery number in the spring of 1970 and was drafted shortly after passing my military physical. At that time, I was applying for Conscientious Objector (CO) status because I did not believe the conflict in Vietnam was a just war and because my practices as a Buddhist led me to a more pacifist position. I persisted in objection to the war, even though this put me into conflict with my father who had been drafted during the Korean War, and despite his experience of prejudice before, during, and after World War II, decided that his obligation to the United States was to report for duty and be sent to Korea. Although I went through the prescribed process for applying for CO status, to make a long story short my CO was turned down at each level of appeal and I was drafted into the United States armed services. I refused induction and because I believed in the United States I spent two years serving my country by first working as surgical orderly and operating room technician in a hospital for one year and for the second year screening children for blood lead poisoning.

However, the FBI did search me out and invited me to what they called an impartial interview. I reported to the FBI office in Boston, and before the interview began I was read my rights (Miranda Act) and then interviewed by an FBI agent. This fellow then filed a report with the United Sates Department of Justice, which had to decide whether or not to prosecute me for breaking a federal law. I tell this story as an example of someone who acted with what they thought was a social conscience to protest an unjust war, but who in the process broke a law and whose case was reviewed for possible prosecution. At this time, my position on the war was still outside the normative values of the majority of the American population, although those values shifted dramatically in the early 1970s. But, let's think for a moment about what happens when very powerful pattern recognition tools are available in technology and information systems and the people who control the usage of these technologies and systems have the power to identify and exert some control over those who have dissident opinions to the social norm or to the controlling power.

In my case of protest against the Vietnam War, the outcome was odd. The Department of Justice decided there had been a procedural error and sent my case back to the local draft board, whose members could then decide whether to draft me again. As far as I can tell, I still have an FBI record, although I've been a little reluctant to request those documents. My point is that with today's technology and information systems it would be extremely easy to identify me as a person with a dissident opinion and to use sophisticated pattern recognition algorithms to decide the level of threat that I posed, and also to decide through computational modeling and decision-support tools what should be done with me. And, this realization while reading Section 4 brought me to the moment when I began to struggle with who defines social deviance and what actions are planned for those who deviate from social norms.

Certainly some of the authors in this book discuss how social deviance can be defined and how such definitions allow us to consider what constitutes unethical or inhumane acts, including breaking of legal codes and laws. The chapters of Section 5, Perspectives on Combating Cybercrime, tackle some of the issues related to social deviance directly. For me, the journey through the chapters of this book evoked a lot of deep feelings, as I think will probably be clear to the readers of this Foreword. Each chapter provided a new insight on the complexities of the human condition. My special child Vanessa certainly deviates from the norm in many of her physical and mental capacities. As a "colored boy" in the Panhandle of Florida and in a small Ohio town, I did not fit easily within the racist norms of the social fabric in the areas where I lived. American citizens of Japanese ancestry were put into relocation camps without due process because of the social norms and hysteria that engulfed the United States at the beginning of World War II. In a protest against a horrifying war that was based on a spiritual practice of Buddhism, I did not fit within the social norms of the United States during the late 1960s although these norms shifted beginning in the early 1970s. In other moments of history, and in other countries, my family members and I might very well have been identified, tracked, and made to disappear. As pattern recognition and knowledge systems increase in their ability to identify, track, discriminate very fine differentials, and allow sophisticated computational modeling of possible outcomes, we have to think very carefully about who controls the underlying technologies and algorithms that drive such systems.

There are two well-known American slogans that I closely paraphrase here.

"Guns don't kill people, people kill people."

"When guns are outlawed, only outlaws will have guns."

The first of these slogans was developed to argue that guns, in and of themselves, do not kill people, although there are odd cases of unprovoked accidental discharge. The second slogan on the other hand was, I think, meant to suggest that government regulation of firearms in the United States was likely to have the undesirable effect of taking weapons out of law-abiding citizens (who, when armed, could defend themselves against criminals) and leave them at the mercy of outlaws who would acquire weapons regardless of the laws restricting weapon ownership. Of course, I simplify here, but I want to end this Foreword with some thoughts about technology and information systems as "neutral" means to an end. Like a weapon, technologies and information systems may be used for good or bad purpose. So, we really ought to constantly examine who is in charge of saying what is good and what is bad?

Interestingly, we can look through lens of different perspectives. From one lens, we can make systems designed with end user designed specifications that allow diverse users better interpretive capacity, regardless of experience and background of the user. Such systems could be applied by the diverse people for the purposes the system was originally designed to do, and for the moment let us assume that the original intent of the design was humanitarian in nature. However, the system could be permuted to other uses. And from another lens, we might see that such permuted versions of the system could be used to target us and those we love in ways that disadvantage us, perhaps slightly or perhaps with dire consequences. Like a gun, we might be holding the system or we might be having the system pointed at us. And like a gun, we should work hard to have a very broad and deep understanding of what happens when we "pull the trigger" that unleashes the power of technology and information systems on people. Even as we move towards natural language systems, we might wonder in advance about why there are already language gradients in the degree languages are informatized. However, if we then extend our exploration of such a gradient, we might worry about what happens when the language developed by a small subpopulation with a specific agenda becomes the norm for systems with natural language interfaces. In particular, a very small percentage of a populations might then be able to exert a powerful positive or negative influence on a large percentage of a population, especially as more and more people need to access information through personal information management systems (computers, cell phones, and so on).

In my reading of history – and I want to own my own biases here – social deviation has often been defined by prevailing social norms and public policy (sometimes encoded in law). The authors of the papers in this book make a very compelling suite of arguments that there are powerful applications that can be used to combat social deviations. And as social deviations are defined within a socio-historical moment, public policies, codes of law, and social norms will allow delineation of what is "illicit appropriation" of information technologies. Furthermore, within a socio-historical moment, public policies, codes of law, and social norms will allow delineation of appropriate responses to illicit appropriation by fostering development and routine usage of technologies and information systems that can combat whatever has been defined as an unwanted social deviation. The target audience of this book is professionals and researchers working in the fields of computer science, as well as students, teachers, instructors, and academics from other related areas, lawyers, people involved in law enforcement and/or humanitarian societies such as Red Cross/Red Half Moon, United Nations, and others generally interested in preserving humanity as well as the Earth and its diverse environments and inhabitants. So, the joy and cautionary note I took away from this wonderful book is the creativity and significant work of the authors and editors as well the deep thinking each of us must pursue to better understand how, when, and why to use the technologies and information systems that are now functional and that will emerge in the future.

Jayshiro Tashiro
University of Ontario Institute of Technology

Jayshiro Tashiro, *PhD, BSN, RND is currently a Professor in the Faculty of Health Sciences at the University of Ontario Institute of Technology (UOIT). His research focuses on telehealth and disease management, the relationships between evidence-based learning and evidence-based practice in healthcare, and assessment of complex competencies within clinical simulations. Beginning in the mid-1990s, Tashiro led research teams in the development and evaluations of virtual clinical simulations that monitor users' choices during treatment of complex patients within the simulations. The principal focus of these monitoring systems has been clinical judgment, with the software conducting automated analysis of choices made by the user while working within a simulation. Funding related to virtual learning environments has been over $10 Million since 1990. At UOIT, Tashiro teaches In the Health Information Management Program and is currently building and evaluating simulation-rich courses that promote interprofessional collaborative patient-centred care. Tashiro also helped establish and now is part of the Management Team for the Health Education Technology Research Unit at UOIT.*

Preface

Technology for Facilitating Humanity and Combating Social Deviations: Interdisciplinary Perspectives emerges as a unique source of information that compiles interdisciplinary perspectives on issues affecting our society, and technological solutions. Whether the answers to those societal problems lay on technology or not can be arguable; and while we would not engage in that debate, we recognize that technology is occupying an increasingly important role in everybody's life, regardless of ethnicity, social status, gender, preferences, nationality, and even technology literacy or Internet connectivity.

A book on technological solutions for society cannot avoid the term "humanity" as it refers to the quality of being humane. Heroic humanitarian deeds take place in different locations across the Globe and in different forms, from law enforcement and emergency response systems to altruist actions by civilians, and from disease control to land field demining. Despite all these efforts, technology is persistently used as either a facilitator or as the arena for crime perpetration, from facilitating illegal distribution of prescription drugs to collection of donations under false pretences. The use of technology for facilitating humanity is a noble action that researchers and practitioners should consider, at least at one point in their careers, part of their mandate.

While humanitarian actions constitute a key piece in the eradication of crimes against humanity, we felt that a barrier to humanity itself is represented by "social deviations." This book endeavoured to put together technological approaches that combat deviant behaviour. We acknowledge that the term "deviation" in its own is a field of study, and even a precise definition of the term can be controversial not only between experts of different areas of social sciences but also from clinical and psychological perspectives. In the interest of avoiding controversies about what is classified as a deviation and what is not, we called the research community for approaches of this nature, including laboratories and research centres of all areas of technology and social sciences, who might be involved in combating social deviations. We received manuscripts from renowned researchers from all around the world with expertise on combating crimes against children, combating hate crimes, humanitarian technologies, moral principles and social responsibility, and different perspectives on fighting cybercrime.

The book opens with a section entitled *Technology for Dealing with Crimes against Children*, featuring two chapters on fighting child pornography. Child pornography is not a new phenomenon. As a species, there has always been a small minority of adults who derive enjoyment from fantasizing about or, worse, acting out on their attraction to children. These adults largely acted alone and rarely formed "communities" because of the difficulties and dangers related to acquiring, disseminating and organizing this illicit and illegal activity. In the 20th century, pornography, in general, became much more common because of the ready availability of photographic equipment. Child pornography, on a personal level, began to manifest itself more readily as photographic evidence of this illegal activity was discovered increasingly more frequently. It appears as though, at least in the United States, postal investigators were able to largely hold the dissemination of child pornography to a minimum. However, digital photography and the Internet at the end of the 20th and the first decade of the 21st century have made child pornography more widely accessible, transferable and organizable. Today, there are innumerable communities whose members share their common interest by distributing or selling child pornography, as well as discussing their experiences and desires with each other. The chapter *Digital Child Pornography: Offender or not Offender* discusses the increasing incidence of child pornography in the Internet, the often difficult legal burden of proof necessary to prosecute crimes and the emergence of forensic computing, which refers to the application of computer science techniques to retrieve digital evidence from electronic devices for legal proceedings. This chapter also provides basic information about computer forensics that can be applied to different common applications including Internet and image browsers, video browsers, instant messaging, file sharing on peer-to-peer networks, etc. The authors then follow up their discussion with a case study of how computer forensics can be used to find "traces" and conclude that for child pornography to decrease and enforcement to be more effective, law officials need to be provided more effective tools and the legal framework needs to facilitate prosecution. And conclude, that the potential traces in the Windows platform can present a useful tool to facilitate prosecution and achieve sentencing, which might serve as a deterrent to this illegal activity. While the sexual exploitation of children is nothing new, the incidence of sexual exploitation of children is increasing globally because of the Internet.

The chapter *Illegal Pornography and IT*, presents first a very ample definition of what child pornography is, including even morphed or virtual reality childlike avatars, mentioning that child pornographers are always testing societal tolerance and the legal underpinnings that are the two measurements as to how much child pornography can be represented on the Internet. The chapter then moves towards framing the problem of child pornography from different points of view, including a discussion from a historical and legal perspective, before moving on to discuss its impacts and how to combat it using information technology (IT). Consequently, the focus of this chapter is on child pornography and IT and the various methods to combat this problem. The advent of peer-to-peer file sharing has made controlling child pornography very difficult to combat because of the particular characteristics of this Internet tool. As far as technology is concerned, the authors specifically mention the development of data sharing systems developed in Canada and the United States, arguing that there is a lack of coordination between agencies as far as sharing information in their databases, and this lack of coordination can be found elsewhere. Following this, the authors then go on to discuss several host-based filtering solutions before moving on to network-based solutions.

The book goes on with a section on *Technological Approaches for Hate Crimes*. Here, *Adult Pornography, Male Peer Support, and Violence against Women: The Contribution of the 'Dark Side' of the Internet* presents a very complete discussion of adult pornography and its relationship to violence against women, in general. As far as the Internet is concerned, the authors argue that there are presently over one million pornographic websites and that thousands more are added every day. This expansion is greatly due to the financial gain pornographers receive upon selling their images and videos on-line to a very large constituency. Anonymously receiving pornographic materials directly in one's home has made it more comfortable for consumers to purchase materials and eliminates the sense of awkwardness people often suffer upon buying pornography from more traditional sources. As a result, the market for pornography has exploded due to the Internet, and is becoming even more common as bandwidth resources increase, thus minimizing the time required to download materials. The authors note that there have been few social scientific studies that have studied the effects of cyberporn and that most of the research to date has been "descriptive and psychological." Furthermore, the authors note that violence against women increases if the male abusers feel support from their male peers. They point out that almost 25% of total search engine requests are for pornographic materials, and thus, if this interest in pornographic materials is supported by on-line social networking, the number of males who can find peer support to "justify" their violence against women is almost limitless. This chapter is interesting in that from a primarily psycho-social point of view, it presents evidence of pornography and physical abuse against women. It then provides a discussion about how the Internet contributes to a growing problem and attempts to draw the reader's attention towards further studying this problem from both a psycho-social and a technological point of view.

Improving the Quality of Life of Older Adults by Preventing and Reducing their Social Discrimination through Interactive Television and Ubiquitous Computing addresses yet another form of hate crime, discrimination based on age. According to the literature research carried out by the authors, elderly population has steadily grown worldwide. Alas, this increase in the number of people with 65 years or older often comes with the loss of close companionship, which can lead in older adults to depression, among other associated mental and physical health problems, decreasing their quality of life. In addition, some aging people are constantly neglected and discriminated by close family and society, and this recurring problem has been found in many countries. The challenge of providing adequate medical, psychological, and social care to elderly people has gained importance in recent years, and a recent body of research suggests that information technology in the form of ubiquitous computing (the seamless integration of information and computing into every day's physical world, so that this technology is available to everyone) can be effectively used to support elderly people's care. This chapter proposes a novel application of ubiquitous computing and an interactive television system embedded in a reactive computing environment that happens in a household, with the objective of supporting communication between elders and medical and social caregivers, and for improving social interaction between elders and their family and friends. The proposed ubiquitous system may serve as well as an automated information gathering system about the state of health of aging people. The chapter also proposes that this type of technology can help decrease neglecting elderly people by providing timely and effective communication and attention between them and their caregivers.

Bullying has always been a social problem that is primarily manifested in schools. Sociologically, bullying often results when a person or persons wish to show social dominance over others and gain acceptance from specific peer groups. The results of bullying, in general, can be very harmful because they may cause anxiety and depression. In severe cases, bulling can even cause suicide. In recent years there have been reports on several famed cases of using electronic media to bully. This practice is called cyberbullying. Relatively few studies have been carried out on the effects of cyberbullying, but much more study is necessary in light of social networks and other tools that permit abuse. *Multiagents System Applied on a Cyberbullying Model for a Social Network*, presents a novel approach that employs multiagents which can improve the understanding of this detrimental social behaviour. The authors state that this is important in that it will be useful to answer queries after gathering information about abusive behaviour. In this way, information regarding abusive behaviour can be gathered and analyzed from either a sociological, psychological or legal point of view, if necessary. In this way, the authors state, vulnerability to cyberbullying can be reduced significantly.

Conversely, while the Internet can be misused to spread racism and commit hate crimes, it also helps foster collaboration between academic peers for open discussions about racism and hate crimes, and ways of preventing them. Virtual learning environments have evolved enormously in the last decades; *Examining Race Hate Crime with Students Using a Cross Cultural International Virtual Learning Envi-*

ronment presents a case study of how virtual worlds can bring different cultural backgrounds together to discuss social issues between students of three universities in three different continents during prolonged periods of time. The scenario they use deals with discussions about a series of talks presented in a conference on racism. To prove their case, the authors cleverly make use of a virtual learning environment to stimulate student participation on sensitive social issues. This study shows how student participation can be overwhelmingly enthusiastic and open even under constraints related to geographical location and social differences between participants.

We included a section on *Humanitarian Technologies* which continue exploring hate crimes but now in ways that deal with the consequences of warfare, immigration, and natural disasters. First, *Humanitarian Demining Action Plan: Humanity and Technological Challenges* deals with one of the most terrible consequences of war, as is the presence of landmines which include antipersonnel (AP) and anti-tank mines, and explosive remnants of war (ERW), including unexploded ordnance (UXO) and abandoned explosive ordnance. They are a major threat to civilian population, often producing a large number of causalities, even long after wars are over. However, the current monetary cost of clearing each mine or ERW is very high, depending on the type of mine, geographical location, and other factors. Humanitarian demining is an important step for the reconstruction of post-war countries, and it requires the entire land area to be free of mines, and hence the need to detect, locate, and remove reliably and safely every single mine and ERW from a targeted ground. For this reason, it is necessary to automate the detection and removal of mines, and to improve the safety of the deminers through the use of efficient, reliable and cost-effective humanitarian mine action equipment (e.g. robots, flexible and intelligent mechanisms, etc.), which also should have a minimum of environmental impact. Such technology should have decision-making capability to locate, mark or neutralize each mine precisely, and to have efficient quality control assurance methods that are reliable and accurate in ensuring no residual mines are left behind within an area declared "clear of mines." It becomes urgent to develop detection (individual mine, and area mine detection), identification and removal technologies and techniques to increase the efficiency of demining operations by several orders of magnitude, to achieve a substantial reduction to the threat of AP mines within a reasonable timeframe and at an affordable cost. This in turn will improve the accuracy, productivity, and safety of the personnel dedicated to remove mines and ERW, not to mention the improvement of safety and well-being of civilian population during the post-war.

A compendium of technology for humanity could not ignore migration issues, especially those that arise by crossing the most transited, and one of the largest, borders in the world. *Using ICT to Foster Human Rights Defense Processes for Mexican Immigrants* describes an initiative that addresses the needs of those who become victims of mistreatment. From a broader perspective, this work could be seen as a general technological approach that can assist people who migrate from any country to another one. The authors of this chapter point out some relevant immigration data that not only support the development of their approach but constitutes a call for action to researchers all over the world, to keep looking at migration problems and continue envisioning viable technological solutions. In particular, the focus of this chapter is on Mexican immigrants in the United States, more specifically from the state of Colima. From a sociological point of view, it is interesting to see a detailed profile of immigrants from that Mexican state, including gender, scholarly level, technical skills, and Internet accessibility. The technological approach proposed in this chapter is a portal specifically targeted to Colima immigrants in the United States, with emphasis on social phenomena that affect the majority of them. This continuing effort has the potential to become one of the primary Internet resources used by Mexican immigrants in the United States, and perhaps evolve and contribute to emerging virtual worlds and serious games for education.

War and conflicts have left millions of unexploded land mines across the globe. *Web-Based Multi-User Distributed and Collaborative Environment Supporting Emergency and Relief Activities* is an awareness chapter that outlines the benefits of using information and communication technologies to improve cooperation during responses to disasters or while providing relief activities. To emphasize the urgency of solutions, the authors give a few examples of land mines and then highlight the need for cooperation and coordination, and present "telecooperation" as a new collaboration paradigm to deal with disasters. Under this urgency, the authors outline the advantages of emerging technologies and the way these can be used to assist during disasters, namely, the World Wide Web, ubiquitous technology, geographical information systems, and global positioning systems. Furthermore, the authors discuss the particulars of these technologies with respect to humanitarian demining, and compile a number of technology requirements in this field.

The Word "deviant" can be applied to a broad spectrum of human activity. Although it is usually used to refer to some sexual component, deviant behaviour can really be used to describe any behaviour that is exaggerated, obsessive or does not meet criteria that is considered by the vast majority of people as within the normal limits accepted as norms for human conduct. The next section of the book comprises two chapters that deal with *Moral Principles and Responsible use of Technology*. The world economic crises that peaked in 2009 is an example of a deviant behaviour by some investment banks in the sense that it was, in part, due to excessive greed, which is considered by many to be a deviant behaviour. The profit motive is considered corporately responsible and implies that a just profit will be earned to compensate for the labor and inputs into any process. However, there is a big difference between profit motive and greed. Because greed was the cause of so much speculation that caused questionable investments that proved to almost send world economy into recession, *'Ethics' and 'Social Responsibility' in Financial Crisis as Parameters of the Financial Environment* becomes extremely timely and relevant. Therefore, this chapter presents a case study that can be used for general reading or even for students in computer security or computer science classes as it presents a case study that blends ethical, legal and technological issues that are applicable to the scenario the authors discuss.

Law, Deviation and Paradigmatic Change: Copyright and its Metaphors deals with the evolution of scientific paradigms in light of illegal file downloading that commonly happens on the Internet. While providing a fascinating compilation of mentalities and paradigms from a historic, sociological point of view, the authors explain how these are the result of metaphors and conceptions which are then extrapolated with the infamous "The Pirate Bay" web site and illegal file downloads. The authors describe a series of concepts as generally perceived in the cyber world, and explain how revealing a concept could be as it is shaped with time and societies across civilizations.

The book closes with a section on *Perspectives on Combating Cybercrime*. This section features chapters that deal with various aspects of cybercrime that have an immediate impact on society and represent a latent threat to humanity. In *The Case of 'Illicit Appropriation' in the Use of Technology* the authors elaborate on the concept of appropriation of technological tools, based on the human-computer interaction (HCI) concept of appropriation. This concept was coined by researcher Alan Dix, and points out that the user of a particular technology appropriates its usage in a way not foreseen or planned by technology designers, and when the user interacts with technology in ways not defined by its designer. However, sometimes this appropriation can be oriented towards activities that can be considered as illicit among a number of social deviations, and in some extreme cases, as criminal. The authors bring about an ample explanation that technology is a "neutral" means to an unspecified number of ends, and that the use of technology to fight upon social deviations is highly dependent on social behaviour and public policies. The authors also argue that technology designers and developers should be aware that any new feature included in technology to stop deviant acts might be used by unscrupulous people to commit those very same acts. The chapter starts describing a review on the concepts of appropriation and user experience, including the concepts of democratic and non democratic states. Secondly, the chapter introduces the concept of illicit appropriation. Thirdly, it shows three case studies of illicit appropriation in democratic, non democratic states and, in order to avoid an excessively state-centric vision of the process of illicit appropriation. The last case study analyzes it in the context of a criminal networked organization and their appropriation of everyday technologies for the commission of deviant acts. In the conclusions, the authors raise awareness directed to interdisciplinary teams and professionals in general, in the sense that it is necessary to understand both technology and social policy to fight upon deviant acts. The authors claim that this would provide an efficient use of technology to facilitate humanity. The chapter is directed to researchers, practitioners and scholars from the fields of law, social policy, human-computer interaction, new technologies design, and other related areas of knowledge.

An intriguing research project on the development and testing of a web filtering system is presented in *Designing Usable Security Feedback for Web Filtering Systems*, with the objective to preventing undesired access to web sites that contain pornographic contents, which also can serve to prevent the access to web pages that contain hate propagation and other social deviations. What is particularly interesting in this chapter is that the authors used data sonification (the use of auditory parameters to represent meaningful information), working along with visual feedback, to alert the user on the possible access of an undesired web page before opening it. The motivation for developing this type of audiovisual system is that combining different forms of feedback it may be possible to increase the likelihood of success of security notifications on the undesired web pages. This chapter also describes a non-exhaustive collection of usability patterns based on the HCISEC design criteria, which are intended to facilitate the design of audiovisual feedback information during the user interface design process of the web filtering system. The authors describe a proof-of-concept, where they successfully tested the usability of the developed audiovisual system to prevent the access to ten of the most visited web pages, according to a popular web page ranking site. The chapter concluded that well-designed auditory and visual feedback can contribute to reduce the risk of accidental pornography views. This chapter is a valuable source of information for researchers, students, and academics from human-computer interaction, information systems, and network security fields.

The last chapter is entitled *Human Factors Required for Building NLP Systems*. It outlines the importance of human factors applied to the design, development, and applications of computer interfaces for natural language processing (NLP). A NLP interface is defined as a system that accepts users' inputs in natural language allowing interaction with some other system, typically a retrieval system, which then results in sufficient responses to input natural language text or query statements. Human factors involves the study of all aspects of the way humans relate to the world around them, with the aim of improving operational performance and safety, through life costs and/or adoption through improvement in the experience of the end user. NLP interfaces have the potential to support the combat of a number of social deviations, such as cyberbullying done over mobile phones using short text messaging (SMS), provided that those NLP interfaces are carefully designed. With NLP, it is possible to automatically analyze the text of an ongoing communication carried out online or on mobile phones and be able to pinpoint threats or hate issues on specific targets. The chapter provides background information on natural language interface styles and processing. In addition, the chapter presents a number of human factors that are involved in designing and building a NLP system. Methods used in incorporating human factors are also discussed. Finally, the chapter describes a case study on a system that can successfully handle spontaneous Arabic SMS text, demonstrating that human factors were key design elements. The chapter makes emphasis on the key elements that are needed to successfully developing NLP systems, for instance, understanding how people encode their thoughts, and finding the right representation to model the concerned domain knowledge. In addition, the chapter clearly highlights the challenges and disadvantages on NLP systems as well.

As technology moves forward, we are witnessing important changes and implications in society. Whether it is moving from "real" to "virtual" or from "face-to-face" to "online," new dimensions of complexity are added, involving new societal aspects sometimes undermined by new technologies. While the future will surely bring new developments and technological applications, we, technology savvy professionals, need to continue developing new ways, better policies, and carrying out actions to use technology for the benefit of humanity while deterring misuses and promoting best practices. Researchers and technology developers must ask how these new technologies might be misused. We believe that emerging technologies such as cloud computing, mobile Internet applications, the evolution of social networking, and multi-sensory human-computer interfaces, among others, will play an important role in facilitating people's daily activities. However, it is likely that some of these technologies will be used as instruments for criminal or socially unacceptable behaviours. The research community must be alert to investigate all these issues in a timely fashion, opening avenues for subsequent editions of this book.

The chapters were selected following a rigorous analysis done by the book editors, and each chapter was double-blind peer-reviewed by at least two experts in the area. This would not have been possible without the valuable help of the Editorial Advisory Board; our sincere appreciation to: Dr. Jihad M. Alja'am, Dr. Raul Aquino-Santos, Dr. Michael E. Auer, Dr. Walter S. DeKeseredy, Dr. Samir Abou El-Seoud, Dr. Patrik Olsson, Dr. Doru Ursutiu, and Dr. Aurora Vizcaino-Barceló. We also thank many other anonymous researchers from around the world who helped with the peer-reviewing process. We also wish to sincerely thank Denise Rowsell for her clerical assistance during the critical final stages in the preparation of this book.

A book that aims at facilitating humanity would not be complete without giving to this cause. Therefore, from the very beginning of the project, we decided to humbly donate all of our revenue generated by this book to UNICEF Canada.

Finally, we would like to thank an anonymous Canadian Foundation, whose name is withheld upon request, for its generous support throughout the different editing stages of this book.

Miguel Vargas Martin
University of Ontario Institute of Technology, Canada

Miguel A. Garcia-Ruiz
University of Ontario Institute of Technology, and University of Colima, Mexico

Arthur Edwards
University of Colima, Mexico

Acknowledgment

We encountered many challenges and had to make sacrifices in the development of this book. I would like to thank my lovely wife for her wholehearted and unconditional support, encouragement, and sincere advice not only throughout this project but in each and every aspect of my personal and professional life. I humbly dedicate all the efforts I put into this book to her. Thank you, Alika.

Miguel Vargas Martin
University of Ontario Institute of Technology, Canada

I would like to thank my colleagues and friends who have supported me in my academic endeavors, and particularly those who contributed to this book. Their assistance and advice have been very valuable. I am thankful to my parents who taught me the value of education. Special thanks are due to my wife Selene, whose endless understanding and support have been profound throughout the difficult times. Without her love and encouragement I am sure that I would not have been able to achieve so much. I also dedicate this book to my son, Miguel Alejandro, for filling my life with joy.

Miguel A. Garcia-Ruiz
University of Ontario Institute of Technology, and University of Colima, Mexico

I would like to thank the many people who have helped me grow professionally. I would also like to acknowledge Stephanie Wilcox and Robert Edwards, two people who are very special to me. However, I would like to especially acknowledge my wife, Marilu, and my children, Elisa and David, who have been so patient with me during the edition of this book.

Arthur Edwards
University of Colima, Mexico

Section 1
Technology for Dealing with Crimes against Children

Chapter 1
Digital Child Pornography:
Offender or not Offender

Frank Y. W. Law
The University of Hong Kong, Hong Kong, China

K. P. Chow
The University of Hong Kong, Hong Kong, China

Pierre K. Y. Lai
The University of Hong Kong, Hong Kong, China

Hayson K. S. Tse
The University of Hong Kong, Hong Kong, China

Kenneth W. H. Tse
The University of Hong Kong, Hong Kong, China

ABSTRACT

Child pornography has become a major cyber crime in recent years. One of the challenging problems in child pornography cases is to distinguish if the subject files were downloaded intentionally or by accident without the knowledge of the computer user. The suspect may admit that he is an erotomania, but argue that the child porn materials were downloaded accidentally while surfing the pornographic web sites. In many jurisdictions, possession of child pornography without user knowledge is not a crime, while the burden of proof is on the prosecution. It is therefore important to identify if the child pornography exists by accident or not. In this chapter, the authors first review the technologies which sustain the prevalence of online child pornography and the recent research on child pornography investigation. Then, the authors present a set of practical investigation techniques. Subsequently, they apply the techniques in a case study with an attempt to distinguish if a suspect is a child pornography offender or just a normal erotomania. This is an important distinction to be made, since a person guilty of child pornography offenses is likely to be punished more seriously under most legal jurisdictions.

DOI: 10.4018/978-1-60960-094-5.ch001

INTRODUCTION

This Internet era has propelled communication and information exchange to the global stage. With an ever-growing penetration rate, the Internet is no longer confined to just homes and offices and is increasingly available wherever and whenever we want it. Technology is constantly changing and the types of computer equipment evolved as well as the capacity of storage media is increasing day by day. The development of technology increases accessibility and the distribution of materials online. These are done by enhancing the ease of possession and dissemination, and decreasing the cost of production and distribution, especially cross-border activities.

The advent of IT technologies not only facilitates people on communication and business, it also provides avenues for savvy criminals in the commission of cyber crimes. In these days, digital crimes have evolved such that digital evidence is found in traditional types of crime. Although the motivations of criminals rarely change, the methods of crime commission do. Mobile phones, emails, memory cards, thumb drives, etc are equipments commonly used by offenders used for crime commission. Child pornography is one of the most prevalent cyber crimes. According to the estimated Internet pornography statistics (Ropelato, 2006), there were 100,000 websites offering illegal child pornography in 2006. A greater number of offenders are now using information technology to organize, maintain and increase the size of their child pornography collections. Furthermore, encryption, P2P networks, light and small storage devices and free online storage are used by offenders to share the illicit materials with high efficiency. It is obvious that when child pornographic materials reach cyberspace, they would be circulated more readily and rapidly.

To cope with the emerging problem of digital child pornography, investigators need to develop the skills and competence required for conducting investigations in the digital environment. They must follow the judicial procedures laid down in their own countries to ensure evidence is admissible. They should always be aware that their investigation might be contested on technical grounds. They are also required to get familiar with the concept of digital evidence and the way to present it at the court of laws.

Echoing the widespread of digital evidence, forensic computing emerges. Forensic computing refers to the application of computer science techniques to retrieve digital evidence from electronic device for legal proceedings. Similar to ordinary forensic science, the analytical results of forensic computing examinations are expected to be reliable, accurate and scientific. However, with the increase of space on digital storage media as well as the strength of encryption, it is envisaged that the process of analyzing and locating digital evidence would become more tedious and time consuming without the help of special digital investigation technique and tools. Though the investigation moved from the physical environment to the digital world, traditional techniques still stand. For example, criminals may claim that they are innocent since they only visit pornographic websites, whereas the child pornographic materials were accidentally downloaded to their computers. In order to prove the knowledge of criminals, digital investigators need to analyze the retrieved digital evidence and distinguish whether the act of criminals is intentional or accidental.

It is obvious that child pornography investigation has become one of the greatest challenges to law enforcement agencies. On one hand, it requires very specialized knowledge in information technology and computer forensics which are not normally possessed by traditional crime investigators. On the other hand, the global reach of the Internet allows criminals to effect their illegal acts in any place they choose. The investigation therefore requires synergy of various parties from private and public sectors. The speed of investigation also becomes one of the critical factors. This requires the harmonization of competencies borne by digital investigators. Knowledge and expertise of investigators varies from region to region. Some of them have advanced techniques while some of them even do not know what information they should look for.

To overcome the deficiency and enhance the effectiveness in digital investigation in the context of child pornography, this chapter discusses a structured approach for performing forensically sound digital investigation on Windows platform computers which may contain child pornography. To understand the digital traces that may be left behind by a computer user, first we will provide an overview of the potential tools that may be used by criminals to access child pornography. The technical challenges that would be faced by digital investigators in the context of investigation will also be discussed. In addition to that, the way to distinguish "intentional" or "accidental" child pornography case will be discussed. Through understanding how latest technologies facilitate the distribution and storage of child pornography, we identify their limitations and propose feasible approaches to enhance the effectiveness and efficiency of child pornography investigation. We will also discuss future challenges in the investigation in the last section.

BACKGROUND

The crime related to child pornography on the Internet has been studied extensively by many researches in many different jurisdictions. Jenkins (2003) introduced various aspects including its organization, operation model and some techniques used by child pornography criminals. Common glossary and language used by the community were also included in this book for reference, which could definitely help investigators in understanding the communications among the offenders.

When we deal with the technical aspects of child pornography investigations, web browser analysis is always a place to look at. Traces including web histories and caches would be left on the computer when a user browses web pages on the Internet. Jones & Belani (2005) provided useful introductions on performing web browser forensics. Their articles discussed the potential traces and the means to find them. Tools assisting the investigator to analyze browser content were also introduced.

The child pornographers are not alone. They communicate with others, who share the similar interests, to share their collections. Instant messengers could provide a handy and effective platform for them to discuss and share. Depending on the instant messaging protocol and software used, there could be various digital traces left on the suspects' computers. Conversation histories are one of the useful traces. Belkasoft. (n.d.). provides a brief overview on investigating instant messenger histories. Morfitt & Valli (2005) have developed a tool for extracting information from the log file of ICQ, which is one of the instant messengers available on the market. Parsonage has discussed the investigation on conversation history of MSN Messenger as well as the successor, Live Messenger, in his online publication (Parsonage, H., n.d.). He has also analyzed artifacts in the MSN protocol to reveal more traces in the publication. In fact, software and protocol artifacts could be useful in digital forensic examination. van Dongen illustrated this by an in-depth analysis of artifacts left by Live Messenger (van Dongen, 2007, June) and pidgin (van Dongen, 2007, December), which is an open source instant messenger client supporting multiple instant messaging protocol including MSN, Google Talk, ICQ and Yahoo! Messenger.

Peer-to-peer (P2P) file sharing networks are another heaven of child pornographers. The anonymity and high file sharing speed of P2P file sharing networks allow them to share their collections in a quick and seemingly safer manner. Researchers have been working on how to perform the forensics investigation in P2P network. Adelstein & Joyce (2007) have developed and presented an automatic tool for extracting data from P2P network including user activities and shared files. Lee & Nam (2007) have also proposed a honeypot technique to monitor and trace illegal files in P2P networks. These tools and techniques could be applied in investigating child pornography activities in P2P network.

Other than instant messaging and P2P networks, child pornographers may also share their collections using removable storage devices like USB flash key. Techniques in tracing USB devices discussed in Luo (2007) could be applied to collect more traces on the criminals' activities. There is recently a new type of USB devices, which is often referred as U3 devices, which split the USB devices into two partitions, one for data storage and another read-only one for application. Tank & Williams (2008) have discussed the impact of these U3 devices on performing forensics analysis.

DEVELOPMENT OF TECHNOLOGIES V CHILD PORNOGRAPHY

Information technology has assisted the transformation of traditional child pornography into a high-tech and complicated crime issue. With the rapid expansion of the Internet nowadays, there are about 1.6 billion online users worldwide (Internet World Stats, 2009, June). Digitalization allows criminals to manufacture and distribute child pornography within a few seconds. The development of communications technology also facilitates criminals to communicate directly with other accomplices and potential victims through emails, instant messaging, newsgroups, chat rooms, bulletin boards, Internet Relay Chats, etc.

Many people thought that cyber world is separated from the physical world. This is obviously not the case because crime on the Internet reflects the crime in the physical world. However, criminals often feel safe in the cyber world because they could hide their identity by using others identities online. There are also a number of tools to assist online users to disguise their true identity. Furthermore, the non-existence of physical evidence minimizes the chance in disclosing the criminal acts to their parents, friends or colleagues. Data encryption also provides a safe solution to protect illicit materials stored inside a computer hard drive. These concepts often lower the alertness of criminals so that they would underestimate the digital trails that are left behind by their acts.

To effectively conduct the investigation, this requires digital investigator having the skills in searching the digital traces that are left behind by criminals. In other words, the ultimate goal of investigation is to find the linkage between the cyber world and the physical world. Before doing so, digital investigator should have the basic knowledge on what's going on in the cyber world. To this end, this section provides an overview to the current tools that could allow offenders to get access to, process or exchange child pornographic materials. By studying the possible means of access, investigators would equip with the knowledge to search for specific digital traces on a computer.

Internet Browser

Internet browser or web browser is a software application installed on an Operating System and is definitely one of the main tools that will be used by computer users for surfing websites on the Internet. Based on the user's input, e.g. such as an Uniform Resource Locator (URL), the browser contacts relevant computer system(s) or Internet server(s) on the Internet, retrieves the user's requested information and depicts the information on the web browser interface. The retrieved information may be website files, text files, image files or video files. Nowadays, there are various kinds of Internet browsers existed in the market and the most famous one are Microsoft Internet Explorer, Mozilla Firefox, Google Chrome, Apple Safari and Opera.

Figure 1. Preview images in thumbnail mode by Windows Explorer

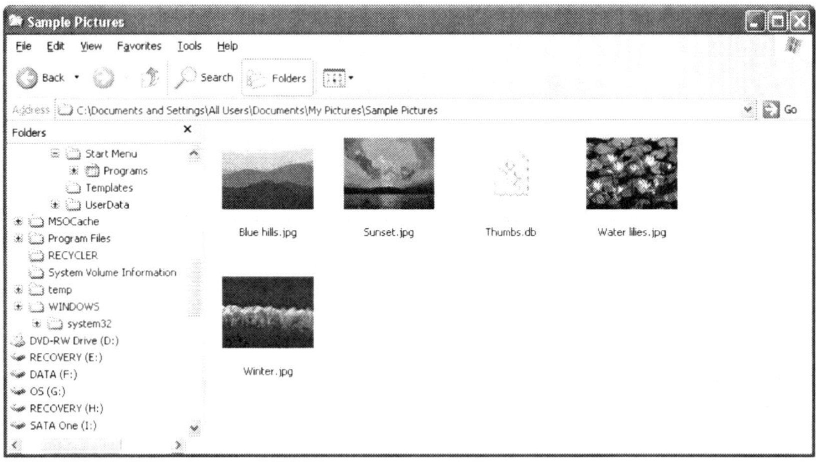

To make the browsers more user-friendly, the application will record the web browsing history and caches of websites by default. This assists fast browsing experience and enhances the ease-of-use of the software. In order to record and analyze users' behavior and web surfing habit, a small file normally refers as "cookie" will often be created within the browser's configuration folder to enhance the user surfing experience.

To access child pornographic information on the Internet, Internet browser is one of the most common applications used by net users. Since Internet browser itself has built-in a number of functions, those functions leave digital traces for digital investigation and assist digital investigator in revealing the following information:

1. The websites that had been visited by the offenders;
2. At what time those websites were visited;
3. What files had been downloaded from the websites;
4. Partial content of the websites which could be built from Internet caches;
5. The Internet surfing habit of the offenders;
6. The username and password which had been used to access the websites.

Image Browser

In order to preview a digital image file, the operating system must have been installed with an image browser or image viewer to transform digital binary data into a human readable picture. Normally, an image browser could handle multiple image file formats and present the images in the context of their pixels, size, date of taken, color depth, photographer details, etc. Recent image browsers normally have the function to organize browsed images into different categories

Some image browsers recorded the thumbnails of images that it had browsed to speed up the image browsing process. Some of them may come along with video file previewing functions to assist user in viewing digital video clips that are stored on the hard drive. Nowadays, image browsers may also be incorporated with basic image editing function so that user could do a simple editing on the images like resizing of images, contrast/brightness adjustment, image cropping, red eye reduction, etc. The most common types of image browser are Microsoft Windows Picture Viewer (Figure 1), ACDsee, Irfanview and fast picture viewer.

Obviously, image browsers are the main application that would be used by offenders to view digital child pornographic files stored at digital media such as hard drive, CD or DVD. With the advance in technologies, more and more functions are added to image browser to enhance user convenience. At the same time, huge amount of meta data about the images that have been viewed by the browser are created. This largely assists the overall digital investigation process. The meta data, normally referred to as EXIF (Exchangeable Image Format) (Exif Version 2.2, 2002) may be able to reveal the following:

1. The name of image files that have been viewed by the browser;
2. The location of image files that have been viewed;
3. The date and time when the image files were viewed;
4. The date and time when the image files were taken;
5. The make and model information of the camera which has been used to make those images.

Video Player

A video player is a kind of software applications that is run on an operating system for the purpose to play back digital video data from files of different formats. A video player is very similar to conventional video cassette recorder and has the functions to play, pause, fast-forward and rewind a video.

With the nature of digital video signal, video player often comes along with video tuning functions to control the aspect ratio, frame size, video modification features such as blurring, softening, sharpening, gray scales and scaling, noise filtering, etc. Furthermore, some of the players are also built-in with video or audio capture function which allows user to capture a specific portion of video from the video file. Common types of video player are Microsoft Windows Media Player, Nero video player, KMPlayer and BS player, etc.

Apart from still images, video provides a more detailed and vivid records of child pornography. Video players are apparently the tool that would be used to play back digital video file that recorded the criminal act relating to child pornography. Similar to image browser, digital video player normally generates meta data which may assist digital investigation. In brief, the meta data of digital video player may contain the following useful information:

1. The name and location of digital file which had been last-played by the video player;
2. The footage image of video that had been created for easy-indexing;
3. The facial index of the people who had been appeared in the video;
4. The date and time when the video was recorded.

Instant Messaging

Instant Messaging (IM) is a form of communication between two or more people based on typed text. The user(s)' computers need to be installed with the IM software which provides an Interface for the user to input the text. The text is then conveyed via a network, sometimes through the server of the IM service provider, to reach another designated user. The most common form is via computer to computer instant text messaging. However, IM can also work with mobile devices, e.g. PDA, Smart phone and Windows Mobile, etc. Due to its convenience, IM has attracted huge number of users, particularly the young generation.

Apart from exchanging text between users, IM software can incorporate voice or video and often allow user to share digital file amongst peers. As a result, IM has provided a faster and often a more convenient form of communication amongst computer users. However, this also created a new platform for child pornography offenders for communicating and sharing child pornography in a more efficient way. Criminals may have found safer to communicate via IM as opposed to telephone because of the difficulties in wiretapping of communication increased in Internet communication. Furthermore, the anonymity and ubiquity of some IM software also make it an attractive medium for criminals. Common IM software includes Windows Messenger, Windows Live Messenger, ICQ, QQ, AOL online and Skype.

IM software, like the other aforementioned software applications, will generate meta data and computer records that may assist digital investigation. The digital traces that may be left behind by an IM software include:

1. The chat history between peers;
2. The file transfer or sharing history between peers;
3. The user profile information of the peers who were on the contact list;
4. The profile image of the peers who were on the contact list;
5. The last chat date and time between peers;

File Sharing on P2P Network

Nowadays, the practice of sharing file online is prevalent and there exists a number of ways to distribute digital files amongst a large number of online users. In the old days, file sharing may be done by establishing a file server on the Internet so that users could obtain a copy of the file by downloading it from the server. The traditional technology that supports file download is known as FTP (File Transfer Protocol). However, this sort of data sharing and distribution is found to be inadequate to cope with the ever-increasing demand for bandwidth and big files, such as movie. Furthermore, there may be problem of Single Point of Failure (SPOF) which largely reduces the availability of the whole file sharing system.

To this end, Peer-to-Peer (P2P) technology is developed to address the problems of traditional client-server file sharing architecture. In a P2P network, each computer can operate autonomously and share resources with other computers without the need of central servers. New peer-to-peer protocols can also enable a computer to upload its downloading data to other computers within the same network. Based on the design or configuration of a particular P2P application, any computer which has downloaded a part or the whole file, can become an uploader of the downloaded data to other participating computers in the network. This distinctive feature not only solves the problems of centralization, but also escalates the overall scalability and flexibility of the network. Common P2P software applications include BitTorrent, uTorrent, Bitcomet, eDonkey, eMule, Kazaa and WinMX, etc.

P2P networks therefore allow large-size files to be distributed among peers within a short interval of time. Moreover, some P2P networks are basically networks of anonymity, without the need of knowing the uploading computers between participating computers. This rapid and decentralized mode of sharing and the level of anonymity have facilitated crimes, especially child pornography, in the cyber world.

Though there are challenges in the context of P2P investigation, the following traces may also exist for digital investigation:

1. The IP address of the peers who were uploading or downloading the digital resource online;
2. The percentage of digital file copy that was owned by the peer;
3. The software tool that had been utilized by peers to upload or download the digital resource;
4. The torrent file which was created by the uploader to share the resource;
5. The storage location of shared resources at the sharer's computer.

Online File Storage Service

Though P2P technology provides a quick way to share large-size digital resources online, online users may still tend to use the traditional way to share resources online. However, the establishment of file sharing server may require a certain degree of expertise and hardware resources. As a result, some websites provide free Internet storage service for Internet user to store digital file, especially large-size file, in order to share with the others online. Online users could access the stored content via web or FTP access. With the free services, the Internet users may be provided with 25 megabytes to a few gigabytes storage size. Though there are terms of service warning the users not to store or share illicit materials online, the detection of illegal materials normally relies on the complaint from public as well as the online users.

Online file storage service provides a great implication in the context of digital investigation because many of the service providers do not keep any log records for their users. Furthermore, the anonymity of its service provides opportunity for criminals to share illegal or pirated materials online. One of the examples is child pornography. Even if the materials exist at the online storage service for a few hours, it could be distributed to huge amount of offenders. Prevalent online file storage service websites included Rapidshare, Badongo, Mediafire, Megaupload, Sendspace, etc.

USB Storage Device

Nowadays, USB storage devices become a popular digital storage media for computer users because of its low cost, high portability and small size. There are several types of USB Storage devices available on the market nowadays, ranging from USB flash drives to external USB hard drives. USB storage devices provide the ability to quickly copy digital data to or from a computer. This replaces the traditional floppy disk or CD-rewriteable in transferring data from one machine to the other. A typical USB thumb drive of 2 GB is around US$10 and its size is usually less than 2 inches.

A lot of people use USB storage devices to help them with flexible working and transferring documents from computer to computer. These devices are easy to conceal and transport and can be disguised as various type of tools like a wrist watch, a Swiss Army knife, a keychain or any number of common devices. With the increase in storage capacity of USB storage devices, some of the device are also able to be installed with an Operating system so that the user can use it as a boot up device for web surfing, instant messaging, picture and video viewing, P2P file sharing, etc. As a result, a threat is posed by such devices as they can also be used to steal important data or store/share illegal materials. It may also contain information such as e-mail messages, Internet browsing history, Instant Messaging chat logs and buddy lists, image/video files and logs. All these can be valuable evidence in an investigation or prosecution.

When a USB storage device is connected to a Windows system, some digital traces will be created at the Windows Registry. They show the type kind of device and the time when the device is plugged into the system. It is because the Windows system requires this information to search for appropriate driver for the device. Such information will also be updated when the device is plugged into the computer system again. Those traces are useful for digital investigators because it could assist in searching for any USB storage device which had been used by the offenders to store data.

Web 2.0

With the introduction of Web 2.0, the world wide web could facilitate interactive information sharing and collaboration between different services. One of the great inventions from Web 2.0 is social networking platforms, e.g. Facebook and MySpace, where Internet users could use it to search or share information in a more effective and efficient manner. Since Web 2.0 concerns the control of information at server side, the web application runs on Web 2.0 does not create much digital traces at the client computer. Computer forensic examiner should bear in mind the implication and the collection of digital traces against the computer user may require the assistance from the service provider.

DIGITAL INVESTIGATION ON CHILD PORNOGRAPHY

The majority of child pornography cases that are investigated by law enforcements normally relates to the possession of child pornography in a digital format, whilst there are cases when child pornography is distributed or sold online.

The prevalence of computers at home will undoubtedly continue to increase, not only due to home entertainment but also as a result of enhanced communications among individuals. Digital materials may be transported on portable disks or handheld mobile devices. Additionally, the data can be coded or disguised in such a way as to hide the true purpose of the material, thereby avoiding detection or enabling it to be transported by other innocent parties.

To thoroughly investigate child pornography related crimes, forensic skills in understanding relevant digital traces that were left behind by criminals are needed. They include Internet browsing history, photo exif data, video player history, file downloading history, signs of photo or video editing, interpreting file timestamps, etc. Apart from common computer forensics knowledge, network investigation skills may also be required if the investigator wants to trace the distributor or online sources where child pornography is disseminated.

Investigators often collect evidence at the scene of crime. After that, they perform certain reconstruction of the events from the traces left behind by criminals to develop leads and decide what evidences may be further needed for investigation. Should the case could be rebuilt from evidences, a more complete picture could be developed and presented at the court of laws. Basically, the process of reconstruction rests with the way to collect and analyze digital evidences through temporal and relational investigation. Though the process of crime scene reconstruction is often time consuming, it can benefit the case as a whole. Sometimes it can shed light on offender's motive and Modus Operandi. It also helps to understand the type of victims that are targeted by the criminal and the place where the criminal come into contact with them. For example, analyzing the Instant Message chats between victim and criminal may reveal information like mode of communication, geographic location, type of spoken languages, profession, interests, personal image, marital status, etc, of the offenders who are involved in the crime. The above information could then be passed to undercover online investigators for further covert online operations.

To this end, this section is to provide some advice for all investigators who may encounter crime cases involved digital child pornography, particularly in the aforementioned digital traces investigation for the purpose of case reconstruction. However, it must be emphasized that there is no simple solution to the problems that may be encountered, but a structured and well-documented approach making full use of available resources and skills often yield good results. This section will start by studying a simple child pornography case scenario to illustrate the ways criminal commits the offence of possession and publishing child pornography. By analyzing the digital traces that may be left behind by the criminal, we follow with the explanation of the investigation process, techniques and tools which are required in the context of investigation.

Case Study

You are a law enforcement officer and received a call from the IT director of ABC Bank. The director reported to you that he had discovered one of his officers' computers stored with child pornography, including both images and videos, on 1st October 2009. Fearing the illicit materials may be distributed amongst the others on the Internet or within the company, the director requested police to launch an investigation against such ABC bank's employee. Initial enquiry by the IT director revealed that the employee subscribed an account from a child pornographic website, which provides child pornography to subscribers at certain fees. Thereafter, he downloaded images and videos from such website to his company computer and view them during his duty time. The employee had shared some of the photos via Windows Live Messenger to his friends. The machine involved is a Windows XP machine.

The documentation and record of the crime are all completed. The employee's machine is also seized for digital investigation. A forensically sound duplication of the computer hard drive has been obtained. We are going to investigate the case in computer forensic context. By reference to the digital traces which we have discussed in the previous section, we summarize below possible digital traces that we may have an interest during digital investigation:

- Internet browsing history;
- Image browsing history and exif data;
- Video browsing history;
- Instant Messaging chat history;
- Digital file time stamp representation;
- USB storage device information.

Part I: Internet Browsing History

Nowadays, most of the computer user activities are closely related to the Internet. Internet browser is the most common utility used by the general computing population for the purpose of web browsing, file uploading or downloading to other web resources. As a result, digital investigation of Internet browsing history constitutes a very large portion of the overall investigation process. Amongst the prevalent Internet browsers, Microsoft Internet Explorer (IE) and Mozilla Firefox are most popular. This section will briefly explain the skills and tools required to reconstruct the Internet browsing histories of such popular Internet browsers.

The main focus of the investigation is definitely the Internet browsing history record generated by the Internet browser application. Before we could spark off the investigation, we need to briefly understand how Internet browser stores their web history.

Internet Explorer basically stores its history data at three different areas. They are 1) Internet browsing history, 2) Temporary Internet files (cache files) and 3) Cookie files. The Internet browsing history contains the website addresses (URLs) and its date and time that the computer user had visited. The Temporary Internet files, usually refers as Internet cache files, are the temporary website files that are created on the users' computer to enhance the Internet browsing speed when the user browses the same website again, i.e. no need for the computer to download the files from the website again. The Cookie files are Internet cookie files that are created by a surfed website for tracking a user's behaviour.

Internet Explorer utilizes a simple binary data structure file called "index.dat" to effectively manage and categorize the information at the above areas. In brief, the "index.dat" file contains information that could be used to link history, cache or cookie files that are created for the application in the context of web browsing. In other words, the "index.dat" file is the heart for digital investigation against IE related activities. The locations of "index.dat" which could be found at Windows XP and Vista machines are listed as follows.

Windows XP:

- \Documents and Settings\<user>\Local Settings\Temporary Internet Files\Content.ie5
- \Documents and Settings\<user>\Cookies

- \Documents and Settings\<user>\Local Settings\History\history.ie5

Window Vista:

- \Users\<user>\AppData\Local\Microsoft\Windows\Temporary Internet
- Files\Content.IE5\index.dat
- \Users\<user>\AppData\Local\Microsoft\Windows\History\History.IE5\index.dat
- \Users\<user>\AppData\Local\Microsoft\Windows\History\History.IE5\MShist01<date> <date>\index.dat
- \Users\<user>\AppData\Roaming\Microsoft\Windows\Cookies\index.dat

There are a number of tools that could be used to analyze the content stored at the "index.dat" file. These include the famous forensic tools Encase, FTK and X-ways forensics. The tools Netanalysis and Pasco are also tailor-made to focus on the analysis of Internet history and cookies. Apart from using the tools to reconstruct the Internet activities, such as web browsing, webmail, online chat, of the users, the tools could assist in analyzing the timestamps of the recorded files. By such analysis, an investigator could conduct a temporal analysis on the Internet activities of the user according to the file accessed, modified and created time. Other than the digital file timestamps and the Internet history files, Windows registry may also provide additional information on the web surfing activities like the last visited URLs by the user at IE. The URLs could be found at the registry hive file "NTUSER.DAT" under the folder Software\Microsoft\Internet Explorer\TypedURLs.

Internet cache normally serves as a temporary storage area for the Internet browser. By examining the content within the temporary Internet cache folder, we may obtain partial content of the website which had been browsed by the user (Figure 2). Furthermore, this cache area can be considered as a buffer area for storing and downloading file from the Internet. Sometimes we may be able to recover files which had been downloaded by the user within the temporary Internet cache files.

Bookmarks, username and password recovery may also be helpful in establishing the intent of user. Unlike Internet cache or Internet history files which may arguably be generated by the machine automatically, bookmarks, username and password are intentionally added by the user to the Internet browser. Together with the Internet history and temporary Internet files findings, this piece of information is relatively useful in showing the knowledge of user for specific website browsing. We set out below the default path for Internet Explorer to store its bookmarks:

Windows XP: \Documents and Settings\<user>\Favorites
Windows Vista: \Users\<user>\Favorites

Figure 2. The Internet history file can tell the URLs of websites which had been visited by the user at specific time

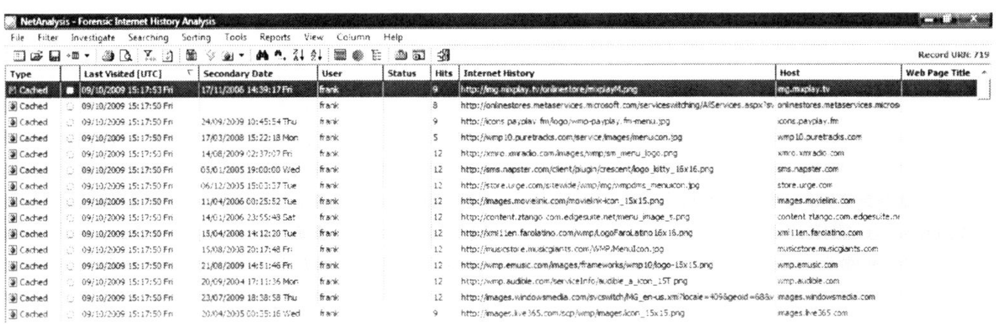

The overall concept of Internet browsing history investigation is to reconstruct the timeline to show the activities done by the Internet user. Such reconstruction could largely assist the investigation because it shows a sequence of activities that were conducted. Particularly in the context of child pornography investigation, the investigation into Internet browsing history could assist the investigator in ascertaining which website that the offender had accessed to child pornographic materials, at what time that website was accessed, the number of occasions such website were visited, the payment request or transaction information as well as the web pages where child pornography could be downloaded to the user's machine.

Part II: Image Browsing History

As mentioned in the previous section, image browser is basically the main tool that is used by computer user to view images stored at the computer. Should the tools be able to record the date and time when the user browsed the images, it could provide a great assistance to investigator as to ascertain approximately the time when the file was downloaded and viewed by the user. Furthermore, if child pornographic photographs were taken by the user himself, the exif data of the photographs may shed light on the tools and locations for taking those photographs.

For the purpose of viewing images, most of the Windows-based users will rely on the Windows file explorer which had incorporated file thumbnail preview and viewing functions. We will focus on analyzing Windows file explorer to see if we could extract out any image browsing history to assist investigation.

When viewing digital image files by file explorer (Windows XP) at thumbnail mode, a hidden compound database file called "thumbs.db" will be created by Windows under the same folder where the images are placed. All of the thumbnail files contained within the thumbs.db have a timestamp indicating when the thumbnails are created. This timestamp provides an indication as to when the files were viewed by the user. In Windows Vista, the thumbs.db is replaced by a number of centralized compound database files in the name of "thumbcache". Various sizes of thumbnails are stored ranging from 32 pixels through to 1024 pixels. The thumbcache file could be located at the following location by default:

\Users\<user>\AppData\Local\Microsoft\Windows\Explorer

There are a number of tools available in the market that could assist the decompilation of thumbs.db and thumbcache files. Furthermore, most of the prevalent forensic tools like Encase, FTK and X-ways forensics had already been equipped with the capability to parse out the stored content within the thumbs.db and thumbcache files.

Apart from Windows file explorer, there are also some software applications developed in the market for the purpose of viewing digital file image. Some image browsers keep records of the images which had been last viewed by the user. Such record may be stored at the Windows registry or at a designated location specified by the application. Moreover, some image browsers, like ACDSee, may be able to categorize the images according to their pixels, sizes, model of camera, date of taken, etc. Some of them even cache the thumbnail of images that had been accessed by the software to assist swift browsing experience. The indexed or cached information are extremely useful for investigators, but the parsing of the information from such application database may require the assistance from the software developer. Figure 3 lists out the information which had been indexed by ACDSee image browser software.

Exif data of image files is a specification for the image file format taken by digital camera. Apart from the image browsing history the exif data may also provide a broad spectrum as to the date and time information of the time when the photos were taken, the digital camera settings

Figure 3. Information indexed by ACDSee image browser software

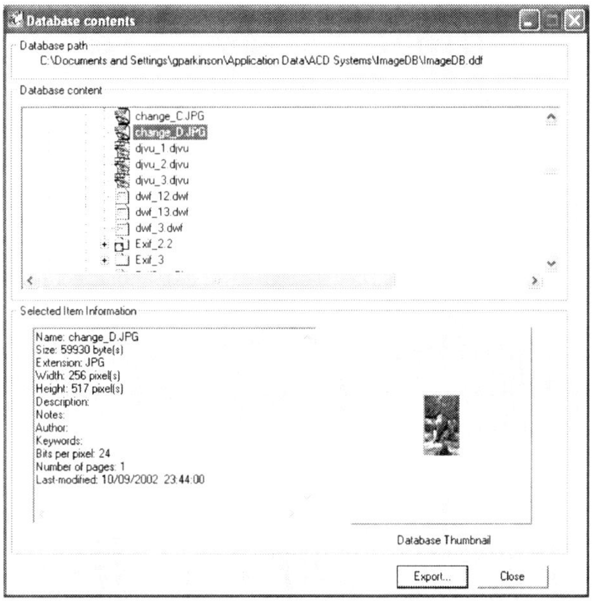

which may include the model and make, the aperture, shutter speed, focal length and ISO speed, as well as the thumbnail of picture which is used to project the image onto the camera's LCD display. Some higher model of digital camera may also incorporate geographical location information at the exif data.

The exif data is considered to be an important digital trace which could provide a lot of hints to investigators to assess whether an offender took the image himself, by what camera, and at which location. The data could assist in the tracing of victims who had been involved in such sexual exploitation activities. These help victims escape from the control of syndicate. Parsing out exif image data is easy and could be done by prevalent image browsing tools like ACDSee and IrfanView. Apart from general image browsing software, computer forensic software or digital image editing software, such as Photoshop, are normally capable of parsing out such information. Figure 4 shows a snapshot of exif data available from a digital photo.

Figure 4. Snapshot of exif data

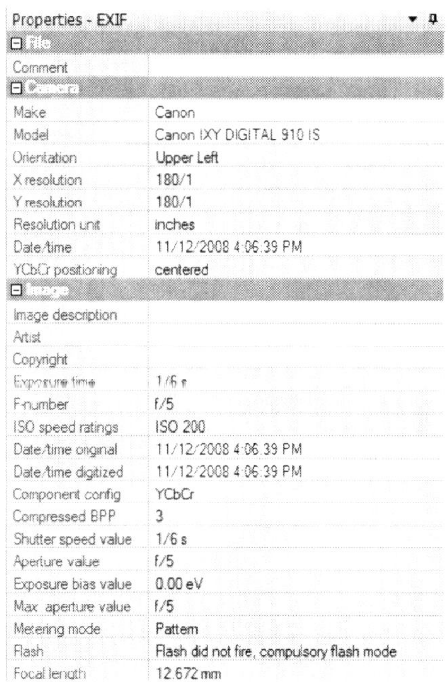

Part III: Video Browsing History

Similar to image browser, video player is basically the main tool that is used by computers users to view digital video footage. The main investigation purpose done on video players is to ascertain the time and the identity of the digital video files which had been viewed by the user at that time. Video players normally keep records regarding the digital video files which had been last viewed by the user. Such records may be stored at the Windows registry or a designated location specified by the application. In order to parse out the relevant digital traces left behind by a video player, the investigator is required to study the digital data that are created by the software application in the context of video playing. Amongst the video playing software, we will analyze Windows Media Player, which is the default Windows video player, as an example to see where we can obtain that information for investigation.

Windows Media Player is the flagship application of Windows and could be used to play and organize digital media files on a computer. Whenever a user opens a digital video file by the application, the recent file list entries of the player will be updated. When the user opens the player again, the recent file list could allow the user to reopen the previous file in a quick way. Such information is not stored together with the application folder but can be found within the Windows registry. The registry key for this information could be recovered from the registry hive file "NTUSER.DAT" (a representation of the key "HKEY_CURRENT_USER" at the Windows registry) which is located at the following location:

Windows XP: \Documents and Settings\<user>\NTUSER.DAT
Windows Vista: \Users\<user>\ NTUSER.DAT

The hive file could be examined by certain registry viewers and the key that we look for is "RecentFileList" under the location *Software\Microsoft\MediaPlayer\Player\RecentFileList*

Figure 5 shows a few video files opened by the player at earlier time. The entry modified time of the key could also suggest the time when the file was opened by the player.

Part IV: Instant Messaging Chat History

The primary use of instant messenger is to assist computer users in the communication with other known contacts by typing text messages. Some of the messengers may even support video or voice chats. According to Microsoft, the latest Windows Live Messenger attracted over 330 million active users per month. With the prevalence of instant messengers, the examination into them for recovery of message conversations is often a necessary skill for investigators.

Since there are so many instant messengers available in the market, it means that we need to understand the application structure of every instant messenger before we could be able to extract relevant chat history and information. Nevertheless, our main focus for examination normally is the recovery of chat conversations. However, due to privacy issue, the preservation of chat conversations is not switched on by default, causing a

Figure 5. Snapshot of registry editor showing a list of video files opened earlier on

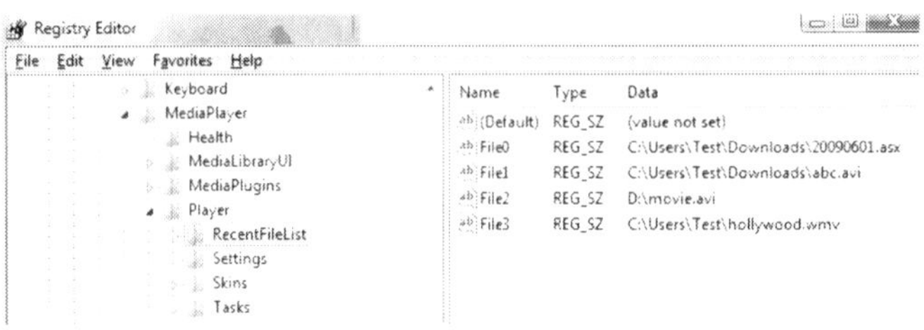

great implication to the digital investigation. Take the Windows Live Messenger as an example, the application does not save any chat conversations. However, a user can enable the automatically save history record function so that the chat contents will be saved to a specific xml format file. The default location of the saved chat contents could be found under the folder \My Documents\My Received Files \PassportName\History or \My Documents\My Chat Logs\.

The default chat history locations for various instant messenger tools under Windows Vista are listed in Figure 6 as reference.

The chat history not only includes text conversation but may also record information about any file transfer between the users. The chat history may be incorporated with timestamp information so that it can reveal the time when the conversation or file sharing was commenced.

Part V: Digital File Timestamp Analysis

With the advance of file system technology, the detailed information on timestamp (modified, accessed and created times) of digital files are now available in various computing environments. They enable computer users to perform analysis on the states and events happened on a machine. This might be of great value as these meta information, if still exists, allow investigators to reconstruct events that were initiated by the computer user.

From the investigative point of view, digital timestamps were influenced and created by human through machine process. There should be specific patterns or trails available for investigators to explain certain phenomena or actions that had been carried out by the user. We set out below some observations of file times on common NTFS file system running on Windows platform.

When a file is freshly established in a file system without any modification, it is in its inherent states and is considered to be intact and is not updated after its creation. In such situation, the three timestamp values are the same. However, should the file be opened or viewed by a computer user, the accessed time changes whilst the modified time and created time still have the same value. This information is very important because it shows that the file may have been opened or viewed by the user on the computer system. The information could assist in proving the computer user has certain degree of knowledge on what type of files being stored on his computer.

If the modified time is found to be equal to the created time, it indicates that the file has neither been modified nor copied from another disk location. This means that the file is intact and has not been updated after its creation. Should the file content be altered by the user, the modified time will be updated and the time should be after the created time, indicating that the user may have modified the image content. However, when a file is copied from one volume to another volume, it would cause the change of modified time to a time before the created time. This is a good indication as to the act of copying of the file by the user.

With the timestamp information, we are able to draw the conclusion that the user of the computer should have certain knowledge of the relevant files. Since different file systems offer different file times behaviors, it is good to use that to show the intention of the user.

Figure 6. Default chat history locations for common instant messenger tools

ICQ	\Users\Documents\ICQ\<ICQ account no.>\ReceivedFiles
AIM	\Users\Documents\AIMLogger\IM log
Yahoo! Messenger	\program Files\Yahoo\messenger\Profiles\<yahoo Account Name>\Archive\Messages

Figure 7. Sample value of the registry key USBSTOR

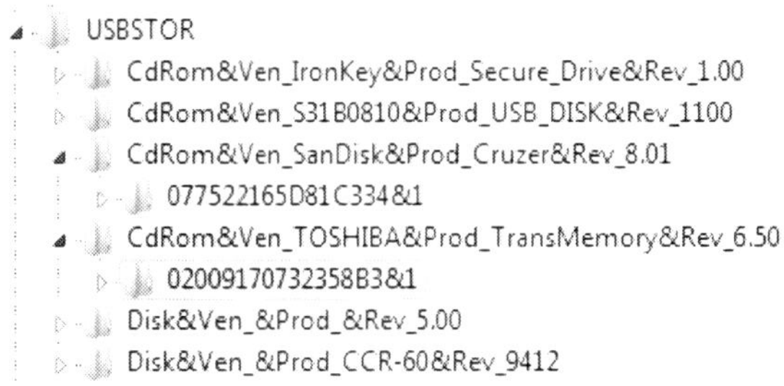

Part VI: USB Storage Device Information

USB storage device is commonly known to be useful for transferring or storing digital files by computer user. With its small size and large capacity, it is used by offenders to hide child pornography or share them amongst peers. When attending a scene of crime, it is of the utmost importance to conduct a proper search of the scene for physical evidence. If a kind of digital storage media is found, it is important to establish a link between the suspect and the media. One of the ways to prove such linkage is by proving that the media has been connected to the suspect's computer.

When a USB storage device is plugged into a Windows system, a key will be created at the Windows registry hive file "SYSTEM". The key is *\Enum\USBSTOR,* which stores a subkey *Disk&Ven_XXX&Rev_XXX* where *XXX* refers to the name and description of the device. Figure 7 shows a sample of this value.

The serial number of USB storage device will also be recorded as a subkey to the above value. In brief, each of the USB storage devices has a unique serial number even under the same manufacturer and production batch. Therefore it can be identify by the system. This serial number could be viewed by software USBview (http://www.ftdichip.com/Resources/Utilities.htm), which is a free utility from Microsoft. The software displays the USB connection tree and shows the USB devices that are currently connected to the computer together with the configuration data (as shown in Figure 8).

Apart from manually examine the relevant registry keys of the Windows system, a tool called USBDeview (http://www.nirsoft.net/utils/usb_devices_view.html) could also be used to list all USB devices that are currently connected to the

Figure 8. Sample snapshot of USB view

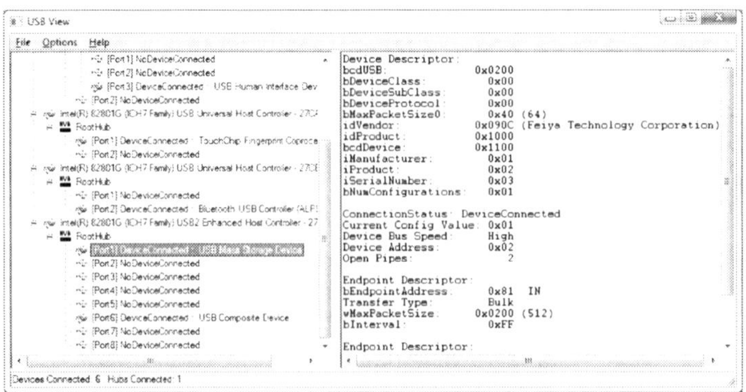

Figure 9. Snapshot of running USBDeview on a running computer

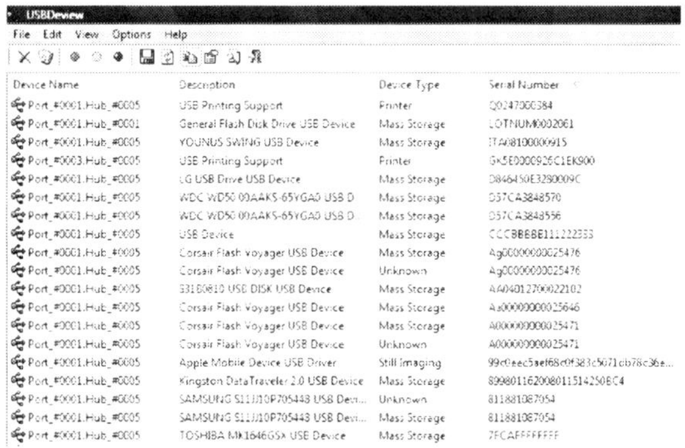

computer as well as all USB devices that were previously used. The program needs to be run on a live machine. The program will extract all relevant USB device registry entries and displays them to the investigator. In order to run the program for the extraction of relevant information for investigation, it is recommended to restore the computer content from the forensically sound image and then run the program on the restored machine. Figure 9 displays the result of running USBDeview on a running computer.

CONCLUSION

Standard methods and procedures facilitate the investigations in digital environment. First of all, evidence has to be gathered in a way that will be accepted by a court of law. If police and judicial authorities follow standard procedures, the evidence so collected will be more likely to be admissible in court. Standard procedures amongst countries in the collection of evidence by investigators will also facilitate the exchange of evidence in international cases. Second, during the investigation of computer systems, every care must be taken to avoid corrupting the data, or causing any other form of damage to the evidence.

The use of standard methods and procedures will reduce, but not eliminate, the risk of damage. In some cases it is inevitable that evidential data will be changed or overwritten during the examination process. Investigators must be aware of the side effects of all operations. They must keep a record of each operation carried out, so that it will be possible to explain their effects afterward.

We note that one of the challenging problems in child pornography cases is to distinguish if the subject files were downloaded intentionally or by accident without the knowledge of the computer user. Apart from manually created user settings, such as typed URLs and bookmarks, which sustain the user has certain knowledge of the files; a large quantity is also an effective indicator. If a person is found to have several hundreds of child pornography files in his hard drive as well as circumstantial evidences from variety of software applications indicated that the files were downloaded intentionally, it is more likely that the files were downloaded on purpose.

On the other hand, as reflected in the latest International Centre for Missing and Exploited Children (ICMEC) report, the control against child exploitation offence is required continually. At the present moment, there were 29 of 187 Interpol member countries have legislations to combat the problem of child pornography. Even if the countries have child pornography legislations, a number of countries do not criminalize the offence of "possession of child pornography".

It is observed that the lack of legislations causes a big implication as to the enforcement to child pornography. Unless most of the countries have respective laws to control the matter, it is expected that the problem of child pornography will be remained at significant level.

In this chapter, we have outlined the potential traces in Windows platform which may assist child pornography investigation. We describe in details the skills and tools for digital evidence recovery and the steps to locate common digital traces for proving an act of browsing, possessing or distributing child pornography. Evidence collected during initial investigation has to undergo further forensics analysis. Although the analysis processes may vary from case to case, we have explained and identified common forensics analysis. Nevertheless, it must be emphasized that there is no simple solution to the problems that may be encountered, but a structured and well-documented approach making full use of available resources and skills often yield good results.

REFERENCES

Adelstein, F., & Joyce, R. A. (2007). File Marshal: Automatic extraction of peer-to-peer data. *Digital Investigation*, *4*(1), 43–48. doi:10.1016/j.diin.2007.06.016

Belkasoft. (n.d.). *Forensic Investigation of Instant Messenger Histories*. Retrieved May 1, 2010, from http://www.forensicfocus.com/ forensic-investigation-of-instant- messenger-histories

Exif Version 2.2 (2002). *Exchangeable image file format for digital still cameras: Exif version 2.2.* Retrieved October 1, 2009, from http://www.digicamsoft. com/ exif22/exif22/html/ exif22_1.htm

Internet World Stats. (2009, June). Retrieved October 1, 2009, from http://www.internetworldstats. com / stats.htm

Jenkins, P. (2003). *Beyond tolerance: Child pornography on the internet*. New York: New York University Press.

Jones, K. J., & Belani, R. (2005). *Web browser forensics*, Part 1 and Part 2. Retrieved October 1, 2009, from http://www.securityfocus.com/ infocus/1827 and http://www.securityfocus. com/ infocus/1832

Lee, H., & Nam, T. (2007). P2P Honeypot to prevent illegal or harmful contents from spreading in P2P network. *The 9th International Conference on Advanced Communication Technology*, 1, Feb 2007, 497-501

Luo, V. C. (2007). Tracing USB Device artefacts on Windows XP operating system for forensic purpose. *In Proceedings of the 5th Australian Digital Forensics Conference*, Dec 2007, 210-218

Morfitt, K., & Valli, C. (2005). After conversation – a forensic ICQ logfile extraction tool. *In Proceedings of the 3rd Australian Computer, Network & Information Forensics Conference.* School of Computer and, Information Science, Edith Cowan University, Perth, Western Australia, 2005, 54-61.

Parsonage, H. (n.d.). *The forensic recover of instant messages from MSN messenger and windows live messenger.* Retrieved October 1, 2009, from http:// computerforensics.parsonage.co.uk/ downloads/ MSNandLiveMessenger ArtefactsOfConversations.pdf

Ropelato, J. (2006). *Internet pornography statistics*. Retrieved October 1, 2009, from http:// internet-filter- review.toptenreviews.com / internet-pornography-statistics.html

Tank, R., & Williams, P. A. H. (2008). The impact of U3 devices on forensic analysis. *In Proceedings of the 6th Australian Digital Forensics Conference*, 199-205 van Dongen, W. S. (2007, June). Forensic artefacts left by Windows Live Messenger 8.0. *Digital Investigation*, *4*(2), 73–87. doi:10.1016/j.diin.2007.06.019

van Dongen, W. S. (2007, November). Forensic artefacts left by Pidgin Messenger 2.0. *Digital Investigation*, *4*(3-4), 138–145. doi:10.1016/j.diin.2008.01.002

ADDITIONAL READING

Adelstein, F., & Joyce, R. A. (2007). File marshal: Automatic extraction of peer-to-peer data. *Digital Investigation*, *4*(1), 43–48. doi:10.1016/j.diin.2007.06.016

Jenkins, P. (2003). *Beyond Tolerance: Child Pornography on the Internet*. New York: New York University Press.

Lee, H., & Nam, T. (2007). P2P Honeypot to Prevent Illegal or Harmful Contents from Spreading in P2P network. *The 9th International Conference on Advanced Communication Technology*, 1, 497-501

Luo, V. C. (2007). Tracing USB Device artefacts on Windows XP operating system for forensic purpose. *In Proceedings of the 5th Australian Digital Forensics Conference*, Dec 2007, 210-218

Tank, R., & Williams, P. A. H. (2008). The Impact of U3 Devices on Forensic Analysis. *In Proceedings of the 6th Australian Digital Forensics Conference*, 199-205

Taylor, M., & Quayle, E. (2003). *Child pornography: An internet crime*. London, UK: Routledge.

KEY TERMS AND DEFINITIONS

Child Pornography: Images or films depicting sexual activities involving a child.

Instant Messaging: A form of real-time text-based communication between two or more people.

Erotomania: A person who has strong sexual desires.

MAC Times: File system metadata which record the times at which certain events (modification, access and creation) pertaining to a computer file occurred most recently.

Online Storage Service: A service with or with a cost for storing files (of any types) online.

Windows Registry: A database that stores settings and information about hardware, operating system software, most non-operating system software, and per-user settings for Microsoft Windows operating systems.

Chapter 2
Child Pornography and IT

Amin Ibrahim
University of Ontario Institute of Technology, Canada

ABSTRACT

The sexual exploitation of children remains a very serious problem and is rapidly increasing globally through the use of the Internet. This chapter focuses on the child pornography and IT, and the various methods to combat this problem. The ease of acquiring IT and digital equipments, the global reach of Internet and freely available peer-to-peer services have made child pornography a very complex issue to undertake. The borderless nature of the Internet and the lack of unified criminal code among nations further escalated the complexity of law enforcement against child pornography.

INTRODUCTION

In our technology savvy society, the sexual exploitation of children takes many forms and various means. An image of a child engaged in real or simulated sexual activities is considered a form of child pornography in the same manner as images depicting parts of child's body with the intention of stimulating sexual arousal or gratification for the viewer of that image. Similar effects are attempted to be achieved in virtual pornography where a morphed or blended artificially created images of children are depicted in sexual activities. Whether these images are limited to a child or involve several children with or without participation of adults, who may or may not be visible to the viewer, they are means of child exploitation and sexualisation of children. While these images are disturbing to most of us, they are used as fulfillment of sexual fantasies by viewers of such images worldwide. As these images are transmitted through electronic means, they are distributed through Internet and test society's tolerance for what is acceptable and not acceptable behaviour involving children of any age.

While child erotica with lewd images of children is not something new, as will be presented in child pornography section to follow, the means of

DOI: 10.4018/978-1-60960-094-5.ch002

transmitting these images has evolved over time. With this change, the challenges to the legal aspect of child pornography and the magnitude of child pornography is getting vast and complex; and how IT is assisting criminals to disseminate this crime and at the same time help law enforcement to mitigate this issue is discussed. Although it is beneficial to discuss the social and psychological effects of child pornography (for more information on psychological effects of child pornography, please refer to Additional Reading section), the main focus of this chapter remains the electronic distribution of images which portray the illegal sexual interaction of children, current detection and prevention methodologies and legal aspects of child pornography.

BACKGROUND

What is Child Pornography?

The exact definition of child pornography differs from country to country, but the core definition of child pornography remains the same everywhere. The Convention on the Rights of Children defines a child as every human being below the age of 18 years unless under the law applicable to the child, majority is attained earlier. Article 34 of the Convention declares that, States Parties undertake to protect the child from all forms of sexual exploitation and sexual abuse. For these purposes, States Parties shall in particular take all appropriate national, bilateral and multilateral measures to prevent: the inducement or coercion of a child to engage in any unlawful sexual activity; the exploitative use of children in prostitution or other unlawful sexual practices; the exploitative use of children in pornographic performances and materials (Child Rights Information Network, 1990, Article 34).

The Canadian statutes and regulations, under criminal code C-46, section 163.1 defines child pornography to include a "photographic, film, video or other visual representation, whether or not it was made by electronic or mechanical means, that shows a person who is or is depicted as being under the age of eighteen years and is engaged in or is depicted as engaged in explicit sexual activity, or the dominant characteristic of which is the depiction, for a sexual purpose, of a sexual organ or the anal region of a person under the age of eighteen years" (Canadian Legal Information Institute, 2008, Bill C-46). On the contrary, the U.S. jurisprudence has established that virtual child pornography without using real children or real images of children is not punishable by U.S. law (Meek-Prieto, 2008, p. 95).

The United Kingdom Protection of Children Act, Section 1 of the Protection of Children Act 1978, as amended by the Criminal Justice and Public Order Act 1994, defines child pornography as, any indecent photographs or pseudo-photographs (an image, whether made by computer graphics or otherwise howsoever, which appears to be a photograph) of a child; including data stored on a computer disc or by other electronic means which is capable of conversion into a photograph (Internet Watch Foundation, 2009).

South Africa Amendment of Section 1 Act 65 defines child pornography as "any image, however created, or any description of a person, real or simulated, who is, or who is depicted or described as being, under the age of 18 years engaged in sexual conduct; participating in, or assisting another person to participate in, sexual conduct; or showing or describing the body, or parts of the body, if such a person in a manner or in circumstances which, within context, amounts to sexual exploitation, or in such a manner that it is capable of being used for the purpose of sexual exploitation" (Government Gazette, 2004, p.2).

From all these, (Maalla, 2009, p. 7) summarized the definition of child pornography on the Internet as "any media file depicting a child engaged in real or simulated explicit sexual activities or lewdly depicting parts of a child's body, with the intention of stimulating sexual arousal and gratification whether there may be one or several children, involved in sexual acts alone or with other children, with or without the participation of adults, who may or may not be visible." Furthermore, she states that these media "might include highly repugnant images involving brutal anal or vaginal rape, bondage, zoophilia, oral-genital sexual relations or other degrading behaviour involving children of any age". She also considered child erotica consisting of images of children posing half-dressed/naked or naked with the emphasis on sexualizing the child as child pornography. Also virtual pornography consisting of morphed or blended artificially created images of children involved in sexual activities as child pornography."

History of Child Pornography

Historically, prior to the invention of digital devices and affordable consumer computing equipment, images of sexual exploitation of children were captured mainly through three mediums: hand drawing, chemical film-based motion video or chemical film-based still frame photography, both of which were often costly to develop and difficult to distribute.

Sketching of images to communicate sexual fantasies among humans has been around for a long time. Child pornography can, reportedly, be traced back to ancient Greeks through England's Libertine movement of the 1600s all the way to the introduction of printing press and photography. Around the beginning of the 20th century, child pornography was widely available in Victorian England (Creighton, 2003, p. 1).

In the late 1960s and early 1970s, the relaxation of laws against the production of all pornography, including child pornography, caused a shift in distribution of child pornography from being an underground activity to being a commercial activity. The production, distribution and possession of child pornography became part of a worldwide commercial industry instead of an illicit activity. As the number of magazines featuring child pornography throughout US and Europe kept increasing, it led to governments introducing legislation against it. As a result, there was a significant decline in its commercial distribution, and an outright disappearance in mainstream magazines.

With the suppression of commercial distribution of child pornography from the late 1970s to the late 1990s, pedophiles and other adults with sexual interest in children started an amateur industry of sharing images of child pornography among themselves. The current distribution of child pornography extends beyond that of a few individuals acquiring and collecting such items for personal reasons, to that of a global underground market extensively driven by supply and demand.

The intense competition among the various digital equipment manufacturers in recent years has led to lower cost devices that are affordable to enter low to medium income homes and revolutionize the photography market. Private chemical film development labs which require expertise, specialized equipment and considerable skill are no longer required to produce professional looking photographs. Thus, those who produce child pornography use readily available and affordable digital imaging equipment to record images of abuse and distribute via personal computers connected to Internet. To make matters worse, the ubiquitous nature of the Internet in particular, has blurred boundaries between commercial and amateur pornography transmission. Now, the distribution child pornography has become a complex issue all over the world via Internet by means of one-to-one messaging, an exchange on a peer-to-peer sharing forum or a commercial trade (Ibrahim, 2009, pp 5 & 6).

The current statistic on the spread of child pornography is alarming. Thousands of new photographs and videos are uploaded on to the Internet every week and hundreds and thousands of searches for images of sexual exploitation of children are carried out every day. Offenders may possess collections of over a million images of sexually exploited children. It is estimated that 200 new images are put into circulation every day (Wortley & Smallbone, 2006, p.12).

The UN Report of the Special Rapporteur on the sale of children, child prostitution and child pornography detailed that the number of sites dedicated to child pornography worldwide is 480,000 identified in 2004 compared to 261,653 in 2001. The number of predators online at any given time is estimated to be 750,000. Since its creation in 1998, and as of the 19th of April 2009, the National Centre on Missing and Exploited Children (NCMEC) in the United States had identified 592,044 out of a total of 681,275 sites as child pornography sites. Furthermore, in 2007 the Internet Watch Foundation (IWF) reported that 2,755 domains contained images of sexual abuse of children. About 80 per cent of those sites were used for commercial purposes and the remaining 20 per cent were for non-commercial. In 2008 the IWF reported that 1,536 domains contained the depiction of sexual abuse of children, 74 per cent of which were for commercial purposes and 26 per cent for non-commercial ones. Some of these images had appalling content: 58% of child sexual abuse domains traced contain graphic images involving penetration or torture of children, of these images, 69% of the children appear to be 10 years old or younger; 24% 6 or under, and 4% 2 or under 2 years old (Maalla, 2009, p. 9).

Legal Aspect of Child Pornography

A major obstacle with the legal aspect of combating child pornography on global Internet stems from the fact that most countries do not have a criminal code pertaining to child pornography. Some countries have chosen to curtail child pornography by criminalizing the possession and distribution of child pornography.

The International Centre for Missing Children conducted a survey concerning child pornography on 184 Interpol Member Countries. The survey tackled issues such as criminalizing possession of child pornography regardless of the intent to distribute, and requiring Internet Service Providers (ISP) to inform on suspected child pornography to law enforcement or other appropriate agencies. The survey concluded that alarmingly only five countries meet their standard. The same study shows that of the Interpol members, 138 countries do not criminalize the possession of child pornography. Furthermore, 122 countries did not have a law to address the distribution of child pornography via computer and the Internet. One shocking finding is that 41 of these countries do not criminalize possession of child pornography, regardless of the intent to distribute (ICMEC Progress Report, 2006, pp 1 & 2).

Even though not all countries have a clear legal standing concerning child pornography, a notable effort has been made in most of developed countries such as the United Kingdom, the United States and Canada. In the United Kingdom under Protection of Children Act and section 160 of the Criminal Justice Act, it is an offence for a person to take, make, distribute, show or possess an indecent photograph or pseudo photograph of someone under the age of 18. This Act further defines "Pseudo-photograph" as an image, whether made by computer-graphics or otherwise howsoever, which appears to be a photograph.

Section 163.1 of the Canadian Criminal Code also states that it is an offence for a person to produce, distribute and sell child pornography. The maximum sentence for such crime is ten years for its production and distribution, and five years for simple possession. Section 163.1 defines "child pornography" as visual representations of explicit sexual activity involving anyone under the age of 18 or depicted as being so; other visual representations of a sexual nature of persons under the age of 18; and written material or visual depictions that advocate or counsel illegal sexual activity involving persons under that age. Moreover Bill C-20 expanded the section 163 definition of "child pornography" provided in the Criminal Code to include written material and audio format whose, "the dominant characteristic of which is the description, for a sexual purpose, of sexual activity with a person under the age of eighteen years that would be an offence under this Act." Therefore, the accused person need not possess written material intended to "advocate or counsel" sexual activity with a person under 18 to be found guilty of a child pornography offence (Canadian Legal Information Institute, 2008, Bill C-20).

In the United States Criminal Code also states that it is an offence for a person to possess any visual depiction, including any photograph, film, video, picture, or computer-generated image or picture, whether made or produced by electronic, mechanical, or other means, of sexually explicit conduct, where the production of such visual depiction involves the use of a minor engaging in sexually explicit conduct; such visual depiction is, or appears to be, of a minor engaging in sexually explicit conduct; such visual depiction has been created, adapted, or modified to appear that an identifiable minor is engaging in sexually explicit conduct; or such visual depiction is advertised, promoted, presented, described, or distributed in such a manner that conveys the impression that the material is or contains a visual depiction of a minor engaging in sexually explicit conduct (Akdeniz, 2008, p. 121).

Mexican Federal Penal code made it an offence for anyone who seeks, compels, facilitates or induces, by whatever means persons under the age of 18 years or person who are unable to understand the meaning of the act or do not have the ability to resist, to perform sexual acts (real or simulated) or lewd exhibitionism or body for sex, with the intention of video recording, photographing, filming, displaying or describing through print ads, transmission of data files in public or private telecommunications, computer systems, electronic or substitutes shall incur a penalty of 7-12 year in prison and 800-2,000 fine. The same penalty is imposed on anyone who reproduces, stores, distributes, sells, buys, leases, exposes, advertises, transmits, imports or exports the materials referred to above (ICMEC Legislative Review 2006, Mexico section).

Although these legislations from various countries might have different interpretation of child pornography, they all have one thing in common which is to protect the rights of children and bring anyone in violation of their right to justice.

Impact of Child Pornography

While the main focus of this chapter is to address the electronic distribution of images which portray the illegal sexual interference of children, current detection and prevention methodologies and legal aspects of child pornography, it is very important to highlight the psychological and sociological impact of child pornography.

Some might argue that the distribution of illegal adult material is a victimless crime where nobody gets hurt and that what is done privately is the person's own business. However, in the case of child pornography, where a child is photographed or videotaped, there is always a victim. The distribution of the material repeats the victimization cycle over and over again even in the future. Similarly, in cases where the face of a child is superimposed on a sexually explicit photo and even though the child never participated in any sexual act, the distribution of such fabricated image may potentially produces a similar effect.

The two ways in which children may be harmed by child pornography include being exposed to child pornography or themselves being filmed or photographed. Children exposed to pornography are at risk of being desensitized and seduced into believing that such pornographic activity is "normal" for children. It can, potentially, provide some sort of conditioning that may adversely affect children's behavior ultimately resulting in learning experiences that links sex to exploitation, force, or violence (Check, 1995, pp. 89-91).

The effect on sexually abused or exploited children is often serious. For instance, children can experience numerous symptoms including emotional withdrawal, anti-social behavior, mood-swings, depression, fear, anxiety, physical symptoms and illnesses. In a study of children involved in sex rings that were sexually abused, 54.8% of those children were used in creating pornographic material. Furthermore, there was a correlation between involvement in pornography and a pattern of identification with the exploiter, along with other symptomatic and deviant behavior (Burgess, Hartman, McCausland & Powers, 1984, Effects of pornography on children section, para 3).

More disturbingly, sexually abused or exploited children may be at high risk of becoming abusers themselves. Worthy of note is the fact that there are those who refute the "cycle of abuse" theory because it is females who are more frequently sexually abused. Yet, in most societies, it is males that are responsible for the majority of child sexual abuse cases. Those who were photographed may take severe vengeful acts. For example, such acts may include burning the house where the pictures are located or stealing records of their exploitation. The media sometimes confuses child victims with perpetrators by publishing pictures of the children and blanking out the faces of offenders. Exploited children who may have enjoyed the attention or who were sexually stimulated bear a particular shame about their participation in pornography. Still, it must be emphasized that whether minors went along with sexual exploitation, enjoyed it, or profited from it, they remain the victims of a criminal and often a destructive act (Healy, 2004, section vii).

The impact of child pornography is not limited to children only; those who investigate such crimes are affected as well. Typically, once the evidence is gathered from the home, the computer is brought to a forensics lab, where analysts must pick through the images looking for evidence of child pornography, determine times of file transfers and verify whether the pictures are of underage children. Investigators carrying out such tasks go through individual and group counseling through their agencies to deal with the mental and emotional toll of viewing the videos, which often include sounds of children crying out in pain and protest (Scott, 2009).

Combating Child Pornography

The trafficking of child pornography through increasingly sophisticated electronic media, including Internet chat rooms, newsgroups, and peer-to-peer networks, has made such images more readily accessible. These technological advances have created more challenges for law enforcement, including requiring effective coordination to combat this crime.

A large amount of resources are needed to conduct child pornography criminals, and almost all are the result of intensive collaboration between law enforcement agencies using integrated systems and extensive communications. One such effective system developed by Microsoft Corporation and the Royal Canadian Mounted Police, allows registered law enforcement agencies to gather and analyze identification information of suspected pedophiles using sophisticated data sharing systems. The Child Exploitation Tracking System, or CETS, was officially launched in April of 2005. Previously, information detailed and gathered was inaccessible to other agencies who may have been tracking the same individuals. This issue was certain to cause a lower arrest rate and entirely ineffective against the wide outreach of the Internet. With the implementation of CETS, agencies across the world can access and update a database of information pertaining to particular persons of interest who may be involved in the production or distribution of child pornography. At present, every law enforcement agency within Canada is utilizing CETS to assist with their investigations (Toronto Police, 2006).

In the U.S., federal agencies have various mechanisms in place to coordinate their efforts to combat child pornography. These mechanisms include the following:

- Monthly meeting of officials from key federal agencies involved in combating child pornography
- Sharing of expertise among federal prosecutors and the consolidation of federal investigative efforts
- Task forces to coordinate federal investigative activities
- A national program to facilitate the sharing of images of exploited children among federal agencies working to combat child pornography
- The federally sponsored National Center for Missing and Exploited Children (NCMEC) to coordinate child pornography tips among federal law enforcement agencies

The 2002 GAO report showed this kind of coordination among several federal agencies led to an increase in the number of child pornography cases prosecuted from 428 to 692 in the years 1998 to 2002 (GAO, 2002, p. 6).

Although most national legislations do not require ISPs, telephone service providers or banking services to report the detection of pornographic sites involving children on their networks or to cooperate with the authorities in that regard, in the United States and Australia there are penalties for ISPs and domain hosts who do not report child pornography sites to the police within a reasonable period of time. The United States Federal Bureau of Investigation commenced *Operation Candyman* by monitoring the electronic mail and chat room interactions of suspected pedophiles. Within several months, agents had assembled and identified a database of thousands of individuals who were then tracked to their homes and workplaces using IP addressing data obtained from service providers (Kevin, 2002).

In South Africa ISPs must take whatever steps necessary to prevent its services from being used to host or distribute child pornography material. All ISPs must notify such activity to the police, as well as the abuser's particulars (name and Internet Protocol (IP) address); the ISP is also obliged to keep a record of that information for use as evidence in legal investigations and proceedings. In Thailand, ISPs are required by law to keep child pornography and data concerning the user of such sites stored for at least 90 days.

CHILD PORNOGRAPHY AND IT

How is Technology Helping Criminals?

The complexity of child pornography relates to the advancement of digital technology; the development of home video equipment and computer technology has revolutionized the international production and distribution of child pornography. Rapidly expanding global access to increasingly inexpensive technologies has transformed child pornography into a sophisticated underground industry. Computer alteration of images, the potential for creating computer generated pornography, encryption, image steganography, and IP spoofing all pose alarming challenges for courts and law enforcement officials throughout the world.

Furthermore, The Internet made the access of child pornography much easier. Anyone who wishes to access such material does not have to find someone in his/her own social circle who can provide such images or somewhere in their locality where such images are available. Instead one can go to online communities where other individuals from different parts of the world come together to create an environment where the use of child pornography and the abuse of children are considered acceptable and even normal.

There is no doubt that new technologies considerably increase the opportunities available to predators by allowing them to stalk, recruit and exploit children anywhere in the world. The United Nations Children's Fund (UNICEF) estimates that there are more than 4 million sites featuring victims who are young minors, including even children under 2. Predators can stalk new victims anonymously in chat rooms and blogs (Maalla, 2009, p. 6)

Interestingly, (Wolak, Mitchell, & Finkelhor 2006, pp 1-5) conducted studies suggest that the number of incidents of youth receiving sexual solicitations from predators has declined between 1999 and 2005. Whereas in 1999 the number of youth receiving sexual solicitation was 1 in 5, that number dropped to 1 in 7 in 2005. This drop can be attributed to youth being more vigilant online when it comes to interacting with strangers. For instance, in 1999, the percentage of youth who use the Internet that interacted with strangers was 40%. This percentage dropped to 34% in 2005.

As online child pornography communities grow, the need for collaboration among several governmental and nongovernmental agencies is essential. The commitment and efforts of many international communities, public authorities, nongovernmental organizations, and the private sector such as ISPs and credit card issuers is beneficial. Such collaborative effort leads to the implementation of measures such as: legislative reforms, dismantling of child pornography networks, reporting facilities for Internet users, filtering and blocking of Internet sites, seizures of pornographic material, arrests, awareness-raising and campaigns.

However, despite these initiatives, there is more and more child pornography on the Internet, becoming what is today a very profitable business with a worldwide market value estimated to be in the billions of dollars. Easy access to new technologies, constant changes in production methods and consumption patterns, in addition to the international dimension of child pornography, all complicate the fight against this plight in both the developed and developing countries. Child pornography continues to be a very topical and alarming issue.

Technology to Fight Child Pornography

Prevention of electronically transmitted illicit pornography involves complex technical measures in relation to detection of child pornography in still images and moving video. Some notable works which are promising have been conducted and others are still in their experimental stage. Of these works, there is none that offers a total solution as of yet; however, the progress that has been made cannot be discounted as each effort is certainly a crucial element to the overall solution.

Tools that are currently available can be divided into two main categories: host-based and network-based solutions. Host-based solutions are mainly installed on personal computers and servers to prevent a user from accessing unwanted material on the Internet. Similarly, network-based solutions reside on a network device such as routers to monitor and prevent the network user from accessing illegal content on the Internet. Although network-based solutions are more complex in implementation than host-based solutions, network-based solutions are also much more difficult to bypass.

Host-Based Solutions

Host-based filtering allows users to install an application which either merges with the operating system or with a primary Internet access application, such as a web browser, to provide protection. Examples of this type of filtering from commercially available products include NetNanny, SafeEyes, CYBERsitter and MaxProtect.

Most host-based solutions are mainly done for commercial purposes and they don't discriminate between "legal" and "illegal" pornography. Even though the perfect Internet filter does not exist in today's marketplace, there are a number of great solutions depending on individuals' need.

Almost all host-based filters are text-based that rely on filtering mechanism based on black-listed URL's and key words. Some have dynamic filtering capabilities and offer an easy-to-use design, making it possible for people with all levels of computer experience to easily install and use the filter to its fullest capacity.

There are also host-based systems that incorporate image analysis tools to filter unwanted images from host systems. One such system is WebGuard which intends to automatically detect and filter adult content from the Internet. WebGuard uses a crawler based system to extract relevant data, combines textual and image content, and the URL name of a site to construct a feature vector. To improve performance, an analysis using a skin colour pixel mode is used (Hammami, 2003, pp 13-57).

Another host-based solution with image analysis tools is BlueBear's Law Enforcement Against Child Exploitation (LACE) – initially deployed on trial basis in 2007. This technology was created to assist Law Enforcement agencies expedite the processing of seized hard drives for known child pornography images. The system uses descriptive elements, such as pixel counts, colour, brightness, size and other patterns to categorize images into predefined lists. It is also capable of detecting and extracting faces from media-evidence files by using automated facial identification to detect facial images of a victim and suspect in previously categorized cases. This technology facilitates and shortens the time taken by investigation officers to sift through hard drives and eliminate the need to repeatedly watch horrendous images (Investigation Software Press, 2008).

Whitehead has also implemented a method of classifying Internet objects using descriptor coefficients, such as name coefficient, text coefficient, image coefficient, audio coefficient, video coefficient, plug-in coefficient, and relational coefficient. The image data is analyzed to determine whether it contains adult content inside an Internet object using predefined skin tone ranges in Hue-Saturation-Value (HSV) colour space. The system divides the number of skin tone pixels within skin tone range in the Internet object image by the total number of pixels in the image and compares the resulting value against an automatically generated or user-defined threshold value. Those images resulting in a ratio less than the threshold proportion of skin pixels are discarded. Images that are equivalent or greater than the threshold proportion of skin pixels are then assigned a nudity coefficient which is equal to the percentage of skin pixels in the image. Factors like the size of the image and encoding type are used to weigh the nudity coefficient of each image in an object (Whitehead, 2001, pp 1-13).

Organizations like Interpol (the world's largest international police organization) use powerful image analysis tools to combat child pornography. This software uses sophisticated image analysis methods to identify criminals even if they get older or change their appearance. By the same token, it is a very useful tool when dealing with child victims because it allows investigators to note if they are looking at the same victim in different videos or images, or the same victim, but older than in the other videos or images (DW – World. DE, 2007).

Network-Based Solutions

The majority of network-based solutions operate by analyzing packet header and packet payload at the network layer. Even though network-based filters are mainly focused towards intrusion detection and prevention systems, or Quality of Service (QoS) features, there are also notable works done to detect and prevent the transmission of child pornography.

One of the network-based solutions towards the prevention of child pornography distribution is NetClean. NetClean blocks child abuse images and videos that are previously identified by law enforcement authorities. The system uses Illegal images from police databases and converts them into unique digital codes. These digital codes are then used to detect illegal images in the network with the aid of automatic image recognition technology (CRIPSYS Data Report, 2008).

Law enforcement agencies also put forward their effort to combat child pornography at the network layer. Wyoming's Internet crimes against children task force, has developed a software that identifies computers, by serial number, that trade child pornography on the Internet to locate the traffickers and obtain search warrants. One main advantage of this system is that it eliminates the use of IP addresses in their investigation. The problem associated with IP addresses is that they are dynamic and subject to frequent change, thus it is difficult to get a conclusive picture of the volume of individual trafficking. This software, used by 1,800 police investigators nationwide, is one of the most powerful new tools being used to combat a growing child porn industry (Koch, 2008).

Another network-based solution, still in research stage is developed by Ibrahim to analyze images based on the actual content of the image. Ibrahim developed a system that can actively monitor data traffic for any child pornography images by analyzing the content of the image. It uses Stochastic Learning Weak Estimators (SLWE) coupled with linear classifiers such as Weighted Euclidian Distance to classify images in two categories: child and non-child pornography images (Ibrahim, 2009, pp 29-52).

One good example is the work of the cyber-forensics lab in Louisiana located at the offices of the states' attorney general and state police. They have created software capable of monitoring child pornography images in real time. In this lab, special investigative units detect computers that are involved in the online exchange of sexually explicit images of children. Recently, the system identified more than 5600 such cases in the state. Despite the great capabilities of the software and the help it offers, the sad reality is that the software finds much more cases than what law enforcement officials are capable of arresting and prosecuting (Scott, 2009).

Besides detection and blocking, another important method of combating child pornography is reporting such incidents. One such tool is NCMEC's CyberTipline which was launched in March of 1998 providing a convenient way to report child sexual exploitation to authorities. There were more than 400,000 reported incidents to CyberTipline in year 2006 alone resulting in arrests and investigations. Furthermore, CyberTipline requires ISPs to report suspected crimes committed against children when such activities are encountered in the course of their services (Wolak, Mitchell, & Finkelhor, 2006, p. viii).

CONCLUSION

The sexual exploitation of children remains a very serious problem and is rapidly increasing globally through the use of technology and the Internet. The ubiquitous nature of the Internet these days, and the affordability and simplicity of digital devices allowed many individuals from different walks of life to access and distribute illegal material. Certain benign technologies have now been misused and abused, and turned into a tool facilitating child pornography.

Host-based and network-based solutions are two types of tools that are helpful in the battle against child pornography. Host-based solutions operate from the user's end by blocking access, to child pornographic resources from the server or the personal computer. Network-based solutions, in the other hand, operate from network device, such as routers, to monitoring and preventing the network's access of such materials.

Besides technological obstacles, there exist legal ones as well. Legal obstacles stem from the fact that some countries do not have a criminal code pertaining to child pornography. There is a pressing need for countries and international organizations to come to a consensus and unify laws to help eradicate child pornography. Furthermore, and as important as technical and legal means, is the need to educate children, parents and everyone involved in this issue. In particular, teens and preteens should be educated on the importance of online vigilance. Such preventive and proactive measures are as effective as the reactive ones.

REFERENCES

Akdeniz, Y. (2008). *Internet child pornography and the law: national and international responses*. Hampshire, England: Ashgate Publishing Limited.

Arellano, N. E. (2008). Winnipeg police combat online child porn with Image matching software. *IT Business online news,* Retrieved July, 13, 2009, from http://www.itbusiness.ca/it/client/en/ Home/News.asp?id=49700&bSearch=True

Burgess, A. W., Hartman, C. R., McCausland, M. P., & Powers, P. (1984, May). Response patterns in children and adolescents exploited through sex rings and pornography. *The American Journal of Psychiatry*, *141*(5).

Canadian Legal Information Institute. (2008). Criminal code of Canada – Bill C-46, *Statues and Regulations of Canada,* R.S.C. 1985, C-46. Retrieved [Aug., 4, 2009] from http://www.canlii.org/en/ca/laws/ stat/rsc-1985-c-c-46/latest/rsc-1985-c-c-46.html.

Check, J. (1995). Teenage training: The effects of pornography on adolescent males. In Lederer, L., & Delgado, R. (Eds.), *The Price We Pay: The Case Against Racist Speech, Hate Propaganda and Pornography* (pp. 89–91). New York: Hill and Wang.

Child Rights Information Network. (1990). *United Nations Convention on the Rights of the Child.* Retrieved August 4, 2009, from http://www.crin.org/docs/resources /treaties/uncrc.asp

Creighton, S. J. (2003). *Child pornography: Images of the abuse of children.* NSPCC Research Department, NSPCC Information Briefings. Retrieved September 13, 2009, from http://wallis.kezenfogva.iif.hu/eu_ konyvtar/projektek/Daphne/www/documents/ projets/2003-017/ChildPornography.pdf

Data Report, CRYPSYS. (2008). Unique software provides child pornography protection: CRYPSYS Data Security is launching NetClean software to prevent the spread of child pornography in the workplace. *My News Desk*, Retrieved August 12, 2009, from www.mynewsdesk.com/se/pressroom/netclean_technologies/pressrelease/download/resource_ attached_pdf_document/215913

DW – World. DE (2007). combating child porn conference focuses on new technology. *Deutsche Welle's online service*. Retrieved August 12, 2009, from http://www.dwworld.de/dw/ article/0,2790025,00.html GAO report (2002). *Combating child pornography: US Government Accountability Office (GAO) Report to the ranking minority member*, Committee on Government Reform, House of Representatives. Washington, DC: United States General Accounting Office.

Government Gazette. (2004). *Republic of South Africa Deeds Films and Publications Amendment Act*. Retrieved September 12, 2009, from http://www.info.gov.za/view/ DownloadFileAction?id=67955

Hammami, M., Chahir, Y., & Chen, L. (2003). WebGuard: Web based adult content detection and filtering system. *IEEE/WIC International Conference on Web Intelligence*, 2, 574-578, 13-17.

Healy, M. (2004). Child pornography: An international perspective. *Computer Crime Research Centre*. Retrieved August 4, 2009, from http://www.crime-research.org/ articles/536/4

Ibrahim, A. (2009). Detecting and preventing the electronic transmission of illegal images. *Unpublished Master's thesis*. Faculty of Engineering and Applied Science, University of Ontario Institute of Technology, Oshawa, Canada.

Internet Watch Foundation. (2009). *Protection of children Act 1978 (England & Wales)*. Retrieved September 12, 2009, from http://www.iwf.org.uk/police/page.22.36.htm

Investigation Software Press. (2008). BlueBear Law Enforcement Services Helps Reduce Your Child Exploitation Investigation Time by Over 85%, *PoliceOne resource for Police and Law Enforcement*. Retrieved September 5, 2009, from http://www.policeone.com/ police-products/ investigation/Investigative-Software/ press-releases/1708908-BlueBear-Law-Enforcement-Services-Helps-Reduce-Your- Child-Exploitation-Investigation-Time-by-Over-85/

Kevin, J. (2002) FBI Arrests 40 in Child Porn Sting, *US Today*. Retrieved September 3, 2009, from http://www.usatoday.com/tech/news/2002/03/18/net-porn.htm

Koch, W. (2008). Software tracks child porn traffickers online. *USA Today,* Retrieved August 12, 2009, from http://www.usatoday.com/news/nation/ 2008-04-15-childporn-side_N.htm

Legislative Review, I. C. M. E. C. (2006). *Legislation across the world on child pornography.* International Centre for Missing & Exploited Children. Retrieved September 5, 2009, from http://www.internationalresourcecentre.org/missingkids/servlet/PageServlet?Language Country=en_X2&PageId=3428

Maalla N. (2009). Promotion and protection of all human rights, civil, political, economic, social and cultural rights, including the right to development: Report of the special rapporteur on the sale of children, child prostitution and child pornography. *United Nations Human Rights Council, Twelve Session, Agenda item 3.*

Meek-Prieto, M. (2008). *Just age playing around? How second life aids and abets child pornography. North Carolina Journal of Law & Technology, 9* (88th ed.). NC: JOLT Online.

Mykytyn, J. (2004). *The Unknown Crisis: Child Pornography on the Internet Anti Child Porn Organization*. Retrieved September 15, 2009, from http://www.antichildporn.org/ whiteppr.htm

Progress Report, ICMEC. (2006). *Child pornography not a crime in most countries*. International Centre for Missing & Exploited Children. Retrieved September 5, 2009, from http://www.icmec.org/en_X1/pdf/ SummerNewsletter-2006formatted.pdf

Scott, R. T. (2009). *Child pornography gets new weapons, The Times-Picayune*. Retrieved February 4, 2010, from http://www.nola.com/crime/index.ssf/2009/11/post_72.html

Toronto Police. (2006). Child Exploitation Tracking System (CETS) fact sheet. *The Toronto Police,* Retrieved August 22, 2009, from http://www.torontopolice.on.ca/ media/text/20060831-cetsfactsheet.pdf

Whitehead A. D & Ryan M. P. (2001). Method and device for classifying internet objects and objects stored on computer-readable media, *Patent Document No: 7383282*, Application No: 09/978,182, Application Date: Oct. 17, 2001.

Wolak, J., Mitchell, K., & Finkelhor, D. (2006). *Online victimization of youth: Five years later*. The Crimes Against Children Research Center, University of New Hampshire. Retrieved January 10, 2010, from http://www.missingkids.com/en_US/publications/NC167.pdf

Wortley, R., & Smallbone, S. (2006). Child Pornography on the Internet. *U.S. Department of Justice Office of Community Oriented Policing Services*, Retrieved September 10, 2009, from http://www.cops.usdoj.gov/files/ric/ Publications/e04062000.pdf

ADDITIONAL READING

Bagley, C. (1991). The long-term psychological effects of child sexual abuse: A review of some British and Canadian studies of victims and their families. *Annals of Sex Research*, *4*, 23–48. doi:10.1007/BF00850138

Beitchman, J. H., Zucker, K. J., Hood, J. E., daCosta, G. A., & Akman, D. (1991). A review of the short-term effects of child sexual abuse. *Child Abuse & Neglect*, *15*(4), 537–556. doi:10.1016/0145-2134(91)90038-F

Brassard, M. R., & Germain, R. (1987). *Psychological maltreatment of children and youth*. New York: Pergamon.

Burgess, A. W., & Clark, M. L. (1984). *Child pornography & sex rings*. New York: The Free press.

Flowers, R. B. (1994). *The victimization and exploitation of women and children: A study of physical, mental, and sexual maltreatment in the United States*. New York: McFarland & Company.

O'Brien, S. (1992). *Child pornography*. Dubuque, IA: KendallHunt Publishing Company.

KEY TERMS AND DEFINITIONS

Child Pornography: Usually refers to images or films depicting sexually explicit activities involving children.

Virtual Child Pornography: A form of child pornography without the use of real children; a computer-generated image (usually cartoon images) depicting children in a sexual act.

Pseudo-Photograph: An image produced manually which is indistinguishable from a real photograph produced using a camera.

Zoophilia: Erotic attraction to or sexual contact with animals.

IP Spoofing: The creation of IP packets with a forged source IP address with the purpose of concealing the identity of the sender or impersonating another computing system.

Image Stenography: The art concealing images in such a way that no one, apart from the sender and intended recipient, suspects the existence of the image. Criminal Code: a compilation of government laws that outline a nation's laws regarding criminal offenses, and the punishments that courts can impose upon offenders when such crimes are committed. Peer-to-peer Network: distributed network architecture is composed of computer systems that make a portion of their resources directly available to other network systems, without the need for central coordination instances.

Section 2
Technological Approaches for Hate Crimes

Chapter 3
Adult Pornography, Male Peer Support, and Violence against Women:
The Contribution of the "Dark Side" of the Internet

Walter S. DeKeseredy
University of Ontario Institute of Technology, Canada

Patrik Olsson
University of Ontario Institute of Technology, Canada

ABSTRACT

It is estimated that there are over a million pornographic sites on the Internet, with as many as 10,000 added every week. In addition to having a major financial impact, adult pornography is strongly associated with various types of violence against women, especially sexual assault. Some studies have found that the contribution of pornography to woman abuse in dating, marriage, and during or after separation/divorce is related to male peer support, which refers to the attachments to male peers and the resources they provide that perpetuate and legitimate woman abuse. The main objective of this chapter is twofold: (1) to review the extant social scientific research on the relationship between violence against women, male peer support, and adult Internet pornography and (2) to suggest new directions in empirical work on the association between these three social problems.

INTRODUCTION

As communications scholar Joseph Walther and his colleagues observed (2001) nearly 10 years ago, "With the expansion of the Internet and new communication technologies, we are witnessing the diffusion of high-end, high-bandwidth multimedia technology for a wide range of people. It is common for many computer-mediated communication (CMC) users to create multi-media World Wide Web sites with graphics and pictures" (p. 105). This statement is still relevant today. Certainly, many such sites are beneficial to corporate executives, small business owners, educators, students, and to a myriad of other people eager

DOI: 10.4018/978-1-60960-094-5.ch003

to enhance their understanding of social, political, cultural, and economic factors that directly or indirectly influence their lives. However, there are also numerous highly injurious features of new information technologies and adult pornography is one major example.

Today, we live in what Jensen (2007) refers to as a "post-*Playboy* world," where defining adult pornography is still subject to much debate. Those who produce adult pornography, consume it, and/or oppose prohibiting it typically refer to harmful, sexually explicit material as erotica. However, there is a big difference between erotica and adult pornography. As Russell (1993) observes, erotica refers to "sexually suggestive or arousing material that is free of sexism, racism, and homophobia and is respectful of all human being and animals betrayed" (p. 3). On the other hand, in adult pornography:

Women are represented as passive and as slavishly dependent upon men. The role of female characters is limited to the provision of sexual services to men. To the extent that women's sexual pleasure is represented at all, it is subordinated to that of men and is never an end itself as is the sexual pleasure of men. What pleases men is the use of their bodies to satisfy male desires. While the sexual objectification of women is common to all pornography, in which women characters are killed, tortured, gang-raped, mutilated, bound, and otherwise abused, as a means of providing sexual stimulation or pleasure to the male characters (Longino, 1980, p. 42).

Although many women consume adult pornography, it is created primarily for generating sexual arousal in heterosexual men (Jensen, 2007). From the standpoint of many feminist scholars (e.g., DeKeseredy, 2009a; Dworkin, 1994),[1] pornography, regardless of whether it appears on the Internet, in stores, on television, in literature, or in other media, is also a variant of hate-motivated violence and it, too, has become "normalized" or "mainstreamed" in North America and elsewhere (Jensen & Dines, 1998), despite becoming increasingly more violent and racist (DeKeseredy, 2009).

While it is beyond the scope of this chapter to graphically describe what appears on contemporary pornographic Internet sites, some brief examples of such violence and racism are necessary. For instance, Doghouse Digital is a company that produced the film *Black Bros and White Ho's*, which offers stereotypical images of "the sexually primitive black male stud" (Jensen, 2007, p. 66). Another example is the interracial film *Blacks on Blondes*, which features a white man in a cage watching black men have sex with his wife (Dines, 2006). An additional common feature of new pornographic films that exist online and elsewhere is painful anal penetration, as well as men slapping women and/or pulling their hair while they penetrate them orally, vaginally, and/or anally (Dines & Jensen, 2008a).

Perhaps "normalized" is an understatement. Pornography is a giant industry and it is estimated that there are over a million pornography sites on the Internet, with as many as 10,000 added every week (Funk, 2006). Note, too, that worldwide pornography revenues from a variety of sources (e.g., Internet, hotel rooms, etc.) recently topped US$97 billion. This is more than the revenues of these world renowned technology companies combined: Microsoft, Google, Amazon, eBay, Yahoo!, Apple, Netflix and Earthlink (Zerbisias, 2008). Another key point to consider is that rare are men who are not exposed to pornographic images and narratives on the Internet (Schwartz & DeKeseredy, 1997). Even if people go out of their way to avoid pornography, it frequently "pops up" on people's monitors while they are working or "surfing the web" for information that has nothing to do with sex (Dines & Jensen, 2008).

To make matters worse, as noted above, what men and boys watch on adult pornographic Internet sites are not simply "dirty pictures that have little impact on anyone." Rather, the images typically endorse "women as second-class citizens" and "require that women be seen as second-class citizens" (Funk, 2006, p. 165). Another challenge to the assertion that "pornography is just fantasy" are quantitative and qualitative data showing that pornography is strongly associated with various types of violence against women (DeKeseredy, 2010; Jensen, 2007), especially sexual assault. In addition, some studies found that the contribution of pornography to woman abuse in dating, marriage, and during or after separation/divorce is related to male peer support (DeKeseredy & Schwartz, 2009), which refers to the attachments to male peers and the resources these men provide that perpetuate and legitimate male-to-female abusive behaviors (DeKeseredy, 1990). For example, many violent, patriarchal men often view pornography in all-male groups and share videos and other media electronically with a "wider circle of friends" via the Internet (DeKeseredy, Schwartz, Fagen, & Hall, 2006; Giordano, 1995).

The main objective of this chapter, then, is twofold: (1) to review the extant social scientific research on the relationship between violence against women, male peer support, and Internet pornography and (2) to suggest new directions in empirical work on the association between these three social problems. Like others who study adult pornography (e.g., Dines & Jensen, 2008a), although we contend that it is strongly correlated with woman abuse, we do not claim that it is the sole or primary cause of male-to-female beatings, sexual assaults, and the like.

BACKGROUND

In Canada, the U.S., Europe, and in other parts of the world, adult pornography became increasingly available in the 1960s and 1970s (Bergen & Bogle, 2000). As time went by, even more people, especially men and boys, could easily access and consume it. For example, in 1995, approximately 43% of U.S. males were exposed to one of seven of the most widely read pornography magazines (Russell, 1998). Obviously, this figure excludes men who viewed, read, or listened to "cyberporn." At the time of writing this chapter, every second, 28,258 Internet users view pornography (Zerbisias, 2008), and the vast majority of them are men (Jensen, 2007). Like the above 1995 statistic, this figure is an underestimate. Whether or not researchers ever obtain an accurate estimate of the percentage of males who consume adult cyberporn, most leading experts in the field agree with Robert Jensen's contention that, "It's become almost as common as comic books were for you and me" (cited in Gillespie, 2008, p. A3).

Consider, too, that one study conducted in Alberta found that one in three boys aged 13 to 14 accessed sexually explicit media content on digital or satellite television, video and DVD and the Internet (Betowski, 2007). These are not innocent users who accidentally come across images, voices, and texts. Nor are they constantly bombarded with such material. Rather they choose to consume and distribute pornography, and unfortunately, some of the consumers will commit criminal acts, including violently attacking intimate female partners (DeKeseredy & Schwartz, 2009).

Over the past 30 years, research into pornography and its potential negative effects on men's behavior has expanded dramatically. The bulk of the studies were psychological in nature and focused on stranger rape or paper-and-pencil measures of attitudes toward women, attitudes toward rapists, and attitudes toward sexual behavior generally. Although there has been a confusing array of mixed findings, most of the research has shown little relationship between erotic nonviolent imagery and behavioral tendencies.

What is disturbing is that a number of psychological laboratory studies show that exposure to more graphic and violent sexual images has changed people's attitudes toward women and rape, as well as increasing male participants' aggressive behavior in the laboratory (Bergen

& Bogle, 2000; Boeringer, 1994; Donnerstein, 1984; Krafka, Linz, Donnerstein, and Penrod, 1997; Russell, 1993). Still, to this day, however, there is as yet relatively little information on the extent to which graphic sexual imagery influences men's sexually aggressive behavior outside the lab setting. Most of the evidence comes to us from artificial lab studies, leading critics to claim that there is little to support any connections with "real world" behavior (Berger, Searles, & Cottle, 1991; Brannigan & Goldenberg, 1987; Schwartz & DeKeseredy, 1998). Certainly, many of the lab scientists who conducted the best known studies have said that politicians have gone too far in suggesting that experimental lab studies are equivalent to the same effects taking place in society (Segal, 1993).

Indeed, there have been some attempts at imaginative alternative methodology. Jensen (1995, 1996), for example, has used personal histories and narrative accounts of men who used pornography. Other researchers have conducted surveys of women to determine how male consumption of pornography affected them. For example, of the 1,638 women who participated in the Canadian National Survey on Woman Abuse in University/College Dating (CNS), 137 (8.4%) stated that they were upset by their dating partners trying to get them to do what they had seen in pornographic media (DeKeseredy & Schwartz, 1998a). This is very similar to the 10% figure that Russell (1990) uncovered from asking a random sample of 930 women in the San Francisco area a similar question. It should be pointed out that, for the most part, Russell interviewed women significantly older than the females who participated in the CNS.

What is more important here is that the CNS found a significant relationship between being upset by men's attempts to imitate pornographic scenes and sexual victimization. Of those who were sexually abused, 22.3% had also been upset by attempts to get them to imitate pornographic scenarios. Only 5.8% of the women who were not victimized reported not being upset by pornography. The relationship also holds for physical violence. Of the female CNS respondents who reported being physically abused in a dating relationship, 15.4% also reported being upset by pornography. Only 4.5% of those who were not physically victimized reported being upset. The CNS data presented here, then, help us to tentatively conclude that pornography plays a major role in the sexual and physical abuse of Canadian women in college and university dating relationships. These findings mirror the abuse reported in surveys by married and formerly married women (Bergen, 1996; Bergen & Bogle, 2000; Harmon & Check, 1989).

More recently, DeKeseredy and Schwartz's (2009) qualitative study of separation/divorce sexual assault in rural Ohio also found a strong relationship between men's consumption of pornography and woman abuse. Sixty-five percent of the male estranged partners of the 43 women interviewed viewed pornography, and 30 percent of the interviewees reported that pornography was involved in sexually abusive events they experienced. This figure supports Jensen's (2007) finding that some violent men use "pornography as a training manual for abuse" (Bergen & Bogle, 2000, p. 231). Thus, some of the men who abused DeKeseredy and Schwartz's (2009) respondents may be graduates of what Bancroft (2002) refers to as "the pornography School of Sexuality" (p. 185). Nevertheless, given that the consumption of pornography is often a secretive event, it is possible that many, if not most of the women in DeKeseredy and Schwartz's sample who stated that their ex-partners did not view pornography were probably unaware of these men's use of pornography. The same can be said about many women who participated in some other studies reviewed here, such as the CNS (Bergen & Bogle, 2000).

Despite a growing body of research, it is still unclear whether pornography of any sort directly causes woman abuse. A long-term and expensive longitudinal design is required to determine whether such a relationship exists (Schwartz & DeKeseredy, 1998). Note, too, that generally, there are some important competing arguments. For example, for men who physically and sexually abuse women, pornography may well be just one more weapon in their arsenal. Hence, a man who cares that his partner would be scared or angry might not expose her to the lessons he learned from a pornographic move, while his abusive friend might try to force his intimate female partner to act out such scenes over her objections (DeKeseredy & Schwartz, 2009). In a somewhat related argument, the same factors that cause a man to abuse women may well also cause him to purchase pornography. In other words, the woman abuse came first, followed by his interest in pornography. In these scenarios, eliminating pornography might not have an effect on the amount of woman abuse, since the men are generally abusive anyway (DeKeseredy & Schwartz, 1998a). However, there is ample empirical evidence that pornography certainly is a component of the problem of woman abuse (Bergen & Bogle, 2000).

Although male consumption of adult pornography is linked to woman abuse, at this point in time, it is unclear whether which type of media plays a stronger role in male-to-female physical and sexual violence. In other words, are men who view, listen to, or read pornography and who abuse women more likely to seek such material online or elsewhere? What Ferguson (1996) stated 14 years ago is still relevant today: "Pornography (both criminal and non-criminal) does exist in Cyberspace, and can be accessed and downloaded by virtually anyone with the appropriate knowledge regarding how to go about obtaining it. In addition sexual victimization can and has occurred" (p. 27). Still, until conclusive evidence is generated and corroborated, it is wrong to conclude that male pornography users who abuse women primarily consume Internet pornography. Nevertheless, according to Dines and Jensen (2008b), "While there are few studies on the effects of Internet pornography, past research suggests that this increase in pornography consumption is implicated in greater levels of male physical and sexual violence against women and children" (p. 366).

THE RELATIONSHIP AMONG PORNOGRAPHY ON THE INTERNET, MALE PEER SUPPORT, AND THE ABUSE OF WOMEN

Issues, Controversies, Problems

A few studies have found the contribution of pornography to woman abuse in intimate heterosexual relationships is related to male peer support (DeKeseredy & Schwartz, 1998b; Schwartz & DeKeseredy, 1998). Further, some men learn to sexually objectify women through their exposure to pornographic media (Funk, 2006; Jensen, 1995), and they often learn these lessons in groups, such as pornographic film showings at university fraternity houses (DeKeseredy & Schwartz, 2009; Sanday, 1990).

Such "strengthening" of male "misogynist bonds" is not a recent phenomenon (Lehman, 2006). As film scholars have documented, cinematic pornography originated in 16 mm silent films, which were:

"usually shown in private all-male 'smokers' in such contexts as bachelor parties and the like. Within such a context, the men laughed and joked and talked among themselves while watching the sexually explicit films about women, who though absent from the audience, were the likely butt of the jokes, laughing, and rude remarks" (Lehman, 2006, p. 4).

Similarly, a study done about 20 years ago uncovered that university fraternity "brothers" also generally went to pornographic theaters in groups. For example, some of the brothers interviewed by Sanday (1990) stated that:

seeing pornography is something to do before their parties start. They want to learn what it's like to "have a two foot dick" and to have a good time together. They never go alone, always together. They go together in order to have a good time, laugh, and make jokes during the movie. They dissociate themselves from the men who go alone to porno movies downtown and sit in seats "with coats and newspapers spread out over their laps" and "jerk off" during the movie. They believe that this is sick, but they don't think "getting off" while reading Playboy privately or enacting a porno fantasy in their house is necessarily sick (p. 129).

Again, we live in a post-*Playboy* world (Jensen, 2007), and increasingly, pornography has moved from theaters to people's homes (Jordan, 2006), including those owned by fraternities. For example, in 2000, a man referred to by DePauw University fraternity brothers as "The Smut Peddler" reported that he used to sell about 100 VHS pornographic tapes to the fraternities per year, but after 15 years in the "business," he witnessed a dramatic decline in sales due to Internet viewing (Claus, 2000). Ten years later, it is logical to assume that this man has probably pursued another "career" due, for the most part, to Internet technology, which provides fraternity brothers easy and constant accessibility that cannot be offered by "The Smut Peddler" (Dines & Jensen, 2008b).

Many people assume that little crime occurs in North American rural communities, an assumption promoted by the media, lay conversations, and even criminological research, which mainly focuses on urban lawbreaking (Donnermeyer, Jobes, & Barclay, 2006). According to newspaper reporter Theresa Boyle (2007), "After all, conventional wisdom holds that the big, bad city is the root of all evil. Small towns are supposed to be peaceful and serene" (p. A19). However, for many rural women, nothing can be further from the truth. There is much woman abuse in rural areas and recent studies show that it is fueled, in part, by male peer support and pornography (DeKeseredy, Schwartz, Fagen, & Hall, 2006). For example, some rural Ohio survivors of separation and divorce sexual assault told DeKeseredy and Schwartz (2009) that their partners consumed pornography with their male friends while drinking excessive amounts of alcohol. One interviewee who experienced this problem described this episode:

They were drinking and carrying on and they had, um, they had a bunch of porno stuff in the garage, and I had walked in and I had started to tear it up. And I was, I was, I thought it was gross. I was mad at it. I was mad at him for being around it. And he just started charging after me, and I started running to my car as fast as I could. And he got into the car and he threw me down in the seat and he just kept punching me, punching me (p. 74).

Regardless of whether they consumed it in groups, 65% of the estranged partners of the 43 women in DeKeseredy and Schwartz's (2009) study viewed pornography, and 30% of the interviewees reported that pornography was involved in sexually abusive events they experienced. Further, there is now evidence suggesting that rural boys consume pornography more than do their urban counterparts, at least in the Canadian province of Alberta (Betowski, 2007). As Scott Bergthold, a U.S. lawyer who helps small towns fight "adult business," told *Los Angeles Times* reporter Stephanie Simon (2004), "Rural communities never thought they'd have to deal with what they perceived to be a big-city problem" (p. 2). Obviously, things have changed, as "hard-core porn" has now "hit the heartland." For example, the Lion's Den chain now has an "adult superstore" in Quaker City, Ohio, population 563 (DeKeseredy & Schwartz, 2009).

Some men, however, abuse female intimates and consume pornography, but do not see it or read it in groups gathered at one particular place. Moreover, they may not directly interact with abusive or sexist peers on a face-to-face basis (DeKeseredy & Schwartz, 1998b). This is not to say that they are not influenced to consume pornography and/or victimize women by male peers. Moreover new cyberspace technology enables men to engage in the on-line victimization of women. This involves men "virtually assaulting," "virtually raping," or "cyberstaking" women who use the Internet (Kendall, 2003; Tucker, Fraser, & Shulruff, 2008). Sadly, there are Internet rape sites (Gossett & Byrne, 2002), and female "virtual victims" experience considerable psychological pain and suffering, as if they had actually been physically assaulted (Ferguson, 1996).

We are also witnessing the emergence of pro-abuse cyberspace male peer support groups (DeKeseredy & Schwartz, 1998b; Kendall, 2003). Although the precise number is unknown, some research shows that many men, most of whom probably never had face-to-face contact with each other, share pornographic material with other men through the Internet (Doring, 2009). Further, there is evidence suggesting that an undetermined number of these consumers and distributors are "part of a broader subculture of sexual deviance that legitimizes various forms of deviant sexuality" (Stack, Wasserman, & Kern, 2004, p. 85). And, some members of this subculture commit deadly violent acts. Consider the following case discussed by Ferguson (1996):

In late October 1996, the body of a woman from Hampton, Maryland, was pulled from a shallow grave outside the trailer of her lover in Lenoir, North Carolina. The incident occurred shortly after the victim had traveled to Lenoir to meet her lover for the first time following an anonymous E-Mail liaison conducted using the pseudonyms "Nancy" and "Slowhand." During this liaison, the two had constructed and participated in several Cyber-sexual scenarios involving sado-masochistic practices, torture and snuff. The victim's lover was charged with her murder, but claims that her death was an accident that occurred while the two were living out the sexual fantasies conceived during their E-Mail liaison (p. 17).

Male peer support theorists DeKeseredy and Schwartz (1998b) contend that the sharing of cyberporn helps create and maintain sexist male peer groups. Further, this sharing reinforces attitudes that reproduce and reconstitute ideologies of male dominance by approvingly presenting women as objects to be conquered and consumed. Such sharing also makes it difficult for users to separate sexual fantasy from reality and assists them in their attempts to initiate female victims and break down their resistance to sexual acts (Dines & Jensen, 2008a). Future research needs to identify the reasons why men join and participate in male subcultures that use adult pornography and how they locate those who are like minded.

It is important to note that there are competing explanations for why people consume Internet pornography, including those offered by psychologists. One such perspective contends that antisocial personality characteristics motivate some people to seek out Internet pornography. According to this account:

The "goodness of fit" of antisocial personality characteristics with antisocial sexual content will, it is speculated, promote a tremendous depth of involvement in antisocial sexual stimuli. Individuals may lose awareness of the constraints of reality regarding enactment of antisocial sexual behavior, and uniquely strong negative effects of antisocial sexual content on the Internet may be seen among those predisposed to access such material (Fisher & Barak, 2001, p. 312).

A related theory is that normal-range people will avoid antisocial, sexually explicit material and will reject such media's messages if encountered (Fisher & Barak, 2001). Psychological theories, such as those briefly described here, may be popular among the general population, but they are not as popular among social scientists today.

Moreover, psychological perspectives that point to some type of psychopathology or mental illness raise some serious questions and criticisms, some of which are described in the next section.

FUTURE RESEARCH DIRECTIONS

Again, there are only a few social scientific studies of the effects cyberporn (Dines & Jensen, 2008b), and most of the research done so far is descriptive and psychological (Stack et al., 2004). Indeed, what is missing from the bulk of the empirical work done so far is what the late C. Wright Mills (1959) refers to as a "sociological imagination." This is hardly a trivial concern, given that there is no reliable evidence linking personality disorders, biological factors, or alcohol/drug abuse to cyberporn use (Stack et al., 2004). Similarly, a large literature shows that most men who abuse female intimates are "less pathological than expected" (Gondolf, 1999, p. 1), with only 10 percent of all incidents of intimate violence resulting from mental disorders (DeKeseredy & Schwartz, 2009; Gelles & Straus, 1988).

If only a handful of men used cyberporn or beat, hit, raped, and killed their intimate female partners, it would be easy to accept non-sociological accounts of their behavior. The reality is, though, that violence against women and cyberporn use are deeply entrenched in our society, and so is patriarchal male peer support. For example, in North America, annually, at least 11% of women in marital/cohabiting relationships are physically abused by their male partners (DeKeseredy, 2009b), and contrary to popular belief, sexual assaults on female college/university students are commonplace. Large- and small-scale surveys consistently show that 25% of North American female undergraduates experience some variation of sexual assault every year (DeKeseredy & Flack, 2007).

Also note that annually there are 72 million visitors to Internet pornography sites and 68 million (25%) of total search engine requests are for pornographic materials (Dines & Jensen, 2008b). Furthermore, there is ample quantitative and qualitative evidence available today to support what Lee Bowker (1983) said more than 25 years ago about all-male patriarchal subcultures of violence:[2]

This is not a subculture that is confined to a single class, religion, occupational grouping, or race. It is spread throughout all parts of society. Men are socialized by other subculture members to accept common definitions of the situation, norms, values, and beliefs about male dominance and the necessity of keeping their wives in line. These violence-supporting social relations may occur at any time and in any place.

Therefore, sociologists ask, given the widespread nature of woman abuse, cyberporn, and male peer support, how can people effectively claim that these problems are committed by "sick" or pathological individuals? Even if this were the case, one would have to spend a great deal of time looking at the social structure of a country that produces more sick or pathological individuals than many other countries (DeKeseredy, in press; DeKeseredy & Schwartz, 1996). Since North America has alarmingly high rates of the three major social problems of central concern to this chapter, individualistic perspectives have little to offer.

Of course, it is incorrect to completely reject individualistic explanations. These points of view, to a certain extent, do help us make sense of criminal acts committed by *some* people. Only an ignorant person would argue that there are not some people who have biological or psychological problems that are factors in their decision to assault women or view degrading, violent pornography. Further, some people have been stopped from committing future crimes through the use of therapy, psychotropic drugs, and other psychologically and biologically informed treatments (DeKeseredy, in press).

Following Mills (1959), the sociological imagination involves attempting to understanding how *personal troubles* are related to *public issues*. Personal troubles are just what you might think. If you are raped, robbed, beaten, or cheated, you have a problem and you have to deal with it. You may need medical attention, comfort from friends or family, financial help, or any of a number of other forms of aid. Sometimes, however, many people are suffering individually from the exact same personal problem at the same time. If a hundred women are raped in one year on university campus, each one of these women has a personal problem, or personal troubles. At the same time, though, something about the broader structural and cultural forces, such as patriarchy or capitalism, Mills would argue, allows for so very many women to be victimized. To be able to look beyond the personal troubles of one or two female students who have been sexually assaulted and see the broader problem of rape on campus and its causes is to possess the sociological imagination.

Consider men who beat their wives or common-law partners. At first glance, a man who assaults the woman he shares an intimate relationship with apparently must be either suffering from life events stress or be mentally ill. Perhaps that seems an adequate explanation for the two or three cases that some readers know well. However, when we look at the 11% or so of women in Canadian marital/cohabiting relationships who are physically abused annually by their male partners, you begin to find "an indication of a structural issue having to do with the institutions of marriage and the family and other institutions that bear upon them" (Mills, 1959, p. 9).

Obviously, more research, regardless of whether it psychological or sociological, on cyberporn's effects on sexual and physical violence is needed (Stack et al., 2004). Empirical work on other negative consequences of cyberporn is also necessary because it is "seamlessly" flowing into our daily lives and is probably causing much damage that we are thus far unaware of. Research done so far on the negative effects reveals an increase in consumer addiction, and studies show that roughly 20% of male cyberporn users admit experiencing negative financial, legal, work-related, relationship, and personal problems related to their on-line activities (Dines & Jensen, 2008b).

Longitudinal research is needed to accurately determine the long-term consequences of Internet pornography use and membership in sexist male peer groups. Many questions are difficult, if not impossible to answer with anonymous, self-report questionnaires given one time, in one place, to one group of men. This "one shot" approach is excellent for documenting the existence of pro-abuse male peer groups and their use of cyberporn; however, it tells us little, if anything about the life course trajectory of people who consume Internet pornography and who are involved in sexist male homo-social networks (DeKeseredy & Schwartz, 1998b).

Over the past 36 years, there have been hundreds of North American studies of the physical, psychological, and sexual abuse of women by their current or former male partners (Brownridge & Halli, 2001; DeKeseredy, in press). Hence, the social scientific community has developed a wealth of knowledge about variations in woman abuse across regions, socioeconomic groups, and racial/ethnic categories. However, to the best of our knowledge, there has only been one sociological study that attempted to discern why some groups are more likely to use cyberporn than are others (Stack et al., 2004). Are we more likely, for example, to find male cyberporn subcultures in eastern or western parts of Canada? Are white men more or less likely than African-American, Aboriginal, or Hispanic males to join and be influenced by pro-abuse male subcultures that distribute and consume cyberporn? These are important empirical questions that can only be answered empirically. Hopefully, researchers will respond to our call to do so in the near future. Social scientific research on cyberporn is in its infancy and much more empirical and theoretical work on a myriad of issues related to this social problem needs to be done.

CONCLUSION

As previously stated, the organic growth of the dark side of the Internet has consequently globalized access to violent and degrading pornographic materials on women and other potentially vulnerable groups in converged online and offline environments. Such media can be diffused to millions of people in only seconds due to faster ways of disseminating digital media productions, and the Internet facilitates access for those seeking violent pornographic content, whether it is legally recognized or not. What used to be rather difficult to access and a secretive phenomenon, due to its degrading nature, is now accessible for larger groups and has subsequently become a significant industry with operations all around the world. Still, before we possibly create moral panics and demand the Internet restrictions, such as censorship, we need to understand that online environments are simply reflections of the offline environments.

The Internet not only facilitates access to previously inaccessible materials, but it has also, to a certain extent, created an environment that normalizes hurtful sexuality. The problem lies in that the users and producers of violent and racist pornographic material are contributing to the normalization of violence and degrading treatment of women to satisfy their needs. For the advanced user, the Internet offers anonymity while searching and accessing material. There are many ways to hide ones IP address by using anonymizing services, VPN-clients, etc., which complicates any legal intervention in cases of unlawful or criminal behavior.

The normalization and acceptance of hard core pornography and use among youth and adults also stems from harsher and more competitive broadcasting conditions where pornography and adult entertainment industry actors are often acknowledged as celebrities. This recent phenomenon has clearly normalized the previous ill-reputed status of the adult entertainment industry and is affecting emerging adults' perceptions of pornography. Certainly, the boundaries have been moved toward a more tolerable direction of violence and deviant behavior than we ever have experienced before among this group. The proliferation of pornography in the current lives of most people is undoubtedly linked to the changing technological context of modern society. (Carroll et al., 2008).

Manuel Castells (2001) asserts that technological systems are socially produced and that social production is culturally informed. The Internet and the online environments are in constant change and new socio-technological inventions are frequently the source for the growth of new enterprises trying to meet the growing demands on the global market. Castells identifies a transformation between the early days and the present situation and distinguishes between producers/users and the consumers/users of the Internet. The producers/users are the ones "whose practice of the Internet feeds directly back into the technological system, while consumers/users are those recipients of applications and systems who do not interact directly with the development of the Internet, although their uses certainly have an aggregate effect on the evolution of the system." The Internet culture, according to Castells, is a collective construction that transcends individual preferences while influencing the practices of people in the culture. Applied to pornography and violence against women, the expansion of the Internet culture might be causing corrupt preferences to influence and strengthen men's patriarchal beliefs, attitudes, and treatment of women.

There has never been a study specifically designed to examine the linkage between adult Internet pornography, male peer support, and violence against women. Nevertheless, preliminary evidence provided in this chapter strongly suggests that the relationship between these three variables is an emerging problem, one that will surely get worse in the near future. Still, to conclusively determine if this is actually the case, at the risk of belaboring the point, much more research is necessary.

What we do know, however, is that, as Katz (2006), among others (e.g., Jensen, 2007) observes:

Mainstream pornography has changed a lot in the past couple of decades. People of a certain age who still associate heterosexual porn with "girlie magazines" and air-brushed photos of big-breasted women shot in soft light on luxurious beds with big pillows would be shocked by the brutality, out-right contempt for women, and racism that is common in today's product (pp. 186-187).

Even though pornography in general is rapidly becoming more degrading, racist, and violent, many people would argue that since an unknown number of men who consume cyberporn and/or other types of pornography never abuse women, the assertion that "porn" is a key determinant of male-to-female violence is refuted. However, as Diana Russell (1998) reminds us:

This is comparable to arguing that because some cigarette smokers don't die of lung disease, there cannot be a causal relationship between smoking and lung cancer. Only members of the tobacco industry and some seriously addicted smokers consider this a valid argument today (p. 150).

Two solutions to the problems caused by cyberporn and other types of pornography often called for are criminalization and abolition. However, proponents of these initiatives are often accused of advocating censorship and trying to eliminate freedom of speech (Funk, 2006; Katz, 2006). Still, why is it that in Canada and elsewhere we have laws and strong reactions against movies showing approvingly the mass execution of Jews by the Nazis in World War II but find it appropriate, or at least a free speech issue, to allow films approvingly showing women being beaten, raped, and degraded in a myriad of ways that are difficult, if not impossible, for the average person to comprehend? Here, following DeKeseredy and Schwartz's (1998a) perspective on pornography, we are not arguing for censorship. Rather, like them, we contend that in a better, equitable society, it would be considered morally reprehensible to view or show pornography, just as it is now for non-documentary films advocating pro-slavery violence and Nazi killings (Schwartz, 1987).

Contrary to popular belief, not all feminists call for censoring adult pornography, even its most violent forms. Gronau (1985), for example, asserts that pornography serves to remind women of the rampant sexist that victimizes and exploits them. For her, if pornography is censored, the evidence of sexism is hidden. Gronau further argues that it is more difficult to mobilize women to fight hidden sexism than it is to fight the obvious and extreme form of sexism manifested in pornography. However, this functional conception of pornography as benefiting women represents a minority view among feminists who have written on the topic (Alvi, DeKeseredy, & Ellis, 2000).

Most people would never dare to praise or support musicians, filmmakers or actors who say hateful things about people's ethnic/cultural backgrounds or spirituality. Still, that cyberporn and other types of pornography constitute a multibillion dollar industry reveals that our media, politicians, and society in general do not find violent, racist, and degrading images of women problematic. As is often said, "What's wrong with this picture?"

There are other initiatives that could be taken to help reduce the harms described in this chapter. For example, since profit is a corporation's "bottom line," many people call for boycotting products offered by pornographers and by companies that distribute their products. Boycotting has a long history and is not a contemporary solution.

Nevertheless, scholars and activists are using new means of boycotting. For example, since violent and racist pornography are "normalized," "mainstreamed," and easily accessible (Jensen & Dines, 1998), some feminist men's groups, such as the Minnesota Men's Action Network: Alliance to Prevent Sexual and Domestic Violence, participate in variations of the Clean Hotel Initiative. This involves encouraging businesses, government agencies, private companies, and so on to only hold conferences and meetings in hotels that do not offer in-room adult pay-per-view pornography (DeKeseredy, 2010). Further, other groups of men and women are joining hands to collectively expose and criticize injurious media coverage of woman abuse (e.g., wife beating) and to boycott companies that profit off of pornography. Robert Jensen (2007) is right to state that, "it's not enough for us to change our personal behavior. That's a bare minimum. Such change must be followed by participation in movements to change the unjust structure and the underlying ideology that supports them" (p. 182).

REFERENCES

Alvi, S., DeKeseredy, W. S., & Ellis, D. (2000). *Contemporary social problems in North American society*. Toronto: Addison Wesley Longman.

Bancroft, L. (2002). *Why does he do that? Inside the minds of angry and controlling men*. New York: Berkley.

Bergen, R. K. (1996). *Wife rape*. Thousand Oaks, CA: Sage.

Bergen, R. K., & Bogle, K. A. (2000). Exploring the connection between pornography and sexual violence. *Violence and Victims*, *15*, 227–234.

Berger, R. J. Searles, P., & Cottle, C.E. (1991). *Feminism and pornography*. New York: Praeger.

Betowski, B. (2007). *1 in 3 boys heavy porn users, study shows*. Retrieved February 23 from http://www.eurekalert.org/pub_ releases/2007-02/uoa-oit022307.php.

Boeringer, S. B. (1994). Pornography and sexual aggression: Associations of violent and nonviolent depictions with rape and rape proclivity. *Deviant Behavior*, *15*, 289–304. doi:10.1080/01639625.1994.9967974

Bowker, L. H. (1983). *Beating wife-beating*. Lexington, MA: Lexington Books.

Boyle, T. (2007). Small towns have higher crime rates. *Toronto Star*, June 29, A1.

Brannigan, A., & Goldenberg, S. (1987). The study of aggressive pornography: The vicissitudes of relevance. *Critical Studies in Mass Communication*, *4*, 289–304. doi:10.1080/15295038709360135

Brownridge, D. A., & Halli, S. S. (2001). *Explaining violence against women in Canada*. Lanham, MD: Lexington Books.

Carroll, J. S. (2008). Pornography and emerging adulthood. *Journal of Adolescent Research*, *23*, 6–30. doi:10.1177/0743558407306348

Castells, M. (2001). *The Internet Galaxy: Reflections on the Internet, Business, and Society*. New York: Oxford University Press.

Claus, M. (2000). Internet changes porn scene at DePauw U. *High Beam Research*, December 12, 1. Retrieved August 26, 2009, from http://www.highbeam.com/ doc/ 1P1-37925753.html.

Daly, K., & Chesney-Lind, M. (1988). Feminism and criminology. *Justice Quarterly*, *5*, 497–538. doi:10.1080/07418828800089871

DeKeseredy, W. S. (2009a). Male violence against women in North America as hate crime. In B. Perry (Ed.), *Hate crimes, volume 3: The victims of hate crime* (pp. 151-172). Santa Barbara, CA: Praeger.

DeKeseredy, W. S. (2009b). Patterns of violence in the family. In Baker, M. (Ed.), *Families: Changing trends in Canada* (6th ed., pp. 179–205). Toronto: McGraw-Hill Ryerson.

DeKeseredy, W. S. (2010). *Contemporary critical criminology*. London: Routledge.

DeKeseredy, W.S. (in press). *Violence against women in Canada*. Toronto, Canada: University of Toronto Press.

DeKeseredy, W. S., Ellis, D., & Alvi, S. (2005). *Deviance and crime: Theory, research and policy*. Cincinnati, OH: LexisNexis.

DeKeseredy, W. S., & Flack, W. F. Jr. (2007). Sexual assault in colleges and universities. In Barak, G. (Ed.), *Battleground criminal justice* (pp. 693–696). Westport, CT: Greenwood.

DeKeseredy, W. S., & Schwartz, M. D. (1996). *Contemporary criminology*. Belmont, CA: Wadsworth.

DeKeseredy, W. S., & Schwartz, M. D. (1998a). *Woman abuse on campus: Results from the Canadian national survey*. Thousand Oaks, CA: Sage.

DeKeseredy, W. S., & Schwartz, M. D. (1998b). Male peer support and woman abuse in postsecondary school courtship: Suggestions for new directions in sociological research. In Bergen, R. K. (Ed.), *Issues in intimate violence* (pp. 83–96). Thousand Oaks, CA: Sage.

DeKeseredy, W. S., & Schwartz, M. D. (2009). *Dangerous exits: Escaping abusive relationships in rural America*. New Brunswick, NJ: Rutgers University Press.

DeKeseredy, W. S., Schwartz, M. D., Fagen, D., & Hall, M. (2006). Separation/divorce sexual assault: The contribution of male peer support. *Feminist Criminology*, *1*, 228–250. doi:10.1177/1557085106288862

Dines, G. (2006). The white man's burden: Gonzo pornography and the construction of black masculinity. *Yale Journal of Law and Feminism*, *18*, 296–297.

Dines, G., & Jensen, R. (2008a). Pornography. In Renzetti, C. M., & Edleson, J. L. (Eds.), *Encyclopedia of interpersonal violence* (pp. 519–520). Thousand Oaks, CA: Sage.

Dines, G., & Jensen, R. (2008b). Internet, pornography. In Renzetti, C. M., & Edleson, J. L. (Eds.), *Encyclopedia of interpersonal violence* (pp. 365–366). Thousand Oaks, CA: Sage.

Donnermeyer, J. F., Jobes, P., & Barclay, E. (2006). Rural crime, poverty, and community. In DeKeseredy, W. S., & Perry, B. (Eds.), *Advancing critical criminology: Theory and application* (pp. 199–218). Lanham, MD: Lexington Books.

Donnerstein, E. (1984). Pornography: Its effect on violence against women. In Malamuth, N., & Donnerstein, E. (Eds.), *Pornography and sexual aggression* (pp. 53–81). Orlando, FL: Academic Press.

Doring, N. (2009). The Internet's impact on sexuality: A critical review of 15 years of research. *Computers in Human Behavior*, *25*, 1089–1101. doi:10.1016/j.chb.2009.04.003

Dworkin, A. (1994). Pornography happens to women. Retrieved August 15, 2009 from http:// www.nostatusquo.com/ ACLU/ dworkin/PornHappens.html.

Ferguson, I. (1996). *A preliminary investigation into offensive and illegal content on the Internet: Deviant criminal pornography*. Ottawa: Department of Justice Canada.

Fisher, W. A., & Barak, A. (2001). Internet pornography: A social psychological perspective on Internet sexuality. *Journal of Sex Research*, *38*, 312–323. doi:10.1080/00224490109552102

Funk, R. E. (2006). *Reaching men: Strategies for preventing sexist attitudes, behaviors, and violence*. Indianapolis, IN: Jist Life.

Gelles, R. J., & Straus, M. A. (1988). *Intimate violence: The causes and consequences of abuse in the American family*. New York: Simon and Schuster.

Gillespie, I. (2008). Nowadays, it's brutal, accessible; pornography. *London Free Press*, June 11, A3.

Giordano, P. C. (1995). The wider circle of friends in adolescence. *American Journal of Sociology*, *101*, 661–697. doi:10.1086/230756

Gondolf, E. W. (1999). MCMI-III results for batterer program participation in four cities: Less "pathological" than expected. *Journal of Family Violence*, *14*, 1–17. doi:10.1023/A:1022843324943

Gossett, J. L., & Byrne, S. (2002). "Click here": A content analysis of Internet rape sites. *Gender & Society*, *16*, 689–709. doi:10.1177/089124302236992

Gronau, A. (1985). Women and images: Feminist analysis of pornography. In Vance, C., & Burstyn, V. (Eds.), *Women against censorship* (pp. 127–155). Toronto: Douglas and McIntyre.

Harmon, P. A., & Check, J. V. P. (1989). *The role of pornography in woman abuse*. Toronto: LaMarsh Research Program on Violence and Conflict Resolution, York University.

Jensen, R. (1995). Pornographic lives. *Violence Against Women*, *1*, 32–54. doi:10.1177/1077801295001001003

Jensen, R. (1996). Knowing pornography. *Violence Against Women*, *2*, 82–102. doi:10.1177/1077801296002001005

Jensen, R. (2007). *Getting off: Pornography and the end of masculinity*. Cambridge, MA: South End Press.

Jensen, R., & Dines, G. (1998). The content of mass-marketed pornography. In Dines, G., Jensen, R., & Russo, A. (Eds.), *Pornography: The production and consumption of inequality* (pp. 65–100). New York: Routledge.

Jordan, Z. (2006). A view at cyberporn and its influence on aggression against women. Unpublished manuscript. Ames, Iowa: Iowa State University.

Katz, J. (2006). *The macho paradox: Why some men hurt women and how all men can help*. Naperville, IL: Sourcebooks.

Kendall, L. (2003). Cyberporn. In Kimmel, M. S., & Aronson, A. (Eds.), *Men and masculinities: A social, cultural, and historical encyclopedia* (*Vol. 1*, p. 193). Santa Barbara, CA: ABC-CLIO.

Krafka, C., Linz, D., Donnerstein, E., & Penrod, S. (1997). Women's reactions to sexually aggressive mass media depictions. *Violence Against Women*, *3*, 149–181. doi:10.1177/1077801297003002004

Lehman, P. (2006). Introduction: "A dirty little secret" – Why teach and study pornography? In Lehman, P. (Ed.), *Pornography: Film and culture* (pp. 1–24). New Brunswick, NJ: Rutgers University Press.

Longino, H. (1980). What is pornography? In Lederer, L. (Ed.), *Take back the night: Women on pornography* (pp. 40–54). New York: William Morrow.

Maidment, M. R. (2006). Transgressing boundaries: Feminist perspectives in criminology. In W.S. DeKeseredy and B. Perry (Eds.), *Advancing critical criminology: Theory and application* (pp. 43-62). Lanham, MD: Lexington.

Mills, C. W. (1959). *The sociological imagination*. New York: Oxford University Press.

Russell, D. E. H. (1993). *Against pornography: The evidence of harm*. Berkeley, CA: Russell.

Russell, D. E. H. (1998). *Dangerous relationships: Pornography, misogyny, and rape*. Thousand Oaks, CA: Sage.

Sanday, P. R. (1990). *Fraternity gang rape*. New York: New York University Press.

Schwartz, M. D. (1987). Censorship of sexual violence: Is the problem sex or violence? *Humanity & Society*, *11*, 212–243.

Schwartz, M. D., & DeKeseredy, W. S. (1997). *Sexual assault on the college campus: The role of male peer support*. Thousand Oaks, CA: Sage.

Schwartz, M. D., & DeKeseredy, W. S. (1998). Pornography and the abuse of Canadian women in dating relationships. *Humanity & Society*, *22*, 137–154.

Segal, L. (1993). Does pornography cause violence? The search for evidence. In P.C. Gibson & R. Gibson (Eds.), *Dirty looks: Women, pornography, power* (pp. 5-21). London: British Film Institute.

Simon, S. (2004). Hardcore porn hits the heartland: Rural superstores are "doing great." *Concord Monitor*. Retrieved on December 7, 2004 from http://www.cmonitor.com/ apps/pbcs.dll/ article?AID=/20041207/REPOSITORY/ 412070332/1014.

Southworth, C., Tucker, S., Fraser, C., & Shulruff, T. (2008). High-tech violence against women. In Renzetti, I. C. M., & Edleson, J. L. (Eds.), *Encyclopedia of interpersonal violence* (pp. 329–330). Thousand Oaks, CA: Sage.

Stack, S., Wasserman, I., & Kern, R. (2004). Adult social bonds and use of Internet pornography. *Social Science Quarterly*, *85*, 75–88. doi:10.1111/j.0038-4941.2004.08501006.x

Walther, J. B., Slovacek, C. L., & Tidwell, L. C. (2001). Is a picture worth a thousand words: Photographic images in long-term and short-term computer-mediated communication. *Communication Research*, *28*, 105–134. doi:10.1177/009365001028001004

Zerbisias, A. (2008). Packaging abuse of women as entertainment for adults: Cruel, degrading scenes "normalized" for generation brought up in dot-com world. *Toronto Star*, January 26, L3.

ADDITIONAL READING

Barak, A., Fisher, W. A., Belfry, S., & Lashambe, D. R. (1999). Sex, guys, and cyberspace: Effects of Internet pornography and individual differences on men's attitudes toward women. *Journal of Psychology & Human Sexuality*, *11*, 63–91. doi:10.1300/J056v11n01_04

Bauserman, R. (1996). Sexual aggression and pornography: A review of correlational research. *Basic and Applied Social Psychology*, *18*, 405–427. doi:10.1207/s15324834basp1804_4

Bergen, R. K., & Bogle, K. A. (2000). Exploring the connection between pornography and sexual violence. *Violence and Victims*, *15*, 227–234.

Biever, C. (2006). The irresistible rise of cybersex: From full-on encounters to online dating with a twist, simulated sex is on the up in mainstream gaming. *New Scientist*, *190*, 30–32. doi:10.1016/S0262-4079(06)61451-6

Bowyer, K. (2000). Pornography on the dean's pc: An ethics and computing case study. *Journal of Information Systems Education*, *11*, 121–126.

Boyle, K. (2000). The pornography debates: Beyond cause and effect. *Women's Studies International Forum*, *23*, 1987–195. doi:10.1016/S0277-5395(00)00077-7

Cate, F. H. (1996). Cybersex: Regulating sexually explicit expression on the Internet. *Behavioral Sciences & the Law*, *14*, 145–166. doi:10.1002/(SICI)1099-0798(199621)14:2<145::AID-BSL238>3.0.CO;2-N

Coopersmith, J. (2006). Does your mother know what you really do? The changing nature and image of computer-based pornography. *History and Technology*, *22*, 1–25. doi:10.1080/07341510500508610

DeKeseredy, W. S. (1990). Male peer support and woman abuse: The current stake of knowledge. *Sociological Focus*, *23*, 129–139.

DeKeseredy, W. S., & Schwartz, M. D. (2009). *Dangerous exits: Escaping abusive relationships in rural America*. New Brunswick, NJ: Rutgers University Press.

Dines, G., Jensen, R., & Russo, A. (Eds.). (1998). *Pornography: The production and consumption of inequality*. New York: Routledge.

Donnerstein, E., Linz, D., & Penrod, S. (1987). *The question of pornography: Research Findings and policy implications*. New York: Free Press.

Dwyer, S. (1995). *The problem of pornography*. Belmont, CA: Wadsworth.

Halavais, A. (2006). *Cyberporn & society*. Dubuque, IA: Kendall/Hunt.

Itzin, C. (Ed.). (1992). *Pornography: Women, violence and civil liberties*. New York: Oxford University Press.

Jensen, R. (2004). Pornography and sexual violence. *VAWnet*, July, 1-8.

Jensen, R. (2007). *Getting off: Pornography and the end of masculinity*. Cambridge, MA: South End Press.

Lane, F. S. (2000). *Obscene profits: The entrepreneurs of pornography in the cyber age*. New York: Routledge.

MacKinnon, C. A., & Dworkin, A. (1997). *In harm's way: The pornography civil rights hearings*. Cambridge, MA: Harvard University Press.

Malamuth, N. M., & Donnerstein, E. (Eds.). (1984). *Pornography and sexual aggression*. Orlando, FL: Academic Press.

Russell, D. E. H. (1998). *Dangerous relationships: Pornography, misogyny, and rape*. Thousand Oaks, CA: Sage.

Schwartz, M. D., & DeKeseredy, W. S. (1997). *Sexual assault on the college campus: The role of male peer support*. Thousand Oaks, CA: Sage.

Schwartz, M. D., & DeKeseredy, W. S. (1998). Pornography and the abuse of Canadian women in dating relationships. *Humanity & Society*, *22*, 137–154.

Segal, L. (1990). Pornography and violence: What the "experts" really say. *Feminist Review*, *36*, 29–41. doi:10.2307/1395107

Seto, M. C., Maric, A., & Barbaree, H. E. (2001). The role of pornography in the etiology of sexual aggression. *Aggression and Violent Behavior*, *6*, 35–53. doi:10.1016/S1359-1789(99)00007-5

Spinello, R. A. (2002). *Regulating cyberspace: The policies and technologies of control*. Westport, CT: Greenwood.

Stack, S., Wasserman, I., & Kern, R. (2004). Adult social bonds and use of Internet pornography. *Social Science Quarterly*, *85*, 75–88. doi:10.1111/j.0038 4941.2004.08501006.x

Uebel, M. (2000). Toward a symptomatology of cyberporn. *Theory & Event*, *3*, 1–50.

Williams, L. (2004). *Porn Studies*. Durham, NC: Duke University Press.

KEY TERMS AND DEFINITIONS

Adult Pornography: Sexually explicit literary, video, or audio material that is violent, racist, and that objectifies and degrades women.

Cyberporn: Sexually explicit literary, video, or audio material distributed by various sectors of the Internet that is violent, racist, and that objectifies and degrades women.

Male Peer Support: The attachments to male friends and the resources they provide that perpetuate and legitimate woman abuse.

Violence Against Women: Any physical or sexual act committed by men with the intent to injure, control, degrade, or kill an intimate female partner.

Misogyny: Hatred of women.

Normalized: Widely accepted in society.

Sociological Imagination: A person's ability to view the relationship between events in their lives and how they are shaped by broader social forces, such as the political economy and culture.

ENDNOTES

1 Following Daly and Chesney-Lind (1988), feminism is defined here as "a set of theories about women's oppression and a set of strategies for change" (p. 502). There are at least 12 types of feminist theory (Maidment, 2006), each of which takes a distinct approach to understanding gender issues, asks different types of questions, and offers different social scientific perspectives on a variety of social scientific topics (DeKeseredy, Ellis, and Alvi, 2005).

2 See DeKeseredy and Schwartz (2009) and Schwartz and DeKeseredy (1997) for in-depth reviews of the extant social scientific literature on the male peer support and woman abuse.

Chapter 4
Improving the Quality of Life of Senior Citizens by Preventing and Reducing their Social Discrimination through Interactive Television and Ubiquitous Computing

Pedro C. Santana
University of Colima, Mexico

Ricardo Acosta-Díaz
University of Colima, Mexico

Juan Contreras-Castillo
University of Colima, Mexico

Pedro Damián-Reyes
University of Colima, Mexico

ABSTRACT

The problem of providing appropriate and effective medical care to the elderly has gained importance in recent years because of the exponential growth of people older than 65 years. This research work proposes using interactive television embedded in a ubiquitous computing environment to help mitigate some of the effects of discrimination and provide health services to older adults living alone in their homes who require timely medical attention.

DOI: 10.4018/978-1-60960-094-5.ch004

INTRODUCTION

The problem of providing appropriate and effective medical care to the elderly has gained importance in recent years due to the exponential growth of people older than 65 years. For example, we can mention the case of the United States, where it is estimated that by the year 2030 there will be over 4 million people older than 85 years of age (Anderson, 1999).

Most of the time, loneliness and depression are associated with old age, which in sum, could result in a reduced quality of life and other health problems (You and Lee, 2006). Senior citizens' loneliness could be one of the consequences of changes that have been happening in the family structure, such as the maturing of children who leave their birth home to form their own families, the reduction of family size, and the migration of family members to other countries.

According to Erickson (1995), over 20% of senior citizens are unable to live alone; therefore, they need other people to take care of them. This represents a major change in their lives and has repercussions on their overall quality of life as they increasingly require special care that can become extremely costly.

BACKGROUND

An aging population is a phenomenon faced by a number of nations. It is estimated that over the next decades the world population will significantly age as a consequence of birth control during the 1950s and 1960s and a worldwide decline in fertility since the 1970s. The percentage of elderly increased from 5.2 in 1950 to 6.2 in 1995; it is projected that by 2050, one out of ten people worldwide will be 65 years or older (Heilig, 1996).

In Mexico, as in other countries, the elderly population is growing faster than ever. According to the Mexican National Council of Population (CONAPO, 2004), Mexico's population will age faster than any other country in the region. In 2005, 7.5% of the Mexican population was 60 years or older; and it is estimated that by 2030 the number will double, reaching 17.5%. Currently, 10% of senior citizens live alone with no family members nearby. The elderly face particular challenges shaping the need for communication with family members living abroad. Their living conditions can be complicated as they are not able to visit or be visited by their families.

Society, industry and government are looking for new technological solutions to support the different needs of the elderly, facilitating and enabling them to cope with their loneliness. For instance, the Mexican government has created a portal named: eMexico. However, the use of technology to provide home care services is usually an unlikely alternative for older adults. Technology is often out of their reach and communication tools are not always designed with the elderly in mind. These elderly go through a natural progression of changes that some consider as age-related barriers that make accessing and adopting current technology more difficult. Other age-related barriers arise from social issues, such as the elderly persons' greater resistance to new ways of doing things (Goodman, Brewster & Gray, 2004). As a consequence of not considering the technological needs of the elderly, this sector of the population continues to perceive technology as complex and will not benefit from what these tools can offer to help them cope with their aged-related needs, such as dealing with their loneliness (Goodman et al., 2004; Newell, 2004).

Social Discrimination

According to Mexico City's human rights commission, the elderly living in this city represents the segment of the population suffering from the most brutal and paradigmatic cases of age discrimination in Mexico (CDHDF, 2009).

Discrimination can be defined as any distinction, exclusion or restriction based on different factors such as ethnicity, national origin, sex, age, disability, socioeconomic status, pregnancy, language, religion, sexual preference, marital status or any other issue that has the effect of impairing or nullifying the recognition, the exercise of rights, and real equality of opportunity for people.

The case of senior citizens is related to discrimination, which is manifested by violence and abandonment, as well as physical, psychological, economic and sexual abuse, among others. Some of these behaviors are due to cultural, economic, and social factors. These behaviors and attitudes can result in a devaluation of the elderly by the generation that unfairly stereotypes them as being unproductive, inefficient, ill, and in general decline.

However, aging should be considered what it is: a natural phenomenon that is part of the human life cycle (CDHDF, 2009).

The legal status of the elderly is quite blurred, especially by the profound and contradictory silence in which they are immersed. For instance, it is stated that people have the right to wealth that all democratic states must ensure, since it is basic. There is talk of civil and political human rights, of economic, social and cultural rights, rights relating to the environment, peace and development, and rights of future generations. Yet, unfortunately, we are also informed on almost a daily basis of how the elderly die alone at home waiting to be accepted into an assisted living setting or a nursing home.

There are reports of elderly people being abandoned by their families or even by their caregivers. The elderly must often simply survive on meager pensions, they receive abuse and mistreatment in the streets, at home, and even by institutions that are supposed to serve and protect them. Unfortunately, all of these situations combine to create a repressive and harsh discrimination of the elderly (Dabove, 2006).

Another form of discrimination is how the elderly are often ignored or ill treated by a society that sometimes attempts to ignore their existence. Society relegates them to the background and public servants often delay proceedings in the case of elderly persons. Senior citizens suffer rejection, abuse and denial of services, as well as physical and emotional exploitation and misappropriation of their assets.

In general, senior citizens are exposed to discrimination in employment, healthcare, and public institutions, including government agencies and transportation systems. It is important to note that some of the basic principles to foster a better quality of life for the elderly include: independence, participation, care, fulfillment, and dignity. Respecting these principles can guarantee an adequate standard of living for the elderly. In this way, society can recognize and appreciate all of the hard work and significant contributions the elderly have made to society, while demonstrating a respect for life and human beings in general.

The Universal Declaration of Human Rights principle of equity states that "*everyone has the rights and freedoms set forth in the Declaration, without distinction of race, color, sex, language, religion, political or other opinion, national or social origin, property, birth or other status. Furthermore, no distinction shall be made based on political, international law or the country or territory under whose jurisdiction a person belongs, whether it is an independent country, as a Trust Territory, non-autonomous or subject to any other limitation of sovereignty*" (United Nations, 2009).

All people should be treated equally regardless their personal or collective differences. Unfair treatment is commonly known as discrimination. Based on the Honduras Documentation Center (CEDOH, 2009), humans have two sets of characteristics: natural or inherent. Examples of inherent characteristics include origin, race, color, and sex, among others. Other characteristics are learned, acquired or modified, including, opinion, religion, culture, language, economic status, etc. There are still other features that can be placed into both groups, for example: ability and motor or mental health, sexual preference and gender identity.

However, discrimination is also treating an individual or a group of people differentially based on their characteristics. This differential treatment can be generated by an individual, organization, or even the government. According to the CEDOH (2009), negative discrimination against others occurs when there is an expressed or implied rejection by act or omission by one part, and this rejection is directed against a natural or acquired characteristic of a human being (age, sex, race, religion, opinion, culture, language, physical appearance, among others). People suffering from discrimination usually belong to a vulnerable group of people with natural or acquired features, based on certain historical circumstances in a society. Therefore, we must strongly emphasize that there is never a legal justification for discrimination.

There are different types of discrimination. However, the United Nations Program on Ageing (UNPA, 2009) and the United Nations Human Rights Commission of the (UNHRC, 2009) state that in some countries elders are discriminated against in the following areas:

1. ***Employment.*** This is one of the clearest signs of discrimination towards elders. A lot of companies exclude seniors. This exclusion includes hiring, firing, promoting, suspending, compensation, benefits, job assignments and training. An example of this type of discrimination is found in Mexico, where newspapers commonly announce age limits in employment ads. Although, in certain circumstances, the notice may not be considered discriminatory because it is a place where old age can be detrimental or even dangerous; however, in most cases, especially in countries that do not have the legislation of the United States or other similar countries, this type of restriction is completely discriminatory.
2. ***Social Security.*** This type of discrimination is associated with the protections "promised" to society's elderly members. The elderly are constantly discriminated against by government institutions created to protect them. In Mexico, the health system has constantly been discriminatory towards older people. Some examples include chronic shortages of medications for the elderly, long waiting periods to see a physician and even rules excluding the elderly from being accepted for organ transplants.
3. ***Shops.*** Private companies in Mexico discriminate against the elderly. For example, some companies do not provide easy access to their buildings by minimally changing the physical infrastructure of their installations. There are no special lifts or ramps to facilitate and assist free movement for them. Furthermore, there are even some regulations that may be considered discriminatory; for example, in Mexico, the elderly often do not have access to credit or medical insurance, due to their advanced age.
4. ***Family.*** This type of discrimination is perhaps one of the most devastating from a human perspective and it can affect them emotionally and psychologically. It occurs when elderly people are relegated or considered a nuisance or a burden for the family. For example, the chronically ill, who require specialized care, often see their rights to property or benefits severely affected.

ICT AS A PROPOSED SOLUTION

The behavior of families can be modified by developing information and communications technologies (ICT). Thanks to these technologies, families can manage their resources more efficiently, freeing time and money for other uses. ICT gives families a more integrated and informational transparent environment, improving their quality of life.

Digitization of Households

Over the years, ICT has promoted changes in the family structure, in society and in schools, however, until now, changes have not been as deep as expected, but has continuously adapted to accommodate new needs resulting from the massive adoption of technology and the consequent changes in society into an increasingly digital environment.

In general, families have decided not to include furniture and spaces designed specifically for ICT, but have adapted existing ones, mostly because of the cost of modifying existing installations. However, the rapid evolution of technologies makes families hesitate to spend money on modifications which may become obsolete in a short period of time.

Schiffman & Kanuk, (1997) identified five characteristics that appear to influence the acceptance of new products by consumers:

1. ***The relative advantage:*** the product attractiveness in terms of quality, price and pertinence that potential customers perceive when compared with existing ones.
2. ***Compatibility:*** the degree to which consumers feel that a potential new product is consistent with their values, needs and practices.
3. ***The complexity of its use:*** the degree to which a new product is difficult to understand or use.
4. ***The facility for testing*** the degree to which a product can be tested by consumers.
5. ***Observability***: the perception of the attributes and benefits of a product.

Additional to the five factors mentioned above, there are others that also affect the diffusion of innovations among consumers. According to the International Network for the Availability of Scientific Publications the communication channels that spread knowledge of innovations, the social system and finally, the level of adoption of any innovation require time to permeate into society and consumers (INASP, 2003). This implies that no matter how revolutionary or valuable a new product is, time is required for these products to be accepted by consumers. Rogers (1962) indicates that the process of adoption of innovations is usually divided into three stages:

1. First, companies begin to offer the new product which is only bought by more innovative consumers since its novel and potential applications are not fully understood.
2. Secondly, the product's use becomes more widespread. As the innovation becomes more common, little by little, people will consider the product's characteristics to be superior to already existing products and begin to replace them.
3. Finally, society becomes fully accustomed to the new product and it becomes a need.

Internet penetration in households has passed its initial phase and is now in a second stage of consolidation. Users no longer are people with high levels of technological knowledge. This stage is characterized by "average" people using technology in mass, although various groups are still underrepresented.

Changes in the Family Environment

One of the main family structure changes that information technology has promoted is the reduction of boundaries between families and society. According to Stolzoff, Shih & Venkatesh (2000) this is primarily due to three reasons:

1. First, family life has become increasingly interconnected with work as ongoing communications by mobile phones and electronic mail break down the separation between work and family.

2. Second, the home has become, to some degree, a commercial area where crowds try to market their goods and services electronically. This phenomenon has changed the way persons shop in traditional places (Hoffman and Novak, 1996).
3. Third, the existence of personal web pages, along with mobile communications, has resulted in ubiquitous homes.

All of this has made the home a place where the changes and tensions arising from the rapid deployment of the digital society are being manifested with particular intensity.

Households consist of a set of technical areas: physical, and sociocultural, which are transformed by the introduction of information technology (Venkatesh, 1996). According to this division, we envision these areas affecting the technological environment as well as the social and cultural context of families.

Technological Environment

In the last two decades, people in Mexico have seen a clear trend towards an increase of household electronic equipment. From the perspective of household equipment, development of ICT has contributed to migration from analog to digital equipment and, in the process, of the interconnection between them, both within the household and the rest of the world. Digital equipment, particularly computers and the Internet, already enjoy a significant presence in homes, although still quite far below that of other more established devices such as televisions and telephones.

We are moving towards an increasingly digital home which is characterized by the integration of four areas under a unified system of information management.

1. **Communications.** It is certainly the core around which the rest of the applications revolve. The high capacity and quality of communications available to the digital home allows such diverse tasks as:
   - Communication via videoconference from any of the terminals available in the house by taking advantage of Internet telephony.
   - Management of information processing equipment such as networked computers, printers, scanners or screens.
   - The control of sensors and actuators that are the home automation network that handles chores such as temperature control or monitoring.
   - Access and purchase of goods and services using electronic commerce.
2. **Work at home.** The Internet connection along with digital home equipment can cover the needs of persons working from their homes, which, in some cases, have become a second office or headquarters.
3. **Entertainment.** The digital home permits the dissemination of high quality audiovisual services throughout the home, including music, video games and video on demand.
4. **Home automation.** New technologies enable comprehensive home management. Electronic services integrated within the house can make chores easier and automate tasks as well as allow remote control of others (Telefonica, 2003).

All of these networks can combine and work together sharing many elements. As a result, all the terminals installed in a house can perform several tasks using the same core data. For example, a person can listen to music from his or her personal music player, which is more satisfying and enjoyable than listening to audio in the living room. If a person desires more privacy, he or she can choose to play music in the bedroom. Amazingly, all of these devices can access the same music files stored in a central storage device.

Social and Cultural Environment

According to the World Internet Report (WIP) although the number of the households using ICT is irrelevant when compared to television and other traditional goods and services, its use is growing rapidly and is considered to be more productive and formative (WIP, 2003). In fact, user perception of new technologies is generally quite positive. They consider most technology to be fun and easy to use. Persons also feel technology promotes creativity. According to Lindlof (1992) and Papera (1996) there are three major reasons why technologies and the Internet are being accepted in homes:

1. **Job.** A large proportion of households obtain personal computers and Internet connections because of job-related needs, either directly because they telecommute from home or simply because they wish to perform additional tasks they do regularly in the office.
2. **Communication.** Many families begin to use the Internet to take advantage of the advanced communications capabilities offered by their computers. One of the most widely used services is undoubtedly e-mail, but instant messaging, IP telephony and videoconferencing are showing very high growth rates.
3. **Information and education.** In many cases, families use ICT to accomplish tasks related to training and education.

The Internet has produced new forms of social interaction, culminating in a historical process that decreases the importance of physical proximity (Castells, 2003). It also promotes individualism; growing up in a neighborhood is becoming less important than establishing social relationships online.

For some people, ICT is detrimental to family ties and general social communication, mostly because it promotes random and anonymous relationships. However, studies suggest that new technologies, far from representing increased isolation, contribute to improving social and family relationships (Howard, Rainie & Jones, 2001; WIP, 2003). A recent study conducted under the auspices of the European Community (Families, 2002) reveals the existence of a wide diversity of family circumstances and ways of working that have been affected by changes linked to the digital society.

Need for Elder Healthcare

As expected, the difficulty of providing health care to elders in an appropriate and effective way has significantly increased in recent years because of the exponential growth of the aged population. This population needs special care. The most common telecare offered to elders today is monitoring against falls (Allegrante, 2003) or malnutrition (McCormack, 1997), as well as checking on whether or not they are taking their medication (Fulmer, 1999). Given the number of patients, this population is perhaps the least served by physicians (Boise, 1999).

Over 20% of people over 85 years of age cannot live alone (Erickson, 1995) and need somebody to monitor or take care of them. As a result, this directly impacts the lives of their families, increases the costs of physician care, and reduces the quality of life of both the elderly and their caregivers.

Several countries are facing problems providing medical services to their growing elderly populations, which, coupled with the rising cost of medical services, may cause some restrictions on public health services in the next decade. One way to address this problem would be to develop effective preventive systems that help people keep themselves as healthy as possible and live without requiring special care.

Medical Care Using Ubiquitous Computing

The main goal of healthcare is to preserve and enhance the quality of life for the patient and those who care for the elderly. This care, reinforced by ubiquitous computing systems, may increase patient independence and provide an integrated environment, especially for the elderly with cognitive disabilities.

The term "ubiquitous medical care" describes the use of ubiquitous computing technologies to provide health services according to the following characteristics:

1. Intelligent environments to provide health care services and general wellbeing.
2. Health care services 24 hours a day using ubiquitous computing.

Ubiquitous Computing

Ubiquitous computing is a term introduced by Mark Weiser to describe his view of the increasing use of computer systems, making them available and yet invisible to the user (Weiser, 1991). His proposal stems from the vision of third generation computing, which has many applications and research teams seeking its development (Weiser, 1993).

The ubiquitous concept generally refers to the non-invasive presence of sensors and transmission capacity that is present in an often unnoticeable way within a specific environment. This technology aims to provide intelligent computer systems that can adapt to user needs.

Ubiquitous computing is generally a background service that is invisible or transparent to the user. The advantage of ubiquitous computing is that it integrates technology into everyday life. Common applications include car navigation systems, timers for different electronic and mechanical devices as well as security systems for homes and offices (McCrory et al, 2000). Very important areas of development include health care, in general, but particularly in disease management and support for independent living (Stanford, 2002, Kidd, C. et al, 1999, Mynatt et al, 2000).

Background

Current developments in sensor technology allow us to obtain health-related information from various sources; for example, embedded sensors that can be carried on our bodies or as part of our clothing. Ubiquitous communication networks based on mobile phones, wireless LANs and other wireless technologies now make it possible to transfer and access personal or medical information from almost anywhere and anytime. Most modern mobile devices now provide ubiquitous computing interfaces for users. The possibilities these technologies offer to provide medical care are extensive and are being implemented progressively as described in Sachpazidis (2002) and Sachpazidis (2001).

In addition to monitoring services and transferring biological information, medical care ubiquitous environmental technology can be used for social interaction. For example, family members living with patients with chronic illnesses may use devices to maintain constant communication with them. The ubiquitous medical care system can be aware of how to help patients better manage their diseases, and provide a means of communication and collaboration between them and the service staff or medical care professionals.

Ubiquitous computing, along with artificial intelligence, can be applied together to help people with cognitive disabilities perform daily activities. People who have cognitive impairments, including young adults with learning disabilities, or seniors who suffer from some form of dementia, such as Alzheimer's, can use intelligent systems to receive health care services. Recently, researchers and industrial partners in ubiquitous computing and artificial intelligence worked together to develop systems able to assist these special population groups (Kautz et al, 2002; Adlam et al, 2001; Mihailidis et al, 2001; Mynatt et al, 2001). However, the use of ubiquitous computing to provide medical care has brought new challenges. In order to handle private information on the health of people, it is important to create and develop systems that are reliable, scalable, secure, capable to maintaining privacy and configurable, among other things.

At the same time, we have to consider that the average user of such systems has no access to the latest technology, so that it should be incorporated into devices for everyday use and it should also be adapted to heterogeneous environments.

AVAILABLE SOLUTIONS

The following sections present a summary of some research projects that implement ubiquitous computing in health care.

Care in Intelligent Environments

Currently, research is being done to create distributed systems that will improve the quality of life of patients and health care providers (Stefan & Andrew, 2003). For example, people with disabilities have serious problems related to independent living that are inherent in their disabilities. To provide support, ubiquitous environments have been developed, such as voice chats, which translate all written text into audio for the visually impaired (Vlad and Felix, 2003). Applications based on cognitive smartphones increase the independence of patients with Alzheimer's or dementia, as well as persons who want to know about important events happening in their area (Sumi et al, 2003). Intelligent environments support the safety and independence of older people with dementia by using a message service or provide activities that remind them of tasks to be performed. Intelligent devices can also monitor the consumption of medications and provide additional services to patients who need to be reminded when to take medications, alert family members when their patient takes their drugs, and indicate whether a medication has expired, among many applications.

Sensors and Monitoring Behavior

Ubiquitous computing technology can be exploited to monitor the activity and behavior patterns of the elderly. Allin et al. (2003) conducted tests to evaluate ubiquitous computing to assess behavioral disturbances in elderly people with dementia. Video cameras, microphones and wearable sensors were used to help monitor patient interpersonal interactions, record changes in daily activities, register the number of times a patient got lost while returning to his or her room after a walk, and to detect physical or verbal abuse among patients. Although there have been significant advances in this area, there are several research questions to be answered about how to manage patient privacy and how to evaluate if ubiquitous computing systems can contribute to reducing patient insecurity.

It has been proven useful to implement systems to observe and understand daily activities performed by individuals. Fishkin et al. (2003) developed Carnac, a system that understands routine activities carried out within the home. Carnac works on objects (utensils, furniture, clothing, etc.) that have been fitted with radio frequency identification (RFID) tags (Finkenzeller, 1998, Patterson et al., 2003). Users who want their activities recorded must use a glove that can read RFID tags, thus allowing the system to observe and record the sequence in which objects are touched and deduce the nature of the performed activities.

Recent advances in wireless networks, ubiquitous computing and sensor technologies offer opportunities to improve the quality of life and health not only for the elderly, but also health care providers (Hayes et al., 2003). However, to develop systems that provide these services, one must identify and understand what changes occur in people with age that are high risk and use models based on stochastic inference to create technologies to make predictions and perform measures according to their needs.

An emerging field of ubiquitous computing involves developing technologies for proactive health care such as systems to detect patients in crisis, systems to detect damage to the health and systems to encourage healthy behaviors (Intelle, 2003).

User Interface and Evaluation

A challenge in this area is to identify real needs that can be solved using ubiquitous computing technology. However, this raises the questions of how to design, install, support and maintain this technology. It is also important to determine how these systems and the use of laboratory devices inclusion will affect hospitals and homes. This requires researchers to define what standards should be used for communications devices within the home and how to transmit data and commands to control medical devices within a hospital (Adlam & Orpwood, 2003).

Most models and devices used in households for health care focus on a small number of physiological parameters (blood glucose, weight, blood pressure, etc.). But for adults with chronic diseases, it is necessary to consider a wider range of parameters, including the social and emotional aspects of health, which have a profound effect on life expectancy, health care and the patient's level of independence.

Probably the most significant barrier facing technology in the area of health care is the development of adaptive interfaces for elderly patients suffering from chronic conditions that affect mobility, knowledge and self-confidence (Jimison, 2003).

Ubiquitous Computing Technologies for Medical Care

The number of computer systems that exist today for health care is extraordinary. Numerous systems exist in which patient medical information is stored and managed, including information about pharmacies, homes, clinics and most importantly, hospitals. But creating and managing repositories of patient medical data is an enormous task.

The ideal strategy would be to create a system that supports the information in any electronic format. There are currently several standards, including the US-based "Health-Level 7" (HL 7), the standard EU-based Health Information System Architecture (HISA) and Denmark's national standard called "G - EPJ." But creating a centralized repository that integrates all formats and is capable of exchanging data with other systems is a daunting and difficult to reach task.

Bardram (2004) is working on the "The Personal Medical Unit" system that uses ubiquitous computing infrastructure for health care, allowing users to store and carry their medical records, synchronize information with other medical systems and monitor devices. However, this system also needs to abide by patient record standards regarding file formats, proprietary brands, and so on.

There has been significant progress in developing flexible electronic devices such as sensors and wireless transmitters that can be incorporated in clothing in what is now known as smart clothing (Park, 2002).

Safari, et al. (2004) have developed a vest used for medical applications which require physiological readings. This vest is controlled by software and can electronically regulate the application of drugs through the skin.

Riisgaard (2004) and her team are working on the "Awarephone,"" which uses mobile phones to provide colleagues remote situational awareness to avoid interruptions. A second project by Riisgaard (2004), known as "The interactive operating theater," allows computing devices to handle special environments using voice or gesture recognition.

Digital devices presently monitor patients with chronic and acute diabetes and monitor blood glucose levels. Also, multiprogramming brain stimulators are available on the market for epilepsy and other neurological disorders and implants; similar applications have been developed in the area of cardiology for the prediction and identification of life-threatening episodes. Laerhoven et al. (2004) are working on the UbiMom project, where hybrid sensor networks combine wearable sensor nodes to monitor patients. In this sense, research is needed in designing practical and reliable interfaces that allows for the fusion obtained by multi-sensors operating in distributed environments.

Tsai et al. (2000) have developed "Micar," a platform for ubiquitous medical care, to provide context-aware authorization to protect patient privacy and data access.

Ambient Intelligence

The goal of ubiquitous computing, also known as pervasive computing (Weiser, 1993), is to integrate information and computing into the everyday physical world so that it is readily available to everyone in a variety of contexts. The idea is to integrate computation into the environment to enable people to move around and interact with computers more seamlessly than they currently do. For instance, some devices are programmed to sense changes in their environment and to automatically adapt and act based on user needs and preferences.

Ambient Intelligence embraces features of several computing areas, such as pervasive or ubiquitous computing (ubicomp), context-aware computing, human-computer interaction (HCI), and artificial intelligence (AI). An intelligent environment is ubiquitous in the sense that it is saturated with heterogeneous computational and wireless communication devices that are naturally integrated to human activity.

The ubiquitous system should be aware of the user's context to provide information and services whenever the user needs them, in a proactive fashion and anticipating the user's needs. Furthermore, the services provided by the environment have to be accessible to diverse and non-specialist users through simple and effortless interactions, that is, the human-computer interaction must be natural (Rodríguez et. al., 2004).

Medical care is one of the most common areas in which ambient intelligence is applied to disease management and support for independent living (Stanford, 2002, Kidd, C. et al., 1999, Mynatt et al., 2000).

Interactive Television

Today, television represents an important part of modern life. Through television, many people know what is happening in the world and view programs for fun and entertainment. Television is a common device with a high household penetration, and has a huge impact on virtually all areas, from information to entertainment and education.

With the advent of interactive digital television (IDTV), viewers have evolved from passive viewers to active participants. The term IDTV refers to television programming with interactive content and digital enhancements. IDTV combines traditional television with interactive digital applications that are developed for use in a television set (Herrero et. al., 2003).

IPTV

IPTV is one of several ways to bring interactive content to users via the TV. IPTV is digital TV distributed over an IP network, instead of a traditional cable network. IPTV does not consist of simply watching videos on websites. Instead, it refers to the way in which information is transmitted.

In IPTV, videos are transmitted to users as IP packets. That is, the TV programming is encoded and converted into IP packets. Then, the IP packets are distributed throughout the network to the end user, which with a decoder (set-top box), converts the digital data into analog television signals.

The IPTV architecture (see Figure 1) consists of the following components:

1. Content Source: A device that receives, encodes, and stores video or other types of content in a database.
2. IPTV service node: A device that receives the video stream and encapsulates it for proper transmission. These nodes enable the delivery of video to the clients.
3. IPTV client: A set-top box providing processing capabilities to the box.

FINAL COMMENTS AND OBSERVATIONS

In order to achieve the functionalities described above, we propose the following architecture as described in Figure 2. The extended architecture includes a XHTML conversion process in order to correctly display the content on a television set. The content is transported in the form of packets using the MHP standard, which also provides a return channel for the interactive applications running on the set-top box at the client location.

The extended IPTV architecture (see Figure 2) consists of the following components:

1. **Multimedia Server (MMS):** We will need a strategy to manage different multimedia documents (video and audio). Therefore, we will require a multimedia server component (MMS) to manage the documents that are transmitted. When the CMS requires retrieving a document, it will send the request to the MMS that will act as a gateway to the actual repository that maintains the files. This proposal opens the door for the elderly to receive special programming designed to help them recognize and react to discrimination. This type of programming also will permit the elderly to interact and form support communities without having to leave the confines of their homes. Group support is used to mitigate or eliminate many emotional and psychological maladies (Hitch, 1994).
2. **Web Content Management System:** The trend of using Content Management Systems (CMS) to manage web content is gaining momentum with the introduction of automated publishing tools that facilitate the publishing process and improves the user experience and usability. We will build our CMS on AJAX and PHP technologies. The Ajax engine will allow users to interact with the

Figure 1. Generic architecture of an IPTV system

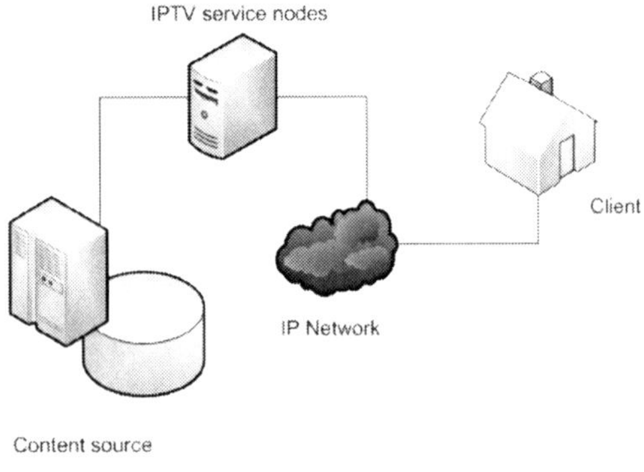

Figure 2. Proposed extended architecture of an IPTV system

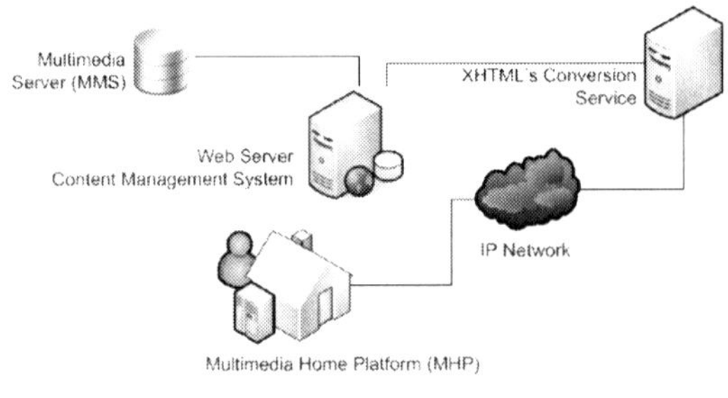

CMS synchronously by using a web-based interface. We also need to implement an API (Application Programming Interface). This API will use the service-oriented computing (SOC) paradigm; thus, we will use services as the fundamental elements for developing applications. This responds to the need to provide a uniform and ubiquitous information distributor for a wide range of computing devices (tablet PCs, PDAs, mobile telephones, or appliances) and software platforms (LINUX or Windows). This flexibility will permit the elderly to select the technology that best fits their needs or personal likes.

3. **XHTML's Conversion Service:** HTML's content, generated by the CMS, will be transformed into standard XHTML data, because low-end set-top boxes can easily manage well-formed XHTML documents. Then, media objects, such as audio and video will be transformed into MHP compliant documents before delivering the content to the client. This conversion service is important to standardize documents so that clients will not have to learn to use different formats. This will reduce the resistance of the elderly to using new technologies and the need for them to require assistance from third parties. The sense of achievement and independence is good for the self confidence and self concept of the users.
4. **IPTV client:** This is a set-top box located in the client side that allows processing interactive applications.

CONCLUSION

The use of a combination of ambient intelligence and interactive television (IPTV) offers the advantage of using the television to bring the elderly together. Although health and monitoring are the most obvious uses for the IPTV architecture we propose, providing the elderly with the capacity to participate with their peers in support groups to discuss discrimination, among other relevant topics, provides this age group with a place where they can identify and obtain support from their peers, thus reducing their isolation and minimizing the devastating effects of exclusion, abuse and other phenomena related to their not being treated with the respect and consideration they deserve.

REFERENCES

Adlam, T., Gibbs, C., & Orpwood, R. (2001). The Gloucester smart house bath monitor for people with dementia. *Physica Medica, 17*(3), 189.

Adlam, T., & Orpwood, R. (2003). Technology, autonomy and cognitive disability. In Proceedings of *UbiComp 2003: Ubiquitous Computing 5th International Conference*. Seattle, WA, USA

Allegrante, J. P. (2003). I*mproving functional recovery after hip fracture*. Retrieved March 8, 2010 from http://clinicaltrials.gov/ct/show/NCT00000436?order=1

Allin, S. J., Bharucha, A., Zimmerman, J., Wilson, D., Roberson, M., Stevens, S., et al. (2003). Toward the automatic assessment of behavioral disturbances of dementia. *UbiComp 2003: Ubiquitous Computing 5th International Conference*. Seattle, WA, USA

Anderson, R. N. (1999). Method for constructing complete annual u.s. life tables. national center for health statistics. *Vital and Health Stat*, Series 2(129).

Bardram, J. (2004). The personal medical unit -- a ubiquitous computing infrastructure for personal pervasive healthcare. *In Proceedings of Ubiquitous Computing 6th International Conference*, Nottingham, UK

Bardram, J., & Bossen, C. (2003). Moving to get ahead: local mobility and collaborativework. In Kuutti, K., Karsten, E. H., Fitzpatrick, G., Dourish, P., & Schmidt K. (Eds), Proceedings of the *Eighth European Conference on Computer Supported Cooperative Work*, p. 355–374, Helsinki, Finland. Amsterdam: Kluwer Academic Publishers.

Bardram, J., Kolbeck, T. A. K., & Nielsen, C. (2003). Supporting local mobility in healthcare by application roaming among heterogeneous devices. In Proceedings of the *Fifth International Conference on Human Computer Interaction with Mobile Devices and Services*. Springer Verlag. 161-176.

Boise, L., Morgan, D. L., Kaye, J., & Camicioli, R. (1999). Delays in the diagnosis of dementia: perspectives of family caregivers. *American Journal of Alzheimer's Disease, 14*(1), 20–26. doi:10.1177/153331759901400101

Castells, M. (2003). *La galaxia internet. reflexiones sobre internet, empresa y sociedad.* Barcelona, Spain: Random House Mondadori.

CDHDF. (2009). *Comisión de Derechos Humanos del Distrito Federal.* Retrieved March 8, 2010 from http://www.cdhdf.org.mx/

CEDOH. (2009). *Nociones fundamentales sobre la discriminación.* Retrieved on 30/04/2010, From http://www.cedoh.org/proyectos/ discriminacion/ assets/Nociones.pdf

CONAPO. (2004). *There are in Mexico 7.9 million elders. Consejo Nacional de Poblacion. Technical Report.* Mexico: Mexico City.

Dabove, M. I. (2006). Discriminación y ancianidad. Reflexiones Filosóficas en torno al sistema jurídico Argentino. *Magazine Telemática de Filosofía del Derecho*, 153-164.

Erickson, P., Wilson, R., & Shannon, I. (1995). Years of healthy life. us department of health and human services, CDC. National Center for Health Statistics, Hyattsville Maryland, *Statistical Notes,* 7.

Families (2002). *families, work and ist: a study of the interactions between family trends and new work methods in the information society.* Project supported under DG-Information Society's IST Programme; Project Co-ordinators: Work Research Centre, Dublin, Ireland. Results of Family Survey, Deliverable No. 3, February 2002, http://www.families-project.com/doc/ families-d3-public.pdf

Finkenzeller, K. (1998). *RFID handbook.* New York: John Wiley and Son, Ltd.

Fishkin, K. P., Kautz, H., Patterson, D., Perkowitz, M., & Philipose, M. (2003). guide: towards understanding daily life via auto-identification and statistical analysis. *UbiComp 2003: Ubiquitous Computing 5th International Conference.* Seattle, WA, USA.

Fulmer, T. T., Feldman, P. H., Kim, T. S., Carty, B., Beers, M., Molina, M., & Putnam, M. (1999). An intervention study to enhance medication compliance in community-dwelling elderly individuals. *Journal of Gerontological Nursing, 25*(8), 6–14.

Goodman, J., Brewster, S., & Gray, P. (2004). Older people, mobile devices and navigation. HCI and the Older Population Workshop. *Paper presented at BCS HCI Workshop 2004.* pp. 13–14. UK: Leeds.

Hayes, T., Pavel, M., Schallau, P. K., & Adami, A. M. (2003). Unobtrusive monitoring of health status in an aging population. *Ubiquitous Computing 5th International Conference, Seattle*, WA, USA.

Heilig, G. K. (1996) World population prospects: Analyzing the 1996 UN population projections. (International Institute for Applied Systems Analysis) *Report WP-96-146* of the IIASA LUC-Project. Retrieved September 1, 2009 from http:// www.iiasa.ac.at/Research/ LUC/Papers/gkh1/ chap1.htm.

Herrero, C., Cesar, P. & Vourimaa, P. (2003). Delivering MHP Applications into a Real DVB-T Network, OtaDigi. *In Proceedings of Telecommunications in Satellite, Cable and Broadcast Service (TELSIKS 2003)*, Serbia – Montenegro, p. 231-234.

Hitch, P. J., Fielding, R. G., & Llewelyn, S. P. (1994, November). Effectiveness of self-help and support groups for cancer patients: A Review. *Psychology & Health*, *9*(6), 437–448. doi:10.1080/08870449408407470

Hoffman, D. L., & Novak, T. P. (1996). Marketing in Hypermedia Computer- Mediated Environments: Conceptual Foundations. *Journal of Marketing*, *60*(3), 211–231. doi:10.2307/1251841

Howard, P., Rainie, L., & Jones, S. (2001). Days and nights on the Internet: The impact of a diffusing technology. *The American Behavioral Scientist*, *45*(3), 383–404.

INASP. (2003). *Income generation: experiences from public library systems in eastern, central and southern Africa*. Oxford, UK: Oxford University Press.

Intill, S. (2003). *Ubiquitous computing technology for just-in-time motivation of behavior change*. S.M. Thesis, Massachusetts Institute of Technology.

Jimison, H. B., Pavel, M., & Pavel, J. (2003). Adaptive interfaces for home health. *Ubiquitous Computing 5th International Conference*, Seattle, WA, USA.

Kautz, H., Fox, D., Etzioni, O., Borriello, G., & Arnstein, L. (2002). An overview of the assisted cognition project. In AAAI-2002 *Workshop on Automation as a Caregiver: The Role of Intelligent Technology in Elder CareMelno Park, CA:* AAAI Press.

Kidd, C. (1999). The aware home: a living laboratory for ubiquitous computing research. *In Proceedings of 2nd International workshop on Cooperative Buildings (CoBuild99). Lecture Notes in Computer Science* 1670, Springer-Verlag, Berlin. p. 191-198.

Laerhoven, V. K., & Lo Benny, P. L. J. W. P., Ng, S. T., King R., Kwan, S., Gellersen, H. W., Sloman, M., Wells, O., Needham, P., Peters, N., Darzi, A., Toumazou, C., & Yang, G. Z (2004). Medical healthcare monitoring with wearable and implantable sensor. *Ubiquitous Computing 6th International Conference*, Nottingham, UK, Proceedings.

Lindlof, T. R. (1992). Computing tales: parents' discourse about technology and family. *Social Science Computer Review*, *10*(3), 291–309. doi:10.1177/089443939201000301

McCormack, P. (1997). Undernutrition in the elderly population living at home in the community: a review of the literature. *Journal of Advanced Nursing*, *26*(5), 856–863. doi:10.1046/j.1365-2648.1997.00415.x

McCrory, A., & Judge, J. (2000). Ubiquitous? pervasive? sorry, they don't compute. *Computerworld*, *34*(12), 44.

Mihailidis, A., Fernie, G. R., & Barbenel, J. C. (2001). The use of artificial intelligence in the design of an intelligent cognitive orthosis for people with dementia. *Assistive Technology*, *13*, 23–39.

Mynatt, E. D., Essa, I., & Rogers, W. (2000). Increasing the opportunities for aging in place. *In Proceedings of CUU Conference*, 1-7. New York:ACM Press.

Nacional de Derechos Humanos, C. *(CNDH)*. (2009). Retrieved March 8, 2010 from http://www.cndh.org.mx/

Newell, A. F. (2004). HCI and older people. HCI and the older population workshop. *In Proceedings of BCS HCI*. pp. 29–30. Leeds, UK.

Papera, S. (1996). *The connected family*. Marietta, GA: Longstreet Press Inc.

Park, S., Mackenzie, K., & Jayaraman, S. (2002). The wearable motherboard: a framework for personalized mobile information processing (PMIP). In *Proceedings of the 39th Annual Design Automation Conference* (New Orleans, Louisiana, USA, June 10 - 14, 2002). DAC '02. New York: ACM.170-174. DOI= http://doi.acm.org/10.1145/513918.513961

Patterson, D. D. F., Kautz, H., & Philipose, M. (2003). Expressive, tractable and scalable techniques for modeling activities of daily living. *Ubiquitous Computing 5th International Conference*, Seattle, WA, USA.

Riisgaard, H. T. (2004). Interacting without keyboard and mouse. *Ubiquitous Computing 6th International Conference*, Nottingham, UK, Proceedings.

Rodríguez, M., Favela, J., Preciado, A., & Vizcaíno, A. (2004). An agent middleware for supporting ambient intelligence for healthcare. Paper Presented at *ECAI 2004 Second Workshop on Agents Applied in Health Care*. Valencia, Spain.

Rogers, E. M. (1962). *Diffusion of innovations*. New York: New York Press.

Sachpazidis, I. (2001). @Home telemedicine. *In Proceedings of Telemed 2001 Conference*. Telematik im Gesundheitswesen, 9th - 10th November 2001, Berlin, Germany.

Sachpazidis, I. (2002). @HOME: A modular telemedicine system Mobile Computing in Medicine. *In Proceedings of the 2 Workshop on mobile computing,* Heidelberg, Germany

Safari, R., Foad, D., & Majid, S. (2004). Reconfigurable fabric vest for fatal heart disease prevention. *Ubiquitous Computing 6th International Conference*, Nottingham, UK, Proceedings.

Schiffman, L. G., & Kanuk, L. L. (1997). *Comportamiento del consumidor*. Mexico City, Mexico: Prentice Hall.

Stanford, V. (2002). Using pervasive computing to deliver elder care. *IEEE Pervasive Computing: Mobile and Ubiquitous Systems*, *1*(1), 10–13.

Stefan, C. & Andrew, Gn (2003). Creating distributed support systems to enhance the quality of life for people with cognitive disabilities. *Ubiquitous Computing 5th International Conference, Seattle, WA, USA.*

Stolzoff, N. C., Shih, E. C., & Venkatesh, A. (2000). *The home of the future: an ethnographic study of new information technologies in the home Project NOAH Center for Research on Information technology and Organizations*. Irvine: University of California.

Sumi, H., Carlos, G., Youssef, K., Choonhwa, L., Hicham, E. Z., & William, M. (2003). External representations in ubiquitous computing design. *Ubiquitous Computing 5th International Conference*, Seattle, WA, USA

Telefonica (2003). *Libro blanco del hogar digital y las infraestructuras comunes de comunicaciones*. Retrieved march 8, 2010 from http://sociedadinformacion.fundacion. telefonica.com/DYC/SHI/TelefonicaSI/Publicaciones/ seccion=1242&activo=1&idioma=es_ES&id=2009100111090001.do

Tsai Tse, M., Jiann-Tsuen, L., Jane, Y., & Hsu, M. (2004). Context-aware authorization for integrated healthcare service. *Ubiquitous Computing 6th International Conference*, Nottingham, UK

UNHRC. (2009). *The United Nations Human Rights*. Retrieved on 30/04/2010, From http://www.un.org/en/rights/

United Nations. (2009). *The Universal Declaration of Human Rights*, Art. 2. Retrieved on 17/10/2009, From http://www.un.org/es/documents/udhr/

UNPA. (2009). *The United Nations Programme on Ageing*. Retrieved on 30/04/2010, From http://www.un.org/esa/socdev/ageing/

Venkatesh, A. (1996). Computers and other interactive technologies for the home. *Communications of the ACM*, *39*(12), 47–54. doi:10.1145/240483.240491

Vlad, C., & Felix, R. (2003). The chatty environment – providing everyday independence to the visually impaired. *Ubiquitous Computing 5th International Conference*, Seattle, WA, USA.

Weiser, M. (1991). The computer for the 21st century. *Scientific American*, *265*(3), 66–75. doi:10.1038/scientificamerican0991-94

Weiser, M. Ubiquitous computing. (1993). *IEEE Computer "Hot Topics"*. October.

WIP. (2003). *Findings from the first World Internet Report*. UCLA Center for Communications Policy.

You, K. S., & Lee, H. (2006). The physical, mental, and emotional health of older people who are living alone or with relatives. *Archives of Psychiatric Nursing*, *20*(4), 193–201. doi:10.1016/j.apnu.2005.12.008

ADDITIONAL READING

Abowd, G. D., & Mynatt, E. D. (2000). Charting past, present, and future research in ubiquitous computing. *ACM Transactions in Computer-Human Interaction*. 7 (1), pp. 29-58. DOI= http://doi.acm.org/10.1145/344949.344988

Hill, J. R., Reeves, T. C., & Heidemeier, H. (2000). *Ubiquitous computing for teaching, learning, and communicating: Trends, issues & recommendations*. Retrieved March 28, 2010, from: http://lpsl.coe.uga.edu/Projects/ AAlaptop/pdf/ UbiquitousComputing.pdf

Johanson, B., Fox, A., & Winograd, T. (2002). The interactive workspaces project: experiences with ubiquitous computing rooms. *IEEE Pervasive Computing / IEEE Computer Society [and] IEEE Communications Society*, *1*(2), 67–74. doi:10.1109/MPRV.2002.1012339

Konomi, S., & Roussos, G. (2007). Ubiquitous computing in the real world: lessons learnt from large scale RFID deployments. *Personal Ubiquitous Comput*. 11 (7), pp. 507-521. DOI= http://dx.doi.org/10.1007/ s00779-006-0116-1

Liu, Z., Wei, B., & Yu, H. (2009). IPTV, towards seamless infotainment. *In Proceedings of the 6th IEEE Conference on Consumer Communications and Networking Conference*. IEEE Press, Piscataway, NJ, p. 201-205.

Qiu, T., Ge, Z., Lee, S., Wang, J., Xu, J., & Zhao, Q. (2009). Modeling user activities in a large IPTV system. *In Proceedings of the 9th ACM SIGCOMM Conference on Internet Measurement Conference* (Chicago, Illinois, USA, November 04 - 06). IMC '09. ACM, New York, NY, pp. 430-441. DOI= http://doi.acm.org/10.1145/ 1644893.1644945

Tokuda, H. (2004). Smart furniture: a platform for context-aware embedded ubiquitous applications. *In Proceedings of the 4th ACM international Conference on Embedded Software* (Pisa, Italy, September 27 - 29, 2004). *EMSOFT '04. ACM,* New York, NY, 2-2. DOI= http://doi.acm.org/10.1145/ 1017753.1017755

Weiser, M. (1993). Some computer science issues in ubiquitous computing. *Communications of the ACM* 36 (7), pp. 75-84. DOI= http://doi.acm.org/10.1145/ 159544.159617

Yang, J., Park, H., Jong, S. G., Lee, S. H., Choi, J. K., & Kim, S. (2009). Interactive control platform for IPTV service in the IMS environment. *In Proceedings of the 11th international Conference on Advanced Communication Technology* - Volume 3. IEEE Press, Piscataway, NJ, pp. 1703-1706.

KEY TERMS AND DEFINITIONS

Ubiquitous Computing: Is a post-desktop model of human-computer interaction in which information processing has been thoroughly integrated into everyday objects and activities. In the course of ordinary activities, someone "using" ubiquitous computing engages many computational devices and systems simultaneously, and may not necessarily even be aware that they are doing so. This model is usually considered an advancement from the desktop paradigm.

Medical Care: The treatment and prevention of illness and disease.

Context-Aware: Originated as a term from ubiquitous computing or as so-called pervasive computing which sought to deal with linking changes in the environment with computer systems, which are otherwise static.

IPTV: System through which digital television service is delivered using the architecture and networking methods of the Internet Protocol Suite over a packet-switched network infrastructure, e.g., the Internet and broadband Internet access networks, instead of being delivered through traditional radio frequency broadcast, satellite signal, and cable television (CATV) formats.

Elders: People of ages nearing or surpassing the average life span of human beings, and thus the end of the human life cycle. Euphemisms and terms for elderly include seniors (American usage), senior citizens (British and American usage) and old people.

Discrimination: Sociological term referring to the treatment taken toward or against a person of a certain group in consideration based solely on class or category.

Interactive Television: Describes a number of techniques that allow viewers to interact with television content as they view it.

ICT: Information and Communication Technology or ICTs allow users to participate in a rapidly changing world in which work and other activities are increasingly transformed by access to varied and developing technologies.

Chapter 5
Multiagents System Applied on a Cyberbullying Model for a Social Network

Alberto Ochoa
University of Ciudad Juárez, Mexico

Julio Ponce
Autonomous University of Aguascalientes, Mexico

Alberto Hernández
Autonomous University of Morelos, Mexico

Felipe Padilla
Autonomous University of Aguascalientes, Mexico

ABSTRACT

The contribution of this chapter is to present a novel approach to explain the performance of a novel Cyberbullying model applied on a Social Network using Multiagents to improve the understanding of this social behavior. This approach will be useful to answer diverse queries after gathering general information about abusive behavior. These mistreated people will be characterized by following each one of their tracks on the Web and simulated with agents to obtain information to make decisions to improve their life's and reduce their vulnerability in different locations on a social network and to prevent its retort in others.

INTRODUCTION

Cyberbullying refers to repeated intimidation, over time, of a physical, verbal or psychological (including indirect and relational bullying) nature of a less powerful person by a more powerful person or group of persons through information and communication technologies, mediums such as mobile phone text messages, emails, phone calls, Internet chat rooms, instant messaging – and the latest trend – social networking websites such as Orkut, MySpace, Hi5, Facebook and video sharing sites like YouTube. Cyberbullying is a fast growing trend that experts believe is more harmful than typical schoolyard bullying. Nearly all of us can

DOI: 10.4018/978-1-60960-094-5.ch005

be contacted via the Internet or our mobile phones. Victims can be reached anytime and anyplace. For many children, home is no longer a refuge from the bullies. "Children can escape threats and abuse in the classroom, only to find text messages and emails from the same tormentors when they arrive home" (Cyberbulling, 2010).

This chapter focuses on the social and cultural implications of cyber technologies. Identity, bullying and inappropriate use of communication are major issues that need to be addressed in relation to communication technologies for the security in the Web use. We present some concepts and tools of Artificial Intelligence that will be used to analyze information to know when someone can be victim of this type of crime, like Text Mining and Multiagent Systems.

First, in section 1 of this chapter, we explain the way to generate a specific behavior with this cyber bullying model on a Social Network, later we discuss social Blockade, Harassment, and Motivational Factors. In next sections, we show the Issues, Problems, and Trends of Cyberbullying (section 2), we approach different concepts related with Artificial Societies and Social Simulation using Multiagent Systems to analyze and model the necessary information to support the correct decisions for the proposed model(section 3)Social Manipulation and Coercion (section 4), and Social exclusion and intimidation (section 5). In section 6 we apply this Model in a social network over the Internet: Orkut (a popular social network in Brazil) and try to explain innovative perspectives of this Model. In section 7 we analyze a case of study: Social Blockade in a Social Networking based on Memory Alpha with 1000 societies represented by agents to demonstrate the concept of cyber bulling. In Section 8, we present a summary of the best practices to avoid cyber bulling by performing automatic text mining over top related Web sites. Finally, in section 9 we provide our conclusions and our future research on this novel topic.

SOCIAL BLOCKADE AND HARASSEMENT

There are different mechanisms of bulling a classmate. Bill Belsey, former president of www.bullying.org and www.cyberbullying.ca, considers next types of cyberbullying: by means of e-mail, cellular phone using text or multimedia messages: by means of instant messages, defamatory web blogs and personal Webs. According to Fante (2005) young people use web blogs, social networks and systems of instantaneous mail to intimidate their classmates, being the diffusion of altered photographies to ridicule to victims one of the preferred methods. These are distributed massively and sometimes indicating the identity of who is put under the humiliation to increase the impact. In the case of the virtual communities, many of them need an invitation to be able to enter to be member of a group, the harassment at school is based on isolating the victims of the humiliations.

Cyberbullying is a form of scholar harassment produced by means of virtual platforms and technological tools, like chats, blogs, fotologs, text messages of cellular phones, e-mail, forums, servers storing videos or photographs, telephone and other electronic media (Alcantara, 2007)

The Cyberbullying being an indirect and non actual form of harassment, the aggressor does not have contact with the victim, does not see his/her face, eyes, pain, suffering, consequently hardly the aggressor will be able to empathize or to wake up compassion on the other. The cyber aggressor obtains satisfaction in the elaboration of the violent act and in the imagination of the damage caused in the other, since is not possible to live it in situ (Hernandez, 2007).

Social Networks

Social network describes a group of social entities and the pattern of inter-relationships among them. What the relationship means varies, from those of social nature, such as values, visions, ideas, financial exchange, friendship, dislike, conflict, trade, kinship or friendship among people, to that of a transactional nature, such as trading relationship between countries. Despite the variability in semantics, social networks share a common structure in which social entities, generically termed actors, are inter-linked through units of relationship between a pair of actors known as: tie, link, or pair. By representing actors as nodes and ties as edges, social network can be represented as a graph (Ponce et al, 2009).

A social network is a social structure made of nodes (which are generally individuals or organizations) that are tied by one or more specific types of interdependency (See Figure 1).

Social Blockade

Also denominated social Isolation, in which one the young person is not allowed to participate, ignoring his/her presence and avoid him/her in the normal activities between friends and classmates (Alcantara, 2009). Sometimes, mainly at the beginning, the majority of the students that compose the class knows the Web on which is making fun of a classmate, and, the victim does not even know it, although the purpose is that sooner of later this one has to find out. In the Web, also voting systems can also be included to choose to the ugliest, to the most idiotic, to the weakest, or contributing another way to stigmatise and to humiliate to the classmates.

Despite being able to denounce one harmful Web page, there is still no guarantee a new Web page won't arise shortly after. This is all taking place before the victim has even had a chance to recover from the social and psychological effects of the previous page.

Harassment

Bullying is a present reality in our scholar centers that contaminates coexistence, producing negative effects not only in those implied directly, but in the totality of the students and teaching staff. The scholar climate is deteriorated seriously, until the point, that for many students to go, daily, to the center supposes a torture. All of this, together with the paper sensationalist that the mass media grant to this problematic, contributes to untie the general anguish in parents, who until recently considered

Figure 1. Social network diagram

the schools like places of peace, security and well-being for the maturity development of their children (Hernandez, 2007). According this author there are two modalities of Cyber Bullying: One that acts as reinforcer of an already started bullying, and that way of harassment among equals by means of Information Technologies without antecedents.

For the first modality, cyber bullying is a more sophisticated way of harassment developed, generally, when traditional ways of harassment becomes unattractive or unsatisfactory. In this case the aggressor is easily identifiable, since it coincides with the actual bullier. The effects of this type of cyber bullying are additive to those the victim already suffers, but also amplify and increase the damages, given to the world-wide opening and generalization of the harassment through the pages Web.

Regarding to the second modality, they are forms of harassment between equals that do not present antecedents, so that without reason it pretends the boy begins to receive forms of harassment through the IT's. Sometimes, after a time to receive this type of harassment, the cyber aggressor decides to complete its work with an actual experience, giving the face. This type of harassment at network presents characteristics of similarity with other forms of harassment, like the fact of being premeditated and a highly deliberated violent conduct or harassment; that based on an asymmetric relation of control and power on the other, but with particular characteristics that differentiate it from other forms of harassment actual and direct.

- Requires of the use and domain of Its.
- Is a form of indirect harassment
- Is an act of violence
- It is an act of camouflaged violence, on which the aggressor is complete unknown, unless he has been harassing actual to the victim before or who decides to be it after the cyber bullying.
- The ignorance of the aggressor superb the impotence feeling.
- It picks up diverse types or forms to indicate the harassment through the TIC's.
- Legal neglect of these forms of harassment, since although it is possible to close the Web, immediately can be opened another one.
- The harassment invades scopes of privacy and pretended security as it is the home, developing a feeling of total lack of protection.
- Harassment becomes public, is opened to more persons rapidly.

Sexual Harassment

On the other hand, sexual harassment also finds in Internet other forms of expression that accompany the actual harassment, for example: continue sending of threatening electronic mails, or the design of a Web page simulating that one woman is a sexual worker that offer her services and that includes her personal data (name, telephone, email, direction, etc.).

Profile of Victims

According to data of (Protégeles, 2009), most of the situations of cyber bullying took place between the 13 and 14 years, a 52% of the total; whereas a 10% of the cases took place before turning the ten years of age. Besides, most of the harassments were committed in the last quarterly of the course; a 35% of April to June, a 32% of October to December, a 13% of January to March and a 10% of July to September.

In the majority of the known cases the pursuers are women (19%) against a 10% of men. In 71% of the cases the gender of the aggressor was not determined.

In 60% of the cases, the girls are those that undergo harassment. The victims undergo cyber bullying through forums or social networks and programs of instantaneous mail in a 45%. In addition, a 12%, through electronic mail and a 19% through cellular phone. According this study, there are minors who undergo harassment by several of these means.

Finally, this study reflects that in almost half of the reviewed cases the situation of harassment takes place almost on a daily basis and that in 80% of the cases the parents finish knowing about the situation, and at the school in 74%.

Motivational Factors Associated with Cyberbullying

Bullying cannot be considered as part of the normal and common adolescent development, several studies have related it to emotional and behavioral problems (Rodríguez et al., 2006).

As part of an ongoing, statewide bullying prevention initiative in Colorado, 3,339 youth in Grades 5, 8, and 11 completed questionnaires in 78 school sites during the fall of 2005, and another 2,293 youth in that original sample participated in a follow-up survey in 65 school sites in the spring of 2006. Questionnaires included measures of bullying perpetration and victimization, normative beliefs about bullying, perceptions of peer social support, and perceptions of school climate. The highest prevalence rates were found for verbal, followed by physical, and then by Internet bullying. Physical and Internet bullying peaked in middle school and declined in high school. Verbal bullying peaked in middle school and remained relatively high during high school. Males were more likely to report physical bullying than females, but no gender differences were found for Internet and verbal bullying. All three types (verbal bullying, physical bullying and Internet Bullying) were significantly related to normative beliefs approving of bullying, negative school climate, and negative peer support (Williams & Guerra, 2007). Calls the attention the existing little monitoring and consequently the few applied corrective measures in the educative centers and at home (Cabezas, 2007).

In the case of women, in a sample from Costa Rica (see (Cabezas, 2007)) it seems that confronting is the main reason for bullying ("to avoid menaces", "to avoid harassment", "she has something of mine", "she criticizes me", "they bother me", among other). In the case of males, motivation is diverse considering intention and pleasure to do it, they provide answers like: "to feel the adrenaline", "just to do it", "nothing special". Bullies have a great necessity to self impose and dominate (Olweus, 1998). Those who bully repeatedly engage in conflicts that they are sure to win because of their superior power, and those who bully are merciless in their tactics (Bitney & Tile, 1997).

Situations like the above mentioned require paying special attention due to the fact that they could develop an antisocial disorder of personality that generates: A general pattern of scorn and violation of the rights of the others, that starts in childhood or at the beginning of the adolescence stage and continues during the adult age" (DSM-IV, 1995, p. 662).

Grouping phenomena is an answer to the established socio-cultural mechanisms that takes to the social contagion, to a weakening of the control, to the division of the responsibility, besides to graduated cognitive changes of the perception of the bullier and the victim, which simultaneously explains, why there are students who are not aggressive but also participate in passive form of the harassment (Olweus, 1998).

ISSUES, PROBLEMS, AND TRENDS OF CYBERBULLYING

Importance of Cyberbullying

In mathematical sociology, interpersonal ties are defined as information-carrying connections between people. Interpersonal ties, generally, come in three varieties: strong, weak, or absent. Weak social ties, it is argued, are responsible for the majority of the embeddedness and structure of social networks in society as well as the transmission of information through these networks. Specifically, more novel information flows to individuals through weak rather than strong ties. Because our close friends tend to move in the same circles that we do, the information they receive overlaps considerably with what we already know. Acquaintances, by contrast, know people that we do not, and thus receive more novel information.

Included in the definition of absent ties, according to Granovetter, are those relationships (or ties) without substantial significance, such as "nodding" relationships between people living on the same street, or the "tie", for example, to a frequent vendor one would buy from. Furthermore, the fact that two people may know each other by name does not necessarily qualify the existence of a weak tie. If their interaction is negligible the tie may be absent. The "strength" of an interpersonal tie is a linear combination of the amount of time, the emotional intensity, the intimacy (or mutual confiding), and the reciprocal services which characterize each tie.

In Granovetter's view, a similar combination of strong and weak bonds holds the members of society together. This model became the basis of his first manuscript on the importance of weak social ties in human life. He submitted his paper to the American Sociological Review in 1969, but it was rejected. Nevertheless, in 1972, Granovetter submitted a shortened version to the American Journal of Sociology, and it was finally published in May 1973. According to Current Contents, by 1986, the Weak Ties paper had become a citation classic, being one of the most cited papers in sociology.

In a related line of research, in 1969 anthropologist Bruce Kapferer, published "Norms and the Manipulation of Relationships in a Work Context" after doing field work in Africa. In the document, he postulated the existence of multiplex ties, characterized by multiple contexts in a relationship. In telecommunications, a multiplexer is a device that allows a transmission medium to carry a number of separate signals. In social relations, by extrapolation, "multiplexity" is the overlap of roles, exchanges, or affiliations in a social relationship (Granovetter, 1973)..

Problems of Cyberbullying

Risk perception is the subjective judgment that people make about the characteristics and severity of a risk. Several theories have been proposed to explain why different people make different estimates of the dangerousness of risks. Three major families of theory have been developed: psychology approaches (heuristics and cognitive), anthropology/sociology approaches (cultural theory) and interdisciplinary approaches (social amplification of risk framework).

Early Theories

The study of risk perception arose out of the observation that experts and lay people often disagreed about how risky various technologies and natural hazards were.

The mid 1960's saw the rapid rise of nuclear technologies and the promise for clean and safe energy. However, public perception shifted against this new technology. Fears of both longitudinal dangers to the environment as well as immediate disasters creating radioactive wastelands turned the public against this new technology. The scientific and governmental communities asked why public perception was against the use of nuclear energy when all of the scientific experts were declaring how safe it really was. The problem, from the perspectives of the experts, was a difference between scientific facts and an exaggerated public perception of the dangers.

A key early paper was written in 1969 by Chauncey Starr. He used a revealed preference approach to find out what risks are considered acceptable by society. He assumed that society had reached equilibrium in its judgment of risks, so whatever risk levels actually existed in society were acceptable. His major finding was that people will accept risks 1,000 greater if they are voluntary (e.g. driving a car) than if they are involuntary (e.g. a nuclear disaster).

This early approach assumed that individuals behave in a rational manner, weighing information before making a decision. Individuals have exaggerated fears due to inadequate or incorrect information. Implied in this assumption is that additional information can help people understand true risk and hence lessen their opinion of danger. While researchers in the engineering school did pioneer research in risk perception, by adapting theories from economics, it has little use in a practical setting. Numerous studies have rejected the belief that additional information, alone, will shift perceptions (Starr C. et al., 1969).

Psychology Approach

The psychology approach began with research in trying to understand how people process information. These early works maintain that people use cognitive heuristics in sorting and simplifying information which lead to biases in comprehension. Later work built on this foundation and became the psychometric paradigm. This approach identifies numerous factors responsible for influencing individual perceptions of risk, including dread, newness, stigma, and other factors (Starr C. et al., 1969).

Heuristics and Biases

The earliest psychometric research was done by psychologists Daniel Kahneman and Amos Tversky, who performed a series of gambling experiments to see how people evaluated probabilities. Their major finding was that people use a number of heuristics to evaluate information. These heuristics are usually useful shortcuts for thinking, but they may lead to inaccurate judgments in some situations – in which case they become cognitive biases.

- The Availability heuristic: events that can be more easily brought to mind or imagined are judged to be more likely than events that could not easily be imagined.
- The Anchoring heuristic: people will often start with one piece of known information and then adjust it to create an estimate of an unknown risk – but the adjustment will usually not be big enough.
- Asymmetry between gains and losses: People are risk averse with respect to gains, preferring a sure thing over a gamble with a higher expected utility but which presents the possibility of getting nothing. On the other hand, people will be risk-seeking about losses, preferring to hope for the chance of losing nothing rather than taking a sure, but smaller, loss (e.g. insurance).
- Threshold effects: People prefer to move from uncertainty to certainty over making a similar gain in certainty that does not lead to full certainty. For example, most people would choose a vaccine that reduces the incidence of disease A from 10% to 0% over one that reduces the incidence of disease B from 20% to 10%.

Another key finding was that the experts are not necessarily any better at estimating probabilities than lay people. Experts were often overconfident in the exactness of their estimates, and put too much stock in small samples of data.

PSYCHOMETRIC PARADIGM

Research within the psychometric paradigm turned to focus on the roles of affect, emotion, and stigma in influencing risk perception. Melissa Finucane and Paul Slovic have been among the key researchers here. These researchers first challenged Starr's article by examining expressed preference – how much risk people say they are willing to accept. They found that, contrary to Starr's basic assumption, people generally saw most risks in society as being unacceptably high. They also found that the gap between voluntary and involuntary risks was not nearly as great as Starr claimed.

Slovic and team found that perceived risk is quantifiable and predictable. People tend to view current risk levels as unacceptably high for most activities. All things being equal, the greater people perceived a benefit, the greater the tolerance for a risk. If a person derived pleasure from using a product, people tended to judge its benefits as high and its risks as low. If the activity was disliked, the judgments were opposite. Research in psychometrics has proven that risk perception is highly dependent on intuition, experiential thinking, and emotions.

Psychometric research identified a broad domain of characteristics that may be condensed into three high order factors: (1) the degree to which a risk is understood, (2) the degree to which it evokes a feeling of dread, and (3) the number of people exposed to the risk. A dread risk elicits visceral feelings of terror, uncontrollable, catastrophe, inequality, and uncontrolled. An unknown risk is new and unknown to science. The more a person dreads an activity, the higher its perceived risk and the more that person wants the risk reduced.

ANTHROPOLOGY/ SOCIOLOGY APPROACH

The anthropology/sociology approach posits risk perceptions as produced by and supporting social institutions. In this view, perceptions are socially constructed by institutions, cultural values, and ways of life.

CULTURAL THEORY

"The Cultural Theory of risk" (with capital C and T). Cultural Theory is based on the work of anthropologist Mary Douglas and political scientist Aaron Wildavsky first published in 1982.

In Cultural Theory, Douglas and Wildavsky outline four "ways of life" in a grid/group arrangement. Each way of life corresponds to a specific social structure and a particular outlook on risk. Grid categorizes the degree to which people are constrained and circumscribed in their social role. The tighter binding of social constraints limits individual negotiation. Group refers to the extent to which individuals are bounded by feelings of belonging or solidarity. The greater the bonds, the less individual choice are subject to personal control. Four ways of life include: Hierarchical, Imndividualist, Egalitarian, and Fatalist.

Risk perception researchers have not widely accepted Cultural theory. Even Douglas says that the theory is controversial; it poses a danger of moving out of the favored paradigm of individual rational choice of which many researchers are comfortable.

INTERDISCIPLINARY APPROACH

Social Amplification of Risk Framework

The Social Amplification of Risk Framework (SARF), combines research in psychology, sociology, anthropology, and communications theory. SARF outlines how communications of risk events pass from the sender through intermediate stations to a receiver and in the process serve to amplify or attenuate perceptions of risk. All links in the communication chain, individuals, groups, media, etc., contain filters through which information is sorted and understood.

The theory attempts to explain the process by which risks are amplified, receiving public attention, or attuned, receiving less public attention. The theory may be used to compare responses from different groups in a single event, or analyze the same risk issue in multiple events. In a single risk event, some groups may amplify their perception of risks while other groups may attune, decrease, and their perceptions of risk (Valdez et al., 2009).

The main thesis of SARF states that risk events interact with individual psychological, social and other cultural factors in ways that either increase or decrease public perceptions of risk. Behaviors of individuals and groups then generate secondary social or economic impacts while also increasing or decreasing the physical risk itself.

These ripple effects caused by the amplification of risk include enduring mental perceptions, impacts on business sales, and change in residential property values, changes in training and education, or social disorder. These secondary changes are perceived and reacted to by individuals and groups resulting in third-order impacts. As each higher-order impacts are reacted to, they may ripple to other parties and locations. Traditional risk analyses neglect these ripple effect impacts and thus greatly underestimate the adverse effects from certain risk events. Public distortion of risk signals provides a corrective mechanism by which society assesses a fuller determination of the risk and its impacts to such things not traditionally factored into a risk analysis.

Trends of Cyberbullying

The main explanation for social loafing is that people feel unmotivated when working with a team, because they think that their contributions will not be evaluated or considered.

According to the results of a meta-analysis study, social loafing is a pervasive phenomenon, but it does not occur when team members feel that the task or the team itself is important. It can occur when the person feels underappreciated within their team or group.

Social loafing occurs in a group situation in which the presence of others causes relaxation instead of arousal. When individuals relax their performance, they are able to fade into the crowd, which is especially appealing to people when they know they are not going to be accountable for their actions or performance. In easier, less demanding tasks, such as singing happy birthday or giving applause, one is likely to exert less effort due to the concept of diffusion of responsibility. This occurs when people think that they can "get a free ride" because someone else will surely pick up the slack.

Social loafing is associated with poor performance on easy tasks. However, people tend to exert more effort on challenging or rewarding tasks. If a group is completing a task for some kind of reward, such as money or a good grade, then members are more likely to try harder. Generally, a greater reward results in more motivation to perform well, and therefore, more effort. People will also work harder when they feel their particular tasks or efforts are indispensable to the group's success (Karau & Williams, 1993).

ARTIFICIAL SOCIETIES AND SOCIAL SIMULATION USING MULTIAGENT SYSTEMS

Social simulation is the modeling or simulation, usually with a computer, of a social phenomena (e.g., cooperation, competition, markets, social networks dynamics, among others). A subset within social simulations is Agent Based Social Simulations (ABSS) which are an amalgam of computer simulations, agent-based modeling, and social sciences.

History and Development

The history of the agent based model can be traced back to Von Neumann, with his theoretical machine capable of reproduction. The device that

Von Neumann proposed would follow detailed instructions to build a copy of it. The concept was then improved by Von Neumann's friend Stanislaw Ulam, also a mathematician, whom suggested that the machine were built on paper, as a collection of cells on a grid. The idea intrigued Von Neumann, who drew it up—creating the first of the devices later named as cellular automata. Another improvement was brought by mathematician (Conway, 2004). He constructed the well-known Game of Life. Unlike the Von Neumann's machine, Conway's Game of Life operated by tremendously simple rules in a virtual world in the form of a 2-dimensional checkerboard.

The birth of agent based model as a model for social systems was primarily brought by computer scientist (Reynolds, 1987). He tried to model the reality of lively biological agents, known as the artificial life, developed the first large-scale agent model, the Sugarscape, to simulate and explore the role of social phenomena such as seasonal migrations, pollution, sexual reproduction, combat, transmission of diseases, and even culture, more recently. Ron Sun developed methods for basing agent based simulation on models of human cognition, known as cognitive social simulation.

Multiagent Systems

The goal of multiagents systems research is to find methods that allow us to build complex systems composed of autonomous agents who, while operating on local knowledge and possessing only limited abilities, are nonetheless capable of enacting the desired global behaviors. We want to know how to take a description of what a system of agents should do and break it down into individual agent behaviors. At its most ambitious, multiagent systems aims at reverse-engineering emergent phenomena as typified by ant colonies. Multiagent systems approaches the problem using the well proven tools from game theory, Economics, and Biology. It supplements these with ideas and algorithms from artificial intelligence research, namely planning, reasoning methods, search methods, and machine learning (Vidal, 2007).

Types of Simulation and Modeling

Social simulation can refer to a general class of strategies for understanding social dynamics using computers to simulate social systems. Social simulation allows for a more systematic way of viewing the possibilities of outcomes.

The history of the agent based model can be traced back to Von Neumann, with his theoretical machine capable of reproduction. The device that Von Neumann proposed would follow detailed instructions to build a copy of it. The concept was then improved by Von Neumann's friend Stanislaw Ulam, also a mathematician, whom suggested that the machine were built on paper, as a collection of cells on a grid. The idea intrigued Von Neumann, who drew it up—creating the first of the devices later named as cellular automata. Another improvement was brought by mathematician (Conway, 2004). He constructed the well-known Game of Life. Unlike the Von Neumann's machine, Conway's Game of Life operated by tremendously simple rules in a virtual world in the form of a 2-dimensional checkerboard.

The birth of agent based model as a model for social systems was primarily brought by computer scientist (Reynolds, 1987). He tried to model the reality of lively biological agents, known as the artificial life, developed the first large-scale agent model, the Sugarscape, to simulate and explore the role of social phenomena such as seasonal migrations, pollution, sexual reproduction, combat, transmission of diseases, and even culture, more recently. Ron Sun developed methods for basing agent based simulation on models of human cognition, known as cognitive social simulation.

There are four major types of social simulation:

1. System level simulation
2. Agent based simulation

3. System level modeling
4. Agent based modeling

A social simulation may fall within the rubric of computational sociology which is a recently developed branch of sociology that uses computation to analyze social phenomena. The basic premise of computational sociology is to take advantage of computer simulations in the construction of social theories. It involves the understanding of social agents, the interaction among these agents, and the effect of these interactions on the social aggregate. Although the subject matter and methodologies in social science differ from those in natural science or computer science, several of the approaches used in contemporary social simulation originated from fields such as physics and artificial intelligence.

System Level Simulation

System Level Simulation (SLS) is the oldest level of social simulation. System level simulation looks at the situation as a whole. This theoretical outlook on social situations uses a wide range of information to determine what should happen to society and its members if certain variables are present. Therefore, with specific variables presented, society and its members should have a certain response to the situation. Navigating through this theoretical simulation will allow researchers to develop educated ideas of what will happen under some specific variables.

AGENT BASED SIMULATION

Agent Based Social Simulation (ABSS) consists of modeling different societies after artificial agents, (varying on scale) and placing them in a computer simulated society to observe the behaviors of the agents. From this data it is possible to learn about the reactions of the artificial agents and translate them into the results of non-artificial agents and simulations. Three main fields in ABSS are agent based computing, social science, and computer simulation. Agent based computing is the design of the model and agents, while the computer simulation is the simulation of the agents in the model as well as the outcomes. The social science is a mixture of sciences and social part of the model. It is the social phenomena that is developed and theorized. The main purpose of ABSS is to provide models and tools for agent based simulation of social phenomena. With ABSS which can explore different outcomes for phenomena where it might not be able to view the outcome in real life. It can provide us valuable information on society and the outcomes of social events or phenomena.

SYSTEM LEVEL MODELING

System level modeling (SLM) aims to specifically (unlike system level simulation's generalization in prediction) predict and convey any number of actions, behaviors, or other theoretical possibilities of nearly any person, object, construct, etcetera within a system using a large set of mathematical equations and computer programming in the form of models. A model is a representation of a specific thing ranging from objects and people to structures and products created through mathematical equations and are designed, using computers, in such a way that they are able to stand-in as the aforementioned things in a study. Models can be either simplistic or complex, depending on the need for either; however, models are intended to be simpler than what they are representing while remaining realistically similar in order to be used accurately. They are built using a collection of data that is translated into computing languages that allow them to represent the system in question. These models, much like simulations, are used to help us better understand specific roles and actions of different things so as to predict behavior and the like.

AGENT BASED MODELING

Agent based modeling (ABM) is a system in which a collection of agents independently interact on networks. Each individual agent is responsible for different behaviors that result in collective behaviors. These behaviors as a whole help to define the workings of the social networking. ABM focuses on human social interactions and how people work together and communicate with one another without having one, single "group mind". This essentially means that it tends to focus on the consequences of interactions between people (the agents) in a population. Researchers are better able to understand this type of modeling by modeling these dynamics on a smaller, more localized level. Essentially, ABM helps to better understand interactions between people (agents) who, in turn, influence one another (in response to these influences). Simple individual rules or actions can result in coherent group behavior. Changes in these individual acts can affect the collective group in any given population.

Agent-based modeling is simply just an experimental tool for theoretical research. It enables one to deal with more complex individual behaviors, such as adaptation. Overall, through this type of modeling, the creator, or researcher, aims to model behavior of agents and the communication between them in order to better understand how these individual interactions impact an entire population. In essence, ABM is a way of modeling and understanding different global patterns.

CURRENT RESEARCH

Agent based modeling is most useful in providing a bridge between micro and macro levels, which is a large part of what sociology studies. Agent based models are most appropriate for studying processes that lack central coordination, including the emergence of institutions that, once established, impose order from the top down. The models focus on how simple and predictable local interactions generate familiar but highly detailed global patterns, such as emergence of norms and participation of collective action. (Macy & Willer, 2002) researched a recent survey of applications and found that there were two main problems with agent based modeling the self-organization of social structure and the emergence of social order. Below is a brief description of each problem:

1. *Emergent structure*. In these models, agents change location or behavior in response to social influences or selection pressures. Agents may start out undifferentiated and then change location or behavior so as to avoid becoming different or isolated (or in some cases, overcrowded). Rather than producing homogeneity, however, these conformist decisions aggregate to produce global patterns of cultural differentiation, stratification, and hemophiliac clustering in local networks. Other studies reverse the process, starting with a heterogeneous population and ending in convergence: the coordination, diffusion, and sudden collapse of norms, conventions, innovations, and technological standards.
2. *Emergent social order*. These studies show how egoistic adaptation can lead to successful collective action without either altruism or global (top down) imposition of control. A key finding across numerous studies is that the viability of trust, cooperation, and collective action depends decisively on the embeddings of interaction.

These examples simply show the complexity of our environment and that agent based models are designed to explore the minimal conditions, the simplest set of assumptions about human behavior, required for a given social phenomenon to emerge at a higher level of organization (see Figure 2).

Researchers working in social simulation might respond that the competing theories from the social sciences are far simpler than those achieved through simulation and therefore suffer the aforementioned drawbacks much more strongly. Theories in social science tend to be linear models that are not dynamic and which are inferred from small laboratory experiments. The behavior of populations of agents under these models is rarely tested or verified against empirical observation.

As we will see in next section, Social Blockade is a process that involves the skills of a number of individuals. The creation of Social Blockade begins with a concept, an argumentation and a situation. For its social modeling, a Dyoram is proposed to display this behavior inside a social networking, that is to say, in this sequence of scenarios, each one of elements who conforms the network according to their roll and society's parameters that they have within the same one. The development of this behavior inside a social networking requires of the concept and diverse similar measures that allow actions of a separated part of a society about a sustained concept in the data. But it is necessary to prioritize the conceptual development and of the categories of the system that conforms the social model, and think simultaneously about the mathematical model. We proposed the next model to represent this behavior (See Figure 3).

Figure 2. Social simulation using a dyoram to represent a social networking

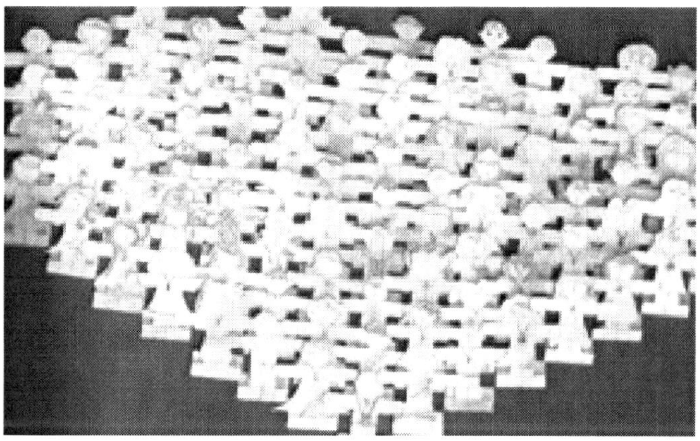

Figure 3. Social blockade model to separate a society of an organized cluster

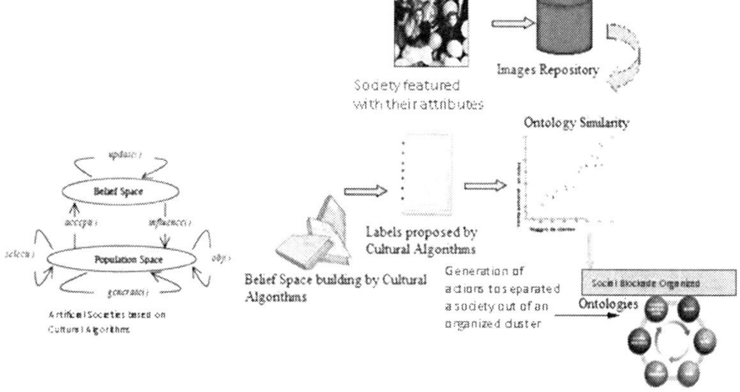

Social Manipulation and Coercion

Cyberbullying is a form of violence, social isolation; public humiliation and malicious gossip have long been the stock in trade of bullies. With the advent of modern communications the bully's reach and powers of social manipulation have been increased exponentially, for intimidate an individual in the attempt to gain power and control over them. The new technologies are advancing more and more every day. People can never know what will be the technologies in the future - a few years ago, in the past we carried a phone with a big size and now – there are small phones with tons of features. Who would have thought? And of course, as the world advances, the bullies do not lag behind and some people said that something needs to be done against cyber-bullies. I would have never thought that it would get to this, but it did. Now, the worst part about this new technique is that it doesn't stop or start with school. You can stay at home and be cyber-bullied from a distance. This activity includes - but is not limited to - threats, intimidation harassment and of course unauthorized publication of personal info in different mediums. So now, not just the whole school will know who the wimps are, but the whole web. This is a lot worse than normal bullying. In any case, there is a good part about this. It will lead to children getting informed on how to stay safe on the web. Some types or cyber violence are showed in Figure 4.

"Cyberbullying is the use of modern communication technologies to embarrass, humiliate, threaten, or intimidate an individual in the attempt to gain power and control over them." (Stutzky, 2010)

Parents are well advised to pay close attention to how this new threat can impact their children. We have already seen too many cases of children subjected to a cyberbullying attack who have been so traumatized that they have committed suicide as a direct result.

As the trial in the first-ever cyberbullying, suicide case began, the prosecutors in the case said the federal court that the suicide of the Missouri teen could have been avoided had she not been tormented online by the woman.

Some signs of cyberbullying (None of these by themselves are a sure indication of being bullied, but several taken together are cause to open a discussion) are the following:

- Unusually long hours on the computer
- Secretive about Internet activity
- Lack of appetite
- Fear

Figure 4. Overview of cyber violence

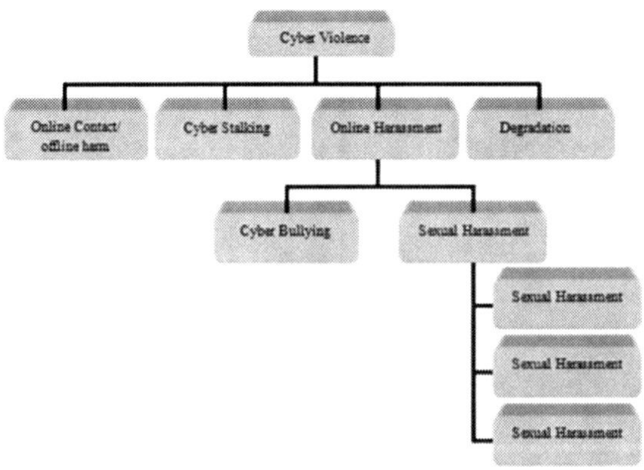

- Closes windows when parents enter his/her room
- Getting behind in school work
- Stomachache

No type of bullying is harmless. In extreme incidents, cyber bullying has led teenagers to suicide. Most victims, however, suffer shame, embarrassment, anger, depression and withdrawal. Cyber bullying is often seen as anonymous, and the nature of the Internet allows it to spread quickly to hundreds and thousands of people. In some cases, it can constitute criminal behavior. The inappropriate use of the Internet, telephones or text messages is a criminal offense. "Bullycide" is the term that has been used to describe suicides caused by relentless bullying. "Cyberbullycide", to coin a phrase, would describe someone driven to suicide following a cyber bullying attack.

Social Exclusion and Intimidation

According to Silva (2001), the intimidation forms that can be presented on bullying are the following:

1) Verbal intimidations (insults, motes, to speak badly of somebody, to seed rumors).
2) Psychological intimidations (threats to provoke fear, to obtain some object or money, or simply to force the victim to make things).

The same author refers regarding social isolation as:

3) Social isolation, preventing the young person to participate, ignoring its presence and not having him in the normal activities between friends and classmates.

According (Aviles, 2006) main types of abusive behavior can be classified on:

- Physical: like pushes, kicks, and aggressions with objects. This type of mistreat occurs with more frequency in the primary school than in the secondary school.
- Verbal: here appear the insults and motes mainly. Contempts in public or being emphasizing and mocking on frequent basis of a physical defect or action are frequent.
- Psychological: are directed actions to mine the self-esteem of the individual and to foment on his/her a sensation of insecurity and fear. The psychological component can be found in all the forms of mistreat.
- Social: they pretend to separately locate to the individual with respect to the group in an bad status and sometimes to make other individuals contributors of this action. These actions are considered indirect bullying.

According to Faustino & Oliveira (2008), cyber bullying, in the virtual community, is characterized, not only for verbal and psychological aggression, besides with diverse insults directed to the victim, as well as the nicknames that are conferred to this person. And also:

- Threats, as in: "I'm waiting to see her in the street to kick her", "Lets go together everybody and give her a .... in the street" and "ahhh and also in her living room... smart loafer... that I will finish her any day....";
- Calumny and defamation: a topic was created by a boy who said that B. had taken a naked photo of a girl in a barbecue, but, when they had asked for showed it, a test of what he was saying, he contradicted the fact to have seen her completely naked and placed doubts in relation to the quality of the photo: "It was not a very clear one, but she wears white panties and with half of it inside her" ...
- Provoking or empowering to commit suicide.

Application of the Proposed Social Model on Orkut (A Popular Social Network in Brazil)

Orkut is a social networks used in Brazil by 13 million users, many create more of a profile, and generate different relationships, to be able to establish communications with people of different life styles, and they doing to believe other users that they are different people.

Nowadays it is demonstrated the feasibility of large-scale, passive deanonymization of real world social networks (Narayanan & Shmatikov, 2009).

Some problems in Orkut that dislike much to their users are: the loss of privacy, the lack of materialization of the relations established through the network. False profiles are created to: make a joke, harass other users, or see who visualizes its profile. Now the users can make denunciations against those false profiles, but without a clear profile of Orkut, the original author (aggressor) can create a new profile.

A serious event in Orkut is the creation of communities with a concept of racism, xenophobia, sales of drugs and paedophilia. Unfortunately, the users who denounce these facts to be eliminated do not reach their objective: the criminals create these data again. Actuality there are a controversy around the use of Orkut by various groups of hate. Some fanatic users, racists and religious, allegedly have a solid following there.

In Orkut exist several hate communities focused on racism, Nazism, and other kinds of specific hate. In 2005, various cases of racism were brought to police attention and reported on in the Brazilian media. According (Racism in Brazil, 2009) in 2006, a judicial measure was opened by the Brazil federal justice denouncing a 20-year-old student accused of racism against those of Black African ancestry and spreading defamatory content on Orkut. In (Racismo na Internet, 2006) the Brazilian Federal Justice subpoenaed Google on March 2006 to explain the crimes that had occurred in Orkut (Ministerio, 2006).

The next numbers about some crimes in Orkut was finding in: (http://www.theregister.co.uk/2008/04/12/google_brazil_pledge/, last quest December 2009). To others Statistics and Tips of cyberbullying in general see I-SAFE website.

- Google is to give Brazilian police access to 3,261 private photo albums on social net-

Figure 5. Traffic on Orkut by country (http://en.wikipedia.org/wiki/Orkut)

Traffic of Orkut on March 31, 2004		Traffic of Orkut on May 13, 2009	
United States	51.36%	Brazil	50%
Japan	7.74%	India	15%
Brazil	5.16%	United States	8.9%
Netherlands	4.10%	Japan	8.8%
United Kingdom	3.72%	Pakistan	6.9%
Other	27.92%	Other	29.6%

working website Orkut, which may contain child pornography.

- Orkut has more than 60 million users, most of them in Brazil. The federal prosecutor for Sao Paulo, in the last two years nearly 90 per cent of the 56,000 complaints in Brazil about net-based paedophilia were linked to the website.
- Until September 2007, all requests for information about users suspected of crimes such as racism or paedophilia were sent to Google US to be examined.
- The private photo albums, considered by some as a safe haven for criminals, were introduced in November 2007, and allow users to block access to the pictures by anyone outside their direct network.

Using the tool of Data Mining denominated WEKA, it was come to develop a denominated Model "Ahankara" of prediction of profiles in users of Orkut, which allows to understand the motivations of this type of profile and determine if the user has a generated Bipolar Syndrome (Ponce et al., 2009).

SOCIAL BLOCKADE IN A SOCIAL NETWORKING REPRESENTED BY AGENTS TO DEMONSTRATE THE CONCEPT OF CYBERBULLYING

A social networking is a social structure that can be represented making use of different types of diagrams. Both more common types are the graph and dyorams. The graph is a collection of objects called vertex or nodes that are connected by lines, edges or arcs. The nodes represent individuals which sometimes are denominated actors and the edges represent the relations that exist between these actors. Consisting of one or more graphs where these graphs conceptualize the network, which is to say, the representation is made mainly on the basis of the relations that exist between the actors who conform the network. In this research, we focused our attention in a practical problem of the Literature related to Modeling of Societies, the social blockade of a society organized for several societies, which allows to understand the behavior that keeps a society with respect to others, the capacity to establish this behavior, allows to establish "the actions to defense" for the given set of societies. The solution to this problem could be given by a sequence of generations of agents, denoted like "community". The agents can only be reassigned a position to blockade with respect to the other societies, determined this according to previous behaviors, shown in each one of them (Ochoa et al., 2007).

Social Blockade in a Social Networking Represented by Multi Agents

From the point of view of the agents, this problem is very complex, on account that the group of agents using strategies to separate another individuals of a society, with respect to the behavior. In the algorithm proposed for the social blockade, the individuals in the space of beliefs (beliefscape) through their better paradigm (BestParadigm) are set to zero to represent the fact that the diversity increases the amount of expectations associated with the separation of a society with respect to other, giving an incentive to the behavior associated with the best paradigm (BestParadigm). For this purpose, we selected 1000 societies described in (Memory Alpha, 2009) and characterized their social behavior with base in seven attributes: emotional control, ability to fight, intelligence, agility, force, resistance, and speed (see Figure 6), these characteristics allow to describe so much to the society as to the individual. The development of the tool was based on our desire to share the intuitive understanding about the treatment of a new class of systems, individuals able to have restlessness, a reserved characteristic of people alive.

Figure 6. Features related with a society in memory alpha (Memory Alpha, 2009)

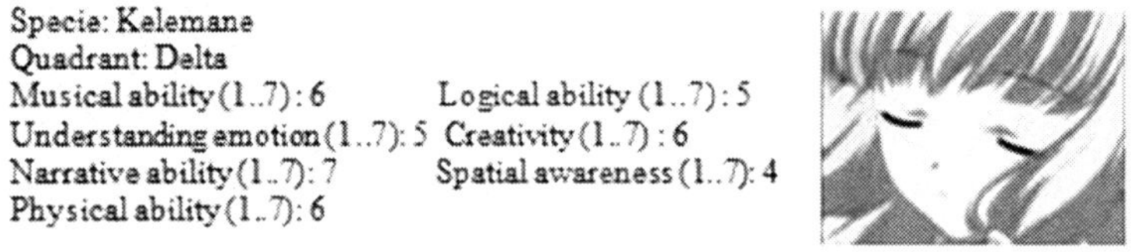

We identify twelve features which increase social blockade: Ethnicity, Use of Technology, Cultural Identity, Commerce, Architecture, Textile Heritage, Language, Environment Protection, Spatial Travels and Exo-collections, Telepathy and Psycho Abilities, The Art of war, Holographic Societies and Living in Class K Planets. Each feature is adjusted according highest interclass similarity and highest intraclass similarity.

We propose seven different similarity equations to organize the separation in societies:

a) Language.

Ordered Symbol and None ordered Symbol and Hamming Similarity to realize disambiguation:

d(Q,C) = b+c

b) Use of Technology.

Simple Matching Coefficient:

$$sim(Q,C) = 1 - \frac{b+c}{n} = \frac{a+d}{a+b+c+d}$$

c) Spatial Location

Euclidean Distance (see Figure 7):

d) Anthropometry

None ordered Symbol to Qualitative Parts and Ordered Symbol to Quantitative parts

e) Dyoram (Social Representation)

Nearest Neighborhood:

$$sim(Q,C) = \sum_{i=1}^{n} f(Qi,Ci) * wi$$

f) Commerce

Strategies Optimist and Pessimist:

If sim(x,y) is a similarity measure.

If has a value of $\alpha = 1 - \alpha \leftrightarrow \alpha > 1/2$, this is optimist.

With $\alpha > 1 - \alpha \leftrightarrow \alpha < ½$, the strategy turn pessimist

g) Social Negotiation

Figure 7. Formula of euclidean distance

$$d(Q,C) = \sqrt{\sum_{i=1}^{n} w_i (q_i - c_i)^2}$$

Asymmetric Similarity Measures:

$$sim(Q, C) = \frac{a}{a + b + c}$$

With this seven specific equations is possible to organize the separation of societies and identify the societies which are involved in this behavior of social blockade.

Experimentation

In order to be able to similar the most efficient arrangement of individuals in a social network, we developed an atmosphere that is able to store the data of each one of the representing individuals of each society; this was done with the purpose of evaluating these societies. One of the most interesting characteristics observed in this experiment was the separation of cultural patterns established by each community. The structured scenes associated with the agents cannot be reproduced in general, since they only represent a little dice in the space and time of the different societies. These represent a unique form and innovating approach of adaptive behavior which explains a computational problem that is separating societies only with a factor associated with his external appearance (genotype), trying to solve a computational problem that involves a complex change between the existing relations. The generated configurations can be metaphorically related to the knowledge of the behavior of the community with respect to an optimization problem (to conform a cluster culturally similar to other societies). The main experiment consisted of detailing each one of the 1000 communities, with 500 agents, and one condition of unemployment of 50 époques (a moment in the history), this allowed us to generate different scenes of Social Blockade in a specific cluster, which was obtained after comparing the different cultural and social similarities from each community, and to determine the existing relations between each one of them (Ustaoglu, 2009). The developed tool represents social blockade, this permits to identify changes over the time respect to other societies (See Figure 8).

Best practices to Avoid Cyber Bullying

Currently, people are not informed about cybersecurity in school, neither about cyber bullying though they should be and now this pops up. We hope people are going to be better informed

Figure 8. Environment constructed by means of the use of Multiagents system (Cultural Algorithms) describing the concept of Social Blockade, each color represented a different society. Green Society generates Social Blockade to the Red Society.

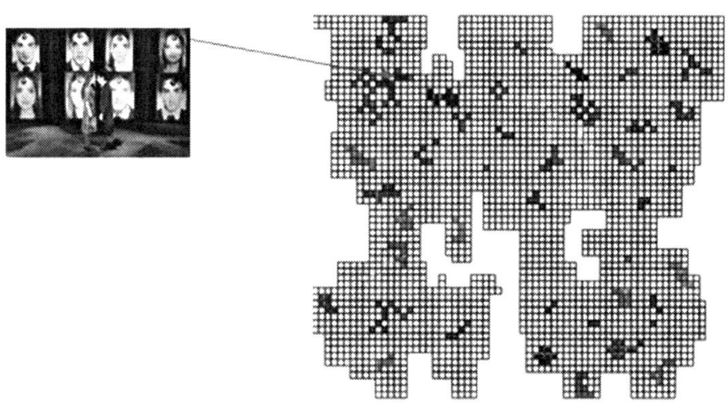

about cyber-security in near future. The situation will change, on this matter there are anti-bullying experts working on it. They're cooperating with a lot of experts to thwart cyber-bullying and all we can say is that this is a great initiative, not only for the victims, but also for cyber-security advance. In this section we present a proposal to answer objectively and in automatically manner the: How to avoid cyber bullying? Question.

Problem at Hands

Is our desire to develop a intelligent tool to provide automatic answers to the question: How to avoid cyber bullying?. After considering several options, we decided to use automatic text mining by means of intelligent agnets to retrieve responses to answer this question from top related Web Sites.

Web Based Text Mining

According to Pérez & Santin(2006), knowledge discovering in databases is a process, a set of stages or tasks, among them are: establishing a relevant problem, selection of proper data to problem solution, exploration and cleaning of data, data processing and transformation, application of modeling techniques, generation and interpretation of models proposed, use of obtained knowledge, and generation of new data by means of its application on the real world. The Web based text mining process we proposed is shown in Figure 9. Determining research objectives is the first stage on this process and determines the success or failure of research, researcher must have a clear idea of which specific information does he want to obtain and the level of depth required to satisfy information requirements (See Figure 9).

Considering objectives, web based data (text) is retrieved from top sites by means of an intelligent agent which explores (executes queries on a web site) and process data (transforms meta data on HTML o XML code on valuable answers for the question), which is stored in a.csv file for further process. Resulting data is transformed and modeled by means of a visual data mining tool (i.e. rapid miner, enterprise miner or WEKA), based on findings a visual model is generated and proved, finally resulting answers become part of new knowledge.

Methodology

Our proposal simplifies this process to three stages:

1. Selection and data base creation from Web Sites using a query with keywords: "How

Figure 9. Web based text mining to answer the question: How to avoid cyber bullying?

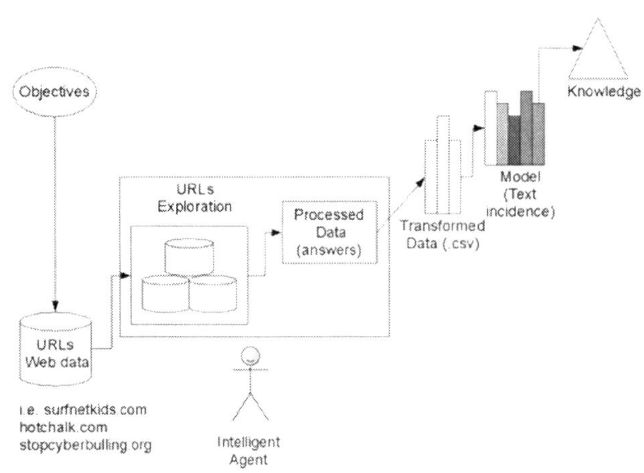

to avoid cyber bullying?" executed on Web search engine.
2. Automatic retrieving of answers using intelligent agents from selected web sites.
3. Text mining of data using a visual tool.

The Experiment

We created an URLs database based on Web Sites retrieved from the Google search engine using the proposed keywords. When URLs were stored, we respected the order provided by Google engine. Each one of the URLs in database was processed by means of our prototype Intelligent Agent written on Java to retrieve automatic answers based on keywords (i.e. Avoid, prevent, don't, refuse, etc.). We limited the number of possible answers to five, once individual URL was processed, we send the id, the url, and the automatically retrieved answers to a.csv file for further analysis using rapid miner.

Preliminary Results and Discussion

On this early stage, our prototype system provided some recommendations to avoid cyber bullying: Do not give/provide private information to anyone (p.e passwords, telephone, address); Never share / post personal information; Report incidents to law enforcements/ school officials / to a trusted adult; Learn to recognize types of cyber bullying; never participate/ never transmit cruel/harmful messages or images; and stop communications with cyber bullies/ tell friends to stop cyber bullying. Despite our system does not provide a complete set of measures to avoid cyber bullying, is summarizes some guidelines to avoid this undesirable practice.

CONCLUSION AND FUTURE RESEARCH

Using Cultural Algorithms, we improved the understanding substantially to obtain the change of "best paradigm", because we appropriately separated the agent communities basing to us on an approach to the relation that keep their attributes. This allowed us to understand that the concept of "Social Blockade" exists with base in the time of époques of interaction with their relationships. This technique allows including the possibility of generating experimental knowledge created by the community of agents for an organized social blockade. The analysis of the level and degree of cognitive knowledge for each community is an aspect that is desired to evaluate for the future work. Understanding the true similarities that have different societies with base in the characteristics that make them contributors of cluster as well as it allows them to keep their own identity, demonstrates that the small variations go beyond phenotype characteristics and are mainly associate to tastes and similar characteristics developed through the time (Ochoa et al, 2009). A new modeling of Artificial Societies can take care of analyzing individually the complexities that each society keeps, without forgetting that they still need methods to understand the original features and particular things for each society.

ACKNOWLEDGMENT

The authors would like to thank to Eliza Coli, and Anna Tordanelli for their insightful comments on the preparation of this chapter.

REFERENCES

Alcantara, M. (2009). El Bulling, Acoso Escolar. Innovación y experiencias Educativas. *CSI-CSIF Revista Digital, 55.*

Aviles, J. M. (2006). Bulling. Intimidación y maltrato entre alumnado. Retrieved May 1, 2010, From: www.educacionenvalores.org/ Bullying-Intimidacion-y-maltrato.html

Bellsey, B. (2009). *Where you are not alone.* Retrieved May 1, 2010, from: http://www.bullying.org

Bitney, J., & Title, B. (1997). *No bullying program: Preventing bullying/victim violence at school (programs directors manual)*. Central City, MN: Hazelden and Johnson Institute.

Cabezas, H. (2007). Detección de Conductas Agresivas "Bullyings" en escolares de Sexto a Octavo Año en Una Muestra Costarricense. *Revista Educación*, *31*(1), 123–133.

Conway, J. (2004). Theory of games to improve artificial societies. *International Journal of Intelligent Games & Simulation*, *2*(1).

Cyberbulling (2010). *What is cyberbulling?* Obtained May 1, 2010, from: http://www.cyberbullying.info/ whatis/whatis.php

DSM-IV. (1995). *Manual diagnóstico y estadístico de los transtornos mentales (DSM-IV)*. Madrid, Spain: Masson, S.A.

Fante, C. (2005). *Fenómeno Bullying. Como prevenir a violência nas escolas e educar para a paz*. Florianopolis, Brasil: Verus.

Google. (2009). *Orkut help forum.* Retrieved May 1 2010 from http://www.google.com/support/forum/p/ orkut/thread?tid=1792d1c990174 1fe&hl=en

Granovetter, M. S. (1973). The strength of weak ties. *American Journal of Sociology*, *78*(6), 1360–1380. doi:10.1086/225469

Hernández, M., & Solano, I. (2007). CiberBullying, Un Problema de Acoso Escolar. *RIED*, 17-36.

I-SAFE. (2010). *Cyber Bullying: Statistics and Tips*. Retrieved May 1 2010, from http://www.isafe.org/channels/sub.php? ch=op&sub_id=media_cyber_bullying

Karau, S. J., & Williams, K. D. (1993). Social loafing: A meta-analytic review and theoretical integration. *Journal of Personality and Social Psychology*, *65*, 681–706. doi:10.1037/0022-3514.65.4.681

Macy, M., & Willer, R. (2002). From factors to actors. [England.]. *Annual Review of Sociology*,28.

Memory Alpha. (2009). *Memory Alpha wiki (Star Trek World)*. Retrieved May 1 2010, from: http://memory-alpha.org/ wiki/Portal:Main

Ministerio (2009). Ministério público pede que Google explique crimes no Orkut (in Portuguese). *Folha Online*. March 10, 2006.

Narayanan, A., & Shmatikov, V. (2009). Deanonymizing social networks. *In IEEE Symposium on Security and Privacy.*

Ochoa, A., et al. (2007). Baharastar – simulador de algoritmos culturales para la minería de datos social. *In Proceedings of COMCEV'2007*.

Ochoa, A. (2008). Social data mining to improve bioinspired intelligent systems. In Giannopoulou, E. G. (Ed.), *Data Mining in Medical and Biological Research*. Vienna, Austria: In-Tech.

Ochoa, A., et al. (2009) Six degrees of separation in a graph to a social networking. *In Proceedings of ASNA'2009*, Zürich, Switzerland.

Olweus, D. (1998). *Bullying at school: What we know and what we can do* [Spanish edition: Conductas de acoso y amenaza entre escolares]. Madrid, Spain: Ediciones Morata.

Pérez, C., & Santín, D. (2006). *Data mining. Soluciones con Enterprise Manager*. Alfa Omega Grupo Editor, S.A. de C.V., Mexico

Ponce, J., Hernández, A., Ochoa, A., Padilla, F., Padilla, A., Alvarez, F., & Ponce de León, E. (2009). Data mining in web applications. In Ponce, J., & Karahoka, A. (Eds.), *Data Mining and Discovery Knowledge in Real Life Application*. Vienna, Austria: In-Tech.

Protégeles. (2009). *Protegeles línea de denuncia*. Retrieved May 1, 2010, from http://www.protegeles.com/

Racism in Brazilian. (2009). *Racism in Brazilian Orkut*. Retrieved May 1, 2010, from http://zonaeuropa.com/20050326_2.htm

Racismo na internet (2006). *Racismo na internet chega à Justiça* (in Portuguese). Estadão. Retrieved July 10, 2007, from http://www.estadao.com.br/tecnologia/ internet/noticias/2006/fev/01/97.htm.

Reynolds, R. (1998). An *introduction to cultural algorithms. Cultural algorithms repository*. Retrieved May 1, 2010, from http://www.cs.wayne.edu/~jcc/car.html

Rodríguez, R., Seoane, A., & Pereira, J. (2006). Niños contra niños: El bullying como transtorno emergente. *Anal Pediátrico*, *64*(2), 162–166. doi:10.1157/13084177

Silva, J. A. (2001). Do hipertexto ao algo mais: usos e abusos do conceito de hipermídia pelo jornalismo on-line. In: A. Lemos, & M. Palacios (orgs.) *Janelas do Ciberespaço - Comunicação e Cibercultura*. Sao Paulo, Brasil: Editora Sulina, pp. 128-139.

Starr, C. (1969). Social benefits versus technological risks. *Science*, *165*(3899), 1232–1238. doi:10.1126/science.165.3899.1232

Ustaoglu, Y. (2009). Simulating the behavior of a minority in Turkish society. *In Proceedings of ASNA'2009*.

Valdez, S. I., Hernández, A., & Botello, S. (2009). Approximating the search distribution to the selection distribution. *In Proceedings of GECCO 2009*, 461-468.

Vidal, J. (2007). *Fundamentals of multiagent systems with NetLogo examples*. Retrieved May 1, 2010, from: http://jmvidal.cse.sc.edu/papers/mas.pdf

Williams, K. R., & Guerra, N. G. (2007). Prevalence and predictors of internet bullying. *The Journal of Adolescent Health*, *41*, S14–S21. doi:10.1016/j.jadohealth.2007.08.018

ADDITIONAL READING

Blumen, W. J. (2005). Cyberbullying: A New Variation on an Old Theme. *Proceedings of Abuse: The darker side of Human-Computer Interaction*. An INTERACT workshop, Rome.

Bontempo, R. N., Bottom, W. P., & Weber, E. U. (1997). Cross-cultural differences in risk perception: A model-based approach. *Risk Analysis*, *17*(4), 479–488. doi:10.1111/j.1539-6924.1997.tb00888.x

Cunningham, S. (1967). The major dimensions of perceived risk. In Cox, D. F. (Ed.), *Risk taking and information handling in consumer behavior*. Cambridge, MA: Harvard University Press.

Kowalski, R. M., Limber, S. P., & Agatston, P. W. (2008). *Cyber bullying: Bullying in the digital age*. Boston: Blackwell Publishing. doi:10.1002/9780470694176

McGrath, H. (2009). *Young people and technology. A review of the current literature (2nd edition). The Alannah and Madeline Foundation*. Retrieved March 25, 2010, from http://www.amf.org.au/Assets/Files/ 2ndEdition_Youngpeopleandtechnology_LitReview_June202009.pdf

McKenna, K. Y. A., & Bargh, J. A. (2000). Plan 9 from cyberspace: The implications of the internet for personality and social psychology. *Personality and Social Psychology Review*, *4*(1), 57–75. doi:10.1207/S15327957PSPR0401_6

Shariff, S. (2009). *Confronting Cyber-Bullying: What Schools Need to Know to Control Misconduct and Avoid Legal Consequences*. New York: Cambridge University Press. doi:10.1017/CBO9780511551260

Stacey, E. (2009). Research into cyberbullying: Student perspectives on cybersafe learning environments. Informatics in Education -. *International Journal (Toronto, Ont.)*, *8*(1), 115–130.

Stutzky, G. R. (2010). *What is Cyberbulling?* Retrieved May 1, 2010, from: http://www.opheliaproject.org /main/ra_cyberbullying.htm

Szoka B. & Thierer A. (2009). Cyberbullying Legislation: Why Education is Preferable to Regulation. *Progress on Point, 16*(12).

Weber, E. U., & Hsee, C. (1998). Cross-cultural differences in risk perception, but cross-cultural similarities in attitudes towards perceived risk. *Management Science*, *44*(9), 1205–1217. doi:10.1287/mnsc.44.9.1205

KEY TERMS AND DEFINITIONS

Artificial Societies: The representation of a group of individuals using Multiagents Systems.

Cyberbullying: Is a type of bullying (embarrass, humiliate, threaten, or intimidate an individual) through information and communication technologies such as mobile phone text messages, emails, phone calls, Internet chat rooms, instant messaging.

Intelligent Agent (IA): Is an autonomous entity which observes and acts upon an environment (i.e. it is an agent) and directs its activity towards achieving goals (i.e. it is rational). Intelligent agents may also learn or use knowledge to achieve their goals.

Intelligent Social Modeling: Technique which permits analyzing diverse social behaviors to its interpretation.

Social Network: Is a social structure made of nodes (which are generally individuals or organizations) that are tied by one or more specific types of interdependency, the relationship means varies, from those of social nature, such as values, visions, ideas, financial exchange, friendship, dislike, conflict, trade, kinship or friendship among people, to that of transactional nature, such as trading relationship between countries.

Social Blockade: Action that try to separate a person or group of persons of a society, with base in a different attribute for example: race, religion, studies or status. This behavior is near to segregation.

Text Mining: Sometimes alternately referred to as text data mining, refers generally to the process of deriving high-quality information from text. Text mining usually involves the process of structuring the input text (usually parsing, along with the addition of some derived linguistic features and the removal of others, and subsequent insertion into a database), deriving patterns within the structured data, and finally evaluation and interpretation of the output.

Social Simulation: Is the modeling, usually with a computer, of a social phenomena.

Chapter 6
Examining Race Hate Crime with Students Using a Cross Cultural International Virtual Learning Environment (VLE)

Julian Buchanan
Victoria University of Wellington, New Zealand (formerly of Glyndŵr University, Wales)

Steve Wilson
Washington University, USA

Nirmala Gopal
University of KwaZulu-Natal, South Africa

ABSTRACT

This chapter explores the use of a Virtual Learning Environment (VLE) to bring together students in criminology and social work from three universities across the globe (the University of KwaZulu-Natal in South Africa; the University of Washington in the United States of America; and Glyndwr University in Wales, United Kingdom), to examine the issue of racism and hate crime. The chapter provides a detailed case study of an online learning environment constructed to facilitate a four week international student conference. The chapter draws strongly upon the student experience and voice, extracted from the anonymous completion of online questionnaires at the end of the conference. The student narrative is supplemented by the experiences of the three authors who facilitated and managed the online learning process. The chapter explores the educational opportunities and limitations in respect of: online learning generally; the cross cultural shared experience; and in terms of engaging students in investigating global social issues.

DOI: 10.4018/978-1-60960-094-5.ch006

INTRODUCTION

Rapid and constantly changing technological advances provide new opportunities for criminals to develop ever sophisticated strategies to plan and manage criminal activities, while also providing law enforcement agencies with new technology to tackle crime. In this spiraling advancement of new gadgets and gizmos, both law breakers and law enforcers need to keep abreast of the latest technological developments. However, technology has also provided new opportunities in respect of crime prevention, collaboration, sharing knowledge, intelligence and ideas. This chapter explores the way in which academics from three continents (Africa, America & Europe) used a Virtual Learning Environment (Blackboard) to collaborate and establish an annual online 'conference' that enabled students from their respected universities to examine the issue of racism and race hate crime. The issue of racism is important to the authors personally and professionally. Interestingly, the three staff reflect a degree of diversity beyond that of geography, Julian Buchanan is a white, English speaking male with Irish/English nationality working in Wales, UK who teaches on criminal justice programmes, Steve Wilson is a Black, middle class, Christian, male living and working in the United States who teaches on social work programmes, while Nirmala Gopal is a South African Indian female living and working in South Africa who teaches criminology.

The issue of racism is important for students to study in all three degree programmes particularly as most students will eventually find employment in the criminal justice and social care sector. Once graduated these students choose professions as police officers, youth justice workers, social workers, drug workers, probation officers, housing officers, etc. The purpose of the conference was to raise students' awareness of the broader issues of discrimination, the abuse of power, stereotyping, hate crime and racism. While the context and culture in three countries is considerably different they all share a history of colonialism and racism. It was therefore felt particularly important to provide students with a historical insight to racism, as well as examine present day understandings of the nature and context of racism in the three countries, and explore the implications for future professional practice as many students will eventually find employment in the human services. The conference had three key learning objectives:

1. To develop an appreciation of the historical and contemporary manifestations of racism today.
2. To develop a critical understanding of the issues of race, inequality and discrimination.
3. To develop an insight into the different issues of diversity, language and power.

Underpinning the conference is the assumptions that unfair discrimination occurs in every society and most, if not all, people will be subject to some form of prejudice at some time in their life. All forms of discrimination warrants concern. Discrimination involves the use of power (by those who are in a position to possess or inherit power), to maintain their privilege and benefits at the expense of others by the perpetuating prejudice and stereotypes to validate their behaviour. Racism is a form of discrimination. Racism then involves the use of power by those who are in a position to possess or inherit power (most commonly white people) to maintain the privilege and benefits at the expense of others (most commonly people who may be seen to belong to a Black and Minority Ethnic group), by perpetuating prejudice and racist stereotypes. As a form of discrimination racism can be particularly damaging, because it is based upon the color of a person's skin, which is not only fundamental to their being and identity, it is something that is always visible, so racism remains an ever present risk which cannot be avoided. This unfair and discriminatory action (racism) is usually upheld by a combination of ignorance, prejudice and stereotyping by those in power. Racism can occur consciously or unconsciously, by action or by inaction, deliberately or accidently – in terms of suffering racism it is the outcome that matters not the intention.

Racism through the transatlantic slave trade has shaped the development and evolution of all three countries (SA, UK & USA) and has arguably been embedded in the very structures of society. Racism is then more than just personal prejudice. Racism involves power that goes beyond personal interaction; it is absorbed within culture and institutionalized within the structures of society. Racism operates at personal, cultural and structural level (Thompson 2006).

In examining a sensitive issue such as racism, the conference (which includes; lectures, videos, audios, PowerPoint presentations and discussions) tries to avoid any reductionism that categorizes or labels students as racists. A person who expresses a view which could be argued to be a 'racist' view should not be labeled a racist. The idea may indeed be racist but it is important in the conference to examine the debate and not the person. An idea can be labeled as perpetuating racism but nobody should be labeled racist without considerable evidence to support the claim. This has never been an issue in the conference as the ethos of the conference is to learn and explore together. The conference is not about correction but exploration. Students are not participating in the conference to absorb politically correct language, but rather they are encouraged to critically explore the issue through examining history, looking at recent events, and considering theoretical frameworks. In doing so staff and students use the discussion forums to critically investigate, think aloud and debate. The conference aims to create androgenic social constructivist strategies to facilitate a student community of learners rather than being based on established and dominant pedagogies found in the traditional Lecture Theatre.

Engaging with students from other countries with different histories, different conceptual frameworks, different use of language and different experiences, enriches the critical exploration of the issue. While there is a global problem of hate crime – the manifestation, motivation and context of such crimes will be different between countries and cultures. This conference was not seeking to provide students with a global legal response to the problem, but rather to engage students to begin to critically explore some shared issues, experiences and trends concerning the hate crime and racism in particular. The creation of a virtual learning environment provided new learning opportunities for staff and students to work collaboratively together to explore this important issue. The different life experiences, difference cultural contexts and indeed different conceptual frameworks for understanding and tackling racism added to the richness of discussion and exploration. The opportunity to be able to study together with students thousands of miles away, with such diverse backgrounds captured students' interest and the different perspectives enhanced learning experience. Indeed, there appears to be educational merit using a cross cultural virtual learning environment to help students to grapple with subject matters such as racism which have historically been fraught with ignorance, prejudice and stereotypes.

THE IMPORTANCE OF CULTURAL COMPETENCE

There are a number of reasons for encouraging students to examine racism and oppression: with increasing social mobility communities are more multi-national and multi cultural; increasingly, daily social interaction necessitates communication and engagement with people from different social backgrounds, different cultures and different nationalities whose first language will not necessarily be English. Students who seek employment in the social care or criminal justice setting have added responsibilities to be sensitive to the increasingly diverse communities they serve. The

United States Surgeon General (US Department of Health, 2001) in a report on mental health provision indicated a need for service providers to understand and respect cultural differences, he stated *'In today's multi-cultural reality, distinct cultures and their relationship to the broader society are not just important for mental health and the mental health system, but for the broader health care system'* (p.1). The report also called for agencies and individuals to provide culturally competent services to the people they serve, that not only recognizes and responds to their own culture, but to also recognizes and responds to the culture of the service users they serve (Johnson et al., 2004).

To be cultural competent a person needs to acquire the necessary knowledge, skills and values to be able to understand and relate sensitively and appropriately to another person from a different background (Betancourt et al., 2003). This 'difference' could be related to a combination of factors such as; race, religion, color, culture or nationality. While it will not be possible for someone to be well-informed on all aspects of diversity, there are transferable skills, shared values and underpinning principles that can help equip staff to be culturally competent. However, cultural competence requires more than just the acquisition of knowledge through cross-cultural training. To become cultural competent involves; learning to see, understand and engage the world from the 'eyes of the other'; developing an appreciation of the nature of discrimination, prejudice and stereotyping; and a commitment to respect and celebrate difference. Insensitive cultural practice has less to do with limited knowledge and much more to do with oppression, stereotyping and prejudice.

Unless agencies and organizations develop cultural competence, disparities in health, social care, justice, housing and employment for Black Minority Ethnic groups will continue. It is important not that we treat everybody the same (sometimes referred to as a non-discriminatory approach), but instead cultural competence requires that we acknowledge difference and treat people differently (sometimes referred to as an anti-discriminatory approach), – in order that they may access services and opportunities equally. Access to resources and opportunities tends to favor the privileged dominant group who hold power. While power is generally held by white middle class males it should be recognized that some groups who may share the same characteristics (white and male) may also find themselves subject to discrimination and marginalization. This does not negate the importance of developing cultural competence but it illustrates the complexity of inequality and discrimination. Tackling the social exclusion and marginalization of poor white (discarded) working class communities (Buchanan & Young, 2000) is important, but it is not the focus of this chapter.

The need to develop culturally competent students is an important strategy to tackle disparities. However, it could be argued that given the threat to heterogeneity from a rapidly expanding globalization (that appears to be creating a homogenous global 'westernized' world), there is an ever-pressing need to educate students to protect, promote and celebrate the richness offered by diversity and difference. There are also pressures towards xenophobia and nationalism characterized in the United Kingdom by the rise in the British National Party and the UK Independence Party which could fuel intolerance, nationalism, racism and hate crime.

While the pressures that might inadvertently nurture racism exist, there are also major changes in global society that could encourage and facilitate greater tolerance, respect and understanding. For example, the constant global information exchange and cultural communication evident in social networking sites such as *Facebook* and *Twitter* has resulted in new networks and social transformations which bring people and ideas together from diverse backgrounds. People from previously hidden and unrepresented communities across the globe are now able to express themselves and converse with anyone in the world who has

Internet access. Without these new online social networking opportunities this range and depth of communication wouldn't have been possible. The Online Race Conference is another example of the network opportunities and collaborations now possible using the digital media.

UNDERSTANDING THE VIRTUAL LEARNING ENVIRONMENT

The Online Race Conference became possible because students at Glyndwr University Wales studying the BA (Hons) Criminal Justice had been using the Blackboard Virtual Learning Environment (VLE) systematically as half of the degree is taught entirely on Blackboard. Online learning is sometimes referred to as e-learning and has been widely embraced by universities in most countries. Dedicated software such as *FirstClass*, *Blackboard*, *WebCT*, *Learnwise* and *Moodle* are widely used to provide a virtual learning environment which can incorporate a wide range of learning methods. The merits or otherwise of e-learning have sometimes been overshadowed by debates upon the merits of different software programmes. What is common across different software programmes is that e-learning is able to afford students a number of key benefits that may otherwise be difficult to secure in traditional lecture based learning:

- Students can access learning at a time that better suits them. This could be particularly important to students who have to juggle learning with other commitments such as employment or for caregivers who look after family or friends.
- Students can access learning at a location of their choice. This could be particularly important in providing access to education for those living in rural areas.
- Students will never miss a lecture or presentation as they can be permanently available online. In face to face lectures students may miss lectures due to other commitments, sickness or transportation difficulties.
- Some students in traditional face to face lectures don't always immediately understand the debates taking place and some can feel overawed asking questions in a large group. In comparison online students have better opportunities to reconsider what they've heard or read, reflect on what is said, and to participate in continued debate and discussion. Their participation in online discussion can be more easily prepared and considered making it easier for less confident students to engage in the learning process.

E-learning has been defined as the process of accessing the internet to engage in a program of formal or informal study. In addition to the traditional lecture which may be in the form of a MS Word document, MS PowerPoint, audio file or video file, students can engaging in synchronous (taking place when all the students are present online at the same time) or asynchronous learning (when students contribute to a discussion 'thread' at various times). Online learning provides enormous scope for a wider range of learning mediums. For example, communication online can include emails, discussion forums, RSS feeds, video conferencing, websites, pdf reports, online questionnaires etc.

At its best the VLE is able to create a community of learners who are not studying in isolation as distant learners, but are engaged in exploring and learning together synchronously and asynchronously in a social and educational network. Whereas, distance learning tends to involve students reading or watching educational material that has been posted to them, and working primarily alone at home. For this reason e-learning is sometimes described as networked learning as opposed to distance learning, although it may well be taking place some distance away from the university campus. When e-learning is supported by face to face traditional learning it is often referred to as Blended Learning.

The experience of being part of a virtual community of learners e-learning should does not seek to replicate the experience of the traditional face to face setting to justify its worthiness. E-learning is different to traditional face to face lectures in the same way that an email or a text message is not the same as a letter. Recognizing and embracing the strengths and weaknesses of new and different forms of learning will enable a better match between the learner, the subject matter, and the most appropriate and best suited learning environment. Online learning provides virtuality which is significantly different from the physicality of traditional campus based learning. The characteristics of the online environment can be disinhibiting, disembodying and anonymizing (Griffiths, 2000).

It is important to recognize that the principles underpinning good learning practices are largely the same whether learning takes place in the lecture theatre or in the virtual world of the e-learning community. E-learning involves the deployment of new technology and opportunities in adult education, as opposed to the adoption of a new educational paradigmatic shift in learning (Madoc-Jones, Buchanan, & Senior, 2003). There is nothing built into e-learning that is likely to make studying online an inherently more positive educational experiences than studying through traditional face to face lectures. Both processes depend largely upon the skills of the educator and how well the distinctive learning opportunities are utilized.

THE STRUCTURE OF THE ONLINE RACE CONFERENCE

At Glyndwr University the Blackboard VLE wasn't used to supplement traditional face to face lectures by providing additional online resources, instead it was used as the sole learning and teaching forum for 50% of the BA (Hons) Criminal Justice. The Social Difference and Inequality module in the second year which runs over 24 weeks requires students to examine issues of discrimination, in particular, looking at inequality, power, stereotyping, race, class, disability, gender, sexual orientation and religion. Since 2004 the issue of racism has been covered by way of a four week Online Race Conference. Following networking opportunities at academic conferences the opportunity to expand the conference to include staff and students from University of Washington and the University of KwaZulu-Natal arose. Getting staff and students together to examine racism, its process and impact within the three countries provided an exciting opportunity. However, due to logistics and resourcing it would never have been possible to physically bring staff and students together in one geographic location, but the VLE provided a relatively easy alternative way of making learning together a real possibility.

The conference was scheduled to run for four weeks and was provided with its own dedicated and customized Blackboard site separate from the BA (Hons) Criminal Justice Blackboard site (see Figure 1). Given the time zone difference with Seattle, USA and the sometimes unpredictable internet speeds in South Africa the conference avoided any attempt to engage in synchronous discussion. The customized Blackboard VLE was simplified so that the content resembled what students could expect to encounter if attending a traditional face to face conference - papers, seminars, informal discussions, meeting places, additional resources and handouts.

The conference entitled 'Racism - Still a Problem?' was based upon three themes spread over the first three weeks of the conference: learning from the past; reflecting upon the present; and theoretical and practical issues. Each week a new set of conference papers and resources were made available. The conference was originally established with UK students in mind so the material tended to have Eurocentric perspective, however, since the collaboration with USA and SA more global perspectives have been incorporated.

The first set of papers 'Learning from the Past' brought students together to understand the nature and context of transatlantic slavery which affected all three continents represented at the conference. This set of papers also provided an insight into the more recent history of the lynching in the USA by so called 'white-supremacists'. In addition to lecture notes and MS PowerPoint presentations these conference papers contained video, audio, music, poetry and web links including: the Liverpool International Slavery Museum, Without Sanctuary Memorial, and Billie Holliday singing 'Strange Fruit'. By appreciating historic aspects of racism, the abuse of power and the way in which such atrocities were passively allowed or even supported by so-called 'ordinary' folk, students were able to develop a more critical insight into division, discrimination and the abuse of power in present society. Discussion board debates suggest that students were surprised and shocked by the inhumanity graphically evident in these conference papers. Some students were shocked by the fact that they'd been previously ignorant of such history in their own country. Another observation from students was surprise at the similarity in patterns of racist abuse across the globe. The following extracts from the Blackboard Discussion Board capture the student response to these first set of papers on slaver and illustrate the way in which students were able to reflect and make connections between the past and present day situations:

People are still trafficked from underdeveloped countries with high levels of poverty to developed nations and exploited in menial jobs, living in squalid conditions with no consideration of safety (the cockle pickers) or worse, in the sex industry. High profile media attention is focused on babies and children "adopted" by wealthy individuals under the guise of giving them a better life. Can being removed from your siblings, wider family, country and culture to live among strangers who do not speak your language be "better" for a child? The use of body organs for transplant is tolerated increasingly obtained from people in dire straits of poverty as if western lives and health needs are superior. Despite efforts to police these inequalities, power and wealth continue to prevail.

As an African American, past slavery continues to have a profound effect in my life, because it is the history of slavery that has robbed African Americans of their origins. This allows for us not to be accepted by both the country that we helped develop, and having no understanding of our past individual orientation. Seemingly lumping all African Americans into one. This feeling of not

Figure 1. The dedicated blackboard virtual learning environment for the race conference

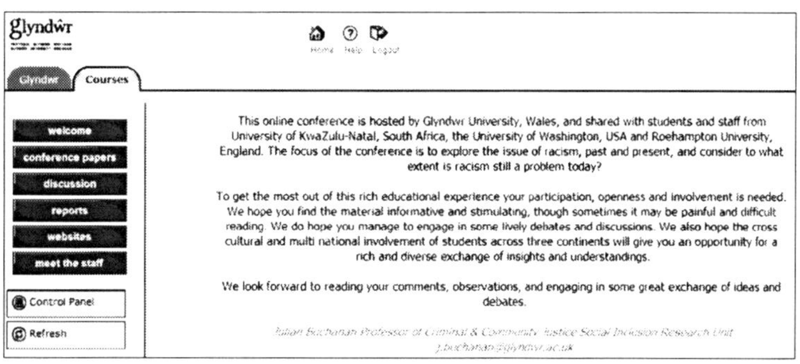

belonging, tugs at the very heart of problems facing African Americans in the United States. It is with this portents that I live my life trying to prove to myself and the dominate culture in America, that I belong here as much as any other. So I continue my daily grind, knowing that I must be on top of my game more than any other American, to disprove any preconceived stereotype, brought on by the construct of slavery.

The Without Sanctuary website that provides a visual legacy or memorial contains photographs and postcards taken as souvenirs at lynching's throughout the United States of America had had a strong impact upon students:

I was taken back with the images that I faced while looking at this website. I could not believe how at ease the majority of onlookers appeared to be, while observing such horrific crimes. Furthermore, to actually have pictures of such scenes on stationary such as post cards. I was shocked to say the least.

Like others I found the image of the spectators to be very disturbing, the expressionless faces and the sheer 'normality' of it all captured in these pictures will haunted me forever. The worst part about it is ... this kind of crime continues even today but is now termed as 'hate crime'. It covers a wide range of issues e.g. disabilities, religion, race, sexual orientation.

In the second week online conference papers moved on from historical events and examined racism in more recent times. This included the murder of Stephen Lawrence and Anthony Walker, the Macpherson Report, deaths in UK police custody and examples of racism in sport. Again a wide range of additional media was used including case studies, newspaper articles, an online questionnaire and various YouTube video clips including poetry by Benjamin Zephaniah, 'Too Black Too Strong', the USA Jena Six and the killing of Oscar Grant. The Discussion Board debates encouraged students to reflect upon what they had learnt and explore why such events happen today. A number of students admitted that before they began the conference they had assumed that racism was largely no longer a problem. Some of these students in their discussions on Blackboard began to reconsider their knowledge and understanding of racism:

I am stunned at the extent of racism today. Before doing this conference, I did not believe the issue was so prevalent. Then again I don't really suppose I have given it much thought tho!

I think it is difficult to be completely oblivious to racism, as it goes on around us so much. I was aware of the Christopher Alder case as I had studied Police powers whilst in college doing Law, back then we watched the police custody video and I found it incredibly disturbing. However, I was unaware how many other similar cases there were. So many cases clearly suggest there is something wrong, whether that is a lack of education within the police forces regarding race, or something deeper set than that, i.e. racist beliefs that are innate to the individuals themselves.

It seems like a lot of these deaths are attributed to the way in which the individual is restrained (asphyxia). I know the police need to use restraint, but in a lot of these cases they seem to have no patience and use immediate brutal force - how none of these officers have been found guilty of an offence is beyond me - even when there are witness statements and videotape footage!!?

The men who died from being restrained by police or prison officers I found very uneasy reading. It is a lot more unsettling that those 'responsible' for public safety and enforcing the law in an appropriate manner may be responsible for deaths such as these is extremely worrying and shocking. The man who died on the police station floor after choking on his own vomit (Christopher Adler) when watching the tape of this it reminded so much of the crowds of people on the postcards etc when watching lynchings. This may seem a rather extreme view but in the cases of the lynching nobody helped and people were left to die. Also very worrying was what could argued is apathy shown by prison offers when Zahid Mubarek wanted to moved away from his overtly racist cell mate which ended in his tragic death.

In the final set of conference papers which looked at theoretical and practical frameworks for practice students examined: statistics of the disproportionate experience of people from Black and Minority Ethnic (BME) groups of the criminal justice system in the UK; the use and abuse of humor to promote racism; the constructs of personal, cultural and structural discrimination (Thompson 2005); notions of discrimination, non-discrimination and anti-discrimination; the abuse of power and privilege; positive action and positive discrimination; and the importance of acknowledging difference so that people can be treated more equally. Again a wide range of multimedia techniques were used including the YouTube video of Brown Eyes Blue Eyes. Discussion Board debates involved students considering a range of policy and practice issues encouraging debates upon: whether we should treat everybody the same; whether it is important to see difference; whether prejudice is something we all suffer at some level; how and why is racism any different; whether it is best for service users and workers to be matched according to color, gender, disability etc; and asking students to explore whether slavery was still a problem in the 21st century. By the third week students progressed beyond knowledge and understanding and began to consider their conceptual framework and explore the implications for policy and practice:

The conference material especially week 3 has helped me realise that I have always tended to treat everybody the same. But now thinking and reading around the subject in much greater detail I have realised that nothing really positive can happen with that. If everybody is 'the same' we cannot understand or celebrate diversity. But not only that we cannot hope to discuss [or tackle] the discrimination that some people may be facing if we do not recognise difference and feelings about it.

I honestly didn't piece together that racism was not just about colour but includes power and prejudice, this subject I found really interesting to read. As I continued to read I found myself answering the questions differently the more I thought about the consequences of what I was thinking, I found it strange how 20 slides can change a person's outlook on life and society.

Reading the PowerPoint and this thread has encouraged me to think of personal experiences. Although when I went to school there were very few BME pupils, there were however a number of Travellers. Looking back they were 'targets' for discrimination by other pupils; however this was by a very small number of individuals and overall because the school had a history of educating members of the travelling community.

The video footage of the Brown Eyes Blue Eyes school experiment had a powerful impact upon a lot of students helping them to see how easily stereotyping, social control and discrimination can take place:

I was shocked how easily the participants, both children and adults, slipped into their roles. When watching the adult version, I expected some to walk out. Just goes to show how easily people can be manipulated and controlled. A very powerful study and very brave of the teacher to address such issues.

The focus of the conference was upon racism but students were also encouraged to make links with other forms of discrimination such as the experiences of women, travelers and people with disabilities. Students frequently make connections with disenfranchised indigenous people participating in the conference. There is a shared experienced with the experience of Native Americans living in the USA, Black and Indian South Africans in South Africa, and first language Welsh speakers in Wales (see Madoc-Jones & Buchanan 2004 for further reading on this issue). It is good to see that students are able to make connections with other forms of oppression and observe similarities in the misuse of power throughout the history of different nations participating in the conference. This student highlighted the legacy of oppression suffered by Welsh speakers in Wales:

During the 18th and 19th century there was a huge effort [by the English] to prevent the [Welsh] language being spoken. Some schools used the "welsh not" as a deterrent for speaking welsh. It was a piece of wood hung around the neck and was passed from culprit to culprit. Children were encouraged to report each other for using Welsh and the child wearing the "not" at the end of the day was punished by the teacher. This method was almost successful in abolishing the Welsh language... it has taken a long long time for the language to recover.

In addition to the conference papers the Blackboard VLE the Discussion Board area contained a Virtual Coffee Bar that provided a forum for staff and students to chat more informally. Additional material was available under the Reports section (see Figure 1) which provided access to a wide range of online reports regarding racism, slavery and broader issues of discrimination. The Websites section provided access to over forty carefully selected websites, while the Meet the Staff section provided a photograph and brief biography on the three staff participating and moderating the online conference.

STUDENT EVALUATION

Most students engaged actively and enthusiastically to the time limited conference material. Not unlike traditional face to face lectures some students remained quiet contributing little by way of discussion but this couldn't be seen as necessarily indicative that they were not involved and not learning. It should be acknowledged that the issue of racism can be difficult to explore with some students reluctant to engage for fear of being misunderstood, exposing their ignorance or being perceived in a negative fashion. Awareness of these difficulties staff were keen to ensure the environment was conducive to learning and were sensitive to such issues.

At the last conference in 2008 the discussion boards (one for each set of conference papers) resulted in a healthy total of 588 contributions from 39 different students out of a possible 43 students. At the end of the 2008 conference all students were asked to complete an anonymous online questionnaire (see Appendix One) to evaluate their learning and experience of the conference. The questionnaire was completed by 21 students (12 from UK, 0 from SA, and 9 from USA). The questionnaire largely involved the students responding to questions with answers on a five point Likert scale but it also included some open ended questions to elicit student comments and feedback. The qualitative student comments provide an additional rich insight to support the quantitative data, however, given the small scale nature of this survey and a response rate of 49% the results should be treated with an appropriate degree of caution.

Given that the online conference replaced four two hour face to face lectures the expectation was that students would devote a similar amount of time (6-10 hours) participating at the online conference. Figure 2 indicates that the majority of student exceeded this:

Had the conference been delivered face to face the vast majority of students wouldn't have been able to attend and those that did would have had to engage with the conference at specified times and locations. The following graph (Figure 3) indicates the way in which the online nature of the conference afforded the students with the possibility of studying not only from a location of their choice (usually the university or their own home), but importantly at a time of their choosing. Interestingly, five students accessed the conference in the early hours of the morning. Evening time proved just as popular as morning or afternoon study with 13 out of the 21 students accessing Blackboard in the evening. The internet is available 24 hours a day, so in theory students in different parts of the world in different time zones should have equal access to participate, however, the Glyndwr University who host the Blackboard VLE 'back up' the computer network between 1.50am and 6.40am BST which meant that students in Seattle USA were unable to access the site between the hours 5.50pm -10.40pm. This did pose additional challenges as one student explained: 'It was a bit inconvenient to have to get up very early to get online before going off to work. The system was shut down during our evening hours'.

Overall the Blackboard Virtual Learning Environment was rated highly by students with 75% (16) of students rating it good or very good (N=19). When asked to explain the benefits of the VLE

Figure 2. Roughly how many hours would you say you have spent on the Blackboard Conference (n=21)

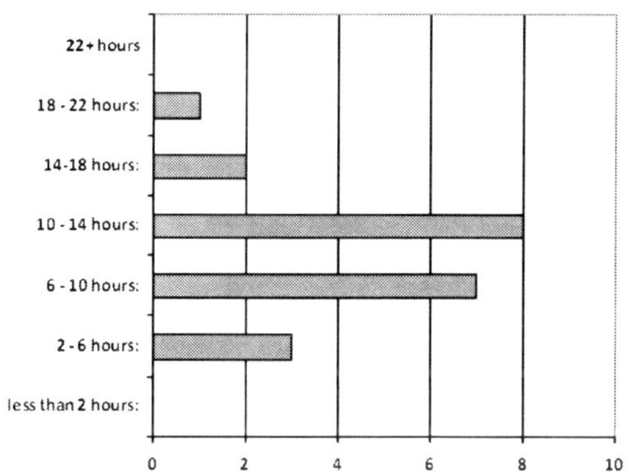

Figure 3. What times of the day/night did you access this conference? Tick all that apply (n= 21 students)

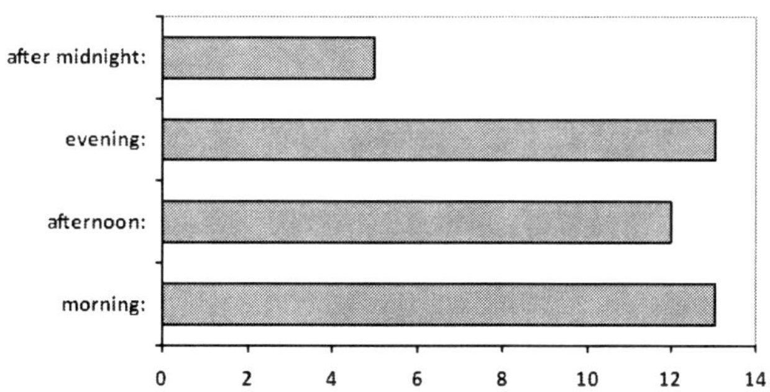

students particularly highlighted access and flexibility as one student explained *'Access when I need it, i.e. when the children are in bed!'* Others mentioned being better able to engage *'equal opportunity to respond'* and *'better able to express myself'*. The VLE tends to be particularly appreciated by quieter students who would otherwise be less vocal in face to face as one student explained: *'You get more time to think about what you want to say, and especially for me I feel more confident about speaking online rather than in a classroom'*. Reflective students also value being able to digest and consider information at a pace that is better suited to their learning style: *'it enables you to do more research of a topic and it enables you to read around a topic before engaging in a debate'*

There was also an appreciation of blended learning as students valued *'a different* [learning] *experience'* and students valued highly being able to explore different cultural perspectives from students studying in different continents: *'It is convenient as I can get online at home, I can interact with people anywhere in the world, I can access online articles and video clips'*. Another student emphasized exposure to alternative perspectives: *'I can converse with others I would not have a chance to converse with otherwise. It is good to hear different perspectives'*. Later the questionnaire specially asked students whether learning with staff and students from other countries made a difference - 91% felt it made a positive difference. Key comments included:

Gained knowledge and insight into their cultures and viewed the subject from a global perspective.

To see different point of views and learn of what goes on in different parts of the world.

I absolutely love the experience of learning with students from other countries. I want to continue to do this in some way during my graduate education.

I wish we'd had time to get to know a bit more about each other before the conference began in earnest.

The whole set up was very good; it was interesting to see other people's views from other countries. One of the most interesting discussions I have been involved in.

When asked about the drawbacks of online learning students missed the nuance of face to face communications. One student highlighted the way in which online discussion lacked the spontaneity and immediacy of face to face debate where '*often a greater debate can be entered into, where people sometimes hold back online in fear of getting it wrong'*. Another student expanded '[You] *don't necessarily get people's initial reaction; you get a thought out censored answer*'. While nuance is constantly evolving in written form - such as capital letters communicating anger or someone shouting, there are major limits to text based communication as another student explained '*much is said within a person's body language / voice that cannot be gathered online*'. The lack of physical contact was clearly a drawback for a lot of students although one student said the only drawback to studying online was '*that you are not meeting others in person*' but the student also suggested '*Maybe someday we can use webcams and have meetings!*'

The main purpose of the conference was to broaden student knowledge in respect of the extent and nature of racism past and present. Students were asked to recall and rate their knowledge of racism before they began the conference using a standard Likert scale from very poor to very good, the question was then repeated asking students to rate their knowledge of racism after completion of the conference. The self assessment results below (Figure 4) indicate good improvements overall:

While the Likert data indicates a clear improvement of knowledge overall in respect of racism the qualitative comments, interestingly provided

Figure 4. Knowledge of racism before and after the conference (n=21)

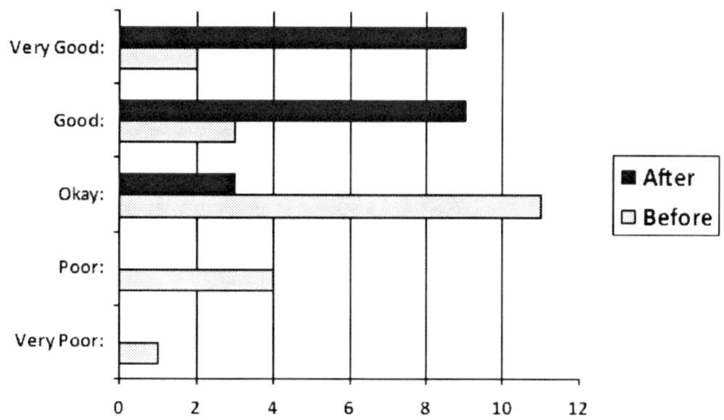

by all 21 students, gave an insight into what particular knowledge students that students had gained. They were asked: What (if anything) have you particularly learnt about racism from doing the conference? Comments were coded into three categories. The first group of students (n=9) emphasized their relative ignorance and naivety towards the issue of racism and valued learning about the historic and present day nature and context of racism. Comments from four students illustrate their acquisition of knowledge and desire to understand more:

I have learnt a lot from the conference, there were sections within the discussion board which I had never heard of or thought about in much detail. It has definitely improved my understanding of the subject.

The fact that it can still have an effect on individuals today and the fact that they suffer so much discrimination within the justice system was surprizing to me.

Just how prevalent it still is today. Was really naive to the extent.

I learned historical facts about racism that I never knew in depth.

Another group of students (n=7) emphasized how they had learnt to understand the problem of racism differently. They gained theoretical and conceptual insights

I have learnt that if you ignore something it is not going to go away. Also doing nothing is almost as bad as contributing to racism.

I think I have a lot left to learn about discrimination and racism. However, this course and materials really opened up my eyes about how and where racism occurs. I never thought about institutionalized racism in high school. It was interesting hearing there is a difference between being non racist and anti racist. Much of the information really got me thinking about this subject in a different way than prior to taking this course.

I think the biggest learning point for me was seeing racism as a mechanism of power.

The remaining students (n=5) expressed comments regarding the diverse and global nature of racism and how discrimination and hate crime impacts other groups. These students valued learning form and about the experiences of others:

It was interesting to learn that students in Wales and the US are fairly similar in our knowledge and experience of racism.

The racism comes is all different forms and is a part of everything in the world, that there are people out there who judge and punish people purely on colour. But also that I might not be able to change the world but I am able to educate people that I know about the effects of racism.

That it is still amongst us throughout the world. Also, it seems that there are people who don't want to discuss racism today.

Students felt they had gained considerably from the opportunity of studying a global social issue such as racism with fellow students from different social backgrounds, different culture, nationality and race, as illustrated by these four students:

Finding out about experiences from people in the other universities.

I enjoy every aspect of the race conference, from meeting new people from around the world to debating different topic.

I enjoyed getting to interact with different global perspectives.

The chance to see differing views from different countries.

Eighty-nine per cent of students felt that an international online conference was a good medium to explore social and cultural issues such as racism. And one student suggested: *'It would be good if we could communicate with other students about issues that occurred in each other's countries throughout the year not only through a 4 week conference, just a thought!'*

DISCUSSION

Glyndwr University is based in North Wales which historically has few people from a Black Minority Ethnic (BME) community, representing only 1% of the population so some students have disappointingly had limited engagement or debate concerning racism. However, students from Wales like students from South Africa and the USA valued the chance to learn from each other. Exposure to different perspectives from across the globe was a highly valued by students, particularly when they discovered new information or learnt alternative ways of seeing. For example discussions regarding race self identification in the three countries illustrated the diversity and indeed complexity of the issue. When exploring appropriate terminology some students wanted to understand why the term 'colored' was considered inappropriate and has been replaced in the UK by the preferred terminology of Black Minority Ethnic (BME). This is relatively easy to explain, however, discussion with staff and students in the USA highlight that in the USA the term 'people of color' is preferred. This emphasis on people and the celebration and uniting of people illustrates the complexity, whereas in South Africa students terms such as Indian and Black are preferred. This sensitive exchange of information regarding appropriate and inappropriate language and the different perspectives held in each of the three countries enabled shared questioning and learning within which students had to grapple to appreciate the underlying principles and events that have resulted in these differences.

The intensive and time limited nature of the conference combined with the attraction of dialogue with students across the globe clearly has many benefits and synergies for the students and the lecturers. The students are actively engaged as adult learners, as researchers sharing experiences and challenging each other to think critically. In South Africa this has been helpful to encourage students to think critically about themselves in a post-apartheid context university and to consider the broader race agenda. Exposure to students from different continents has helped students from each country to become more open to ideas and challenge and learn from each other. The opportunity to read, consider and discuss the experience and storied from other students has been educationally stimulating and at times challenging of beliefs and values. This is particularly helpful for issues such as racism and maybe easier to explore in written format rather spoken word.

There is much debate about the strengths and weakness of different learning environments such as whether to use open source *Moodle* or the commercial software *Blackboard*, however in my experience much more important than a pre-occupation with the 'tools of the trade' is to concentrate upon the quality of teaching and appreciate what significantly contributes to a successful VLE. However, the technology has provided higher education new opportunities to engage different social groups, to develop new IT and online skills, to engage students in new learning environments with better quality of study materials with easier access for searching and retrieval. As Sharma (2006) highlighted:

'Technology has given us a new way to communicate and come together despite boundaries of distance, time zones and cultures. We can use it in a way that furthers human communication and learning from each other, rather than becoming a slave of technological tools.' (p.81)

When seeking to provide a successful educational online learning experience there is a risk of focusing too much upon the technological opportunities. Technology will not make a poor teacher a good teacher, nor will it transform a boring lecture into an interesting lecture. Online learning environments are too often left to become a repository from which the student downloads, reads and digests knowledge. Dialogue, engagement and exploration are key so that as Boyer (2006) explains: *'the learning landscape demands additional support for these online tools to move beyond the simple transfer of information to deep, meaningful learning that alters perspectives'* (p.338) To be successful the online learning process should utilize a range of varied teaching methods designed with the specific learners in mind, to engage the tutor and student in collaboration and construction of knowledge (Bach et al 2007). The online lecturer then becomes much more of a facilitator or moderator managing and steering the student learning experience (Sharma, 2006). This contrasts to the classical pedagogic approach to education. Literally taken pedagogy refers to the process of teaching children. In these terms the learning process mirrors that of adult to child rather than between two adults. The teacher has control and responsibility for deciding what will be learned, when and how it will be learned. In contrast the social constructivist approach is androgenic based upon adult learners who are mature, independent and responsible for their engagement and participation in learning (Madoc-Jones, Buchanan, & Senior, 2003). This social constructivist approach is based upon the principle that students will learn not only from what they read and discover but importantly they will learn through the experiences of others (Boyer 2006) and this principle underpins the approach to learning in the Online Race Conference.

SOLUTIONS AND RECOMMENDATIONS

1. While the 'MacDonaldization' of everyday life threatens to undermine and marginalize the richness of difference, diversity and distinctiveness, the internet can to some extent provide an antidote by bringing together new networks for individuals from very different backgrounds that can communicate, listen and learn from each other. The challenge is for these networks to create heterogeneity

rather than homogeneity – this should be more than possible provided the networks are not based upon power, competition or profit.

2. Successful online learning needs to be based on androgenic social constructivist strategies to facilitate a student community of learners rather than being based on established and dominant pedagogies that are already well served in the traditional Lecture Theatre. While technology afford tremendous opportunities, the success of the Online Race Conference is based much more upon good preparation, a well managed and facilitated VLE, underpinned by a androgenic social constructivist philosophy to provide a safe and collaborative adult learning environment. Successful facilitation requires timely substantial considered responses to students' contributions to encourage further exploration, investigation and discussion. Throughout the conference, students were encouraged to engage in reflection and dialogue with fellow students as well the lecturers.
3. Unresolved issues remain such as limited internet access in some countries that serves to marginalize some countries and communities. Engagement at the conference by South African students was at times been hampered by limited or poor internet access. The conference was originally established exclusively for students in the UK and despite adaptations the material is still too Eurocentric. Despite these shortcomings the experience of students participating in the online conference suggest there is a real opportunity for international global collaborative amongst academics and students to utilize VLE in a managed structured way to help educate students on a wide range of shared concerns that infringe and diminish the quality of life for others.
4. Despite running collaboratively for three years the conference is still at an early developmental stage and more work is needed to ensure the conference material better reflects a shared curriculum that incorporates the history and experience of all three countries. It is a challenge to coordinate and ensure the conference takes place at a time that is suitable for all and is embedded within the different degree programme structures.
5. The Online Race Conference has demonstrated fairly well that it is possible to involve and integrate students from across the globe to study and learn together in an online cross cultural environment to examine key global social issues such as racism. Other social issues that have a global concern could also usefully be explored through international online conferences such as disability, gender, domestic violence, child abuse, drug misuse, etc The use of a VLE provides an exciting relatively easy to manage (compared to organizing a geographically based conference) educational opportunity for staff and students that appears effective and cheap.

CONCLUSION

The evidence, albeit small scale, suggests that students learning regarding the complexity, extent and nature of racism has (for those who completed the online questionnaire), been successful. Although entitled an Online Race Conference the conference has also facilitated much broader learning regarding issues of diversity, discrimination, power, stereotyping and cultural competence. Extending the learning over a four week period appears to give the student the necessary time needed to 'mull over' and reflect on the learning. When issues often fraught with ignorance, prejudice and misunderstanding are covered quickly there is a risk that students resist learning as it may sometimes require a review of deep-seated held assumptions. For staff it has also raised discussions regarding adult learning, social constructivist approaches, teaching methods. Subjects such as racism which are deeply embedded personally, culturally and structurally are not easy to examine but arguably educational programmes that are equipping students with knowledge, skills and values needed to work human service organizations have a responsibility to provide an educational opportunity to examine racism. This cross-cultural examination of an important issue, by a generation of individuals who are familiar with technology, social networking and virtual learning, provides new opportunities and models for learning and teaching methods.

ACKNOWLEDGMENT

We would like to thank the students from Glyndŵr University, Wales; University of Washington and University of KwaZulu Natal who engaged enthusiastically and openly in the International Online Race Conference and for completing the anonymized questionnaires that have informed this chapter.

REFERENCES

Betancourt, J. R., Green, A. R., Carrillo, J. E., & Annaneh-Firempong, O. (2003). Defining cultural competence: a practical framework for addressing racial/ethnic disparities in health and health care. *Public Health Reports*, *118*(4), 293–302.

Boyer, N. R., Maher, P. A., & Kirkman, S. (2006). Transformative learning in online settings: The use of self-direction, metacognition, and collaborative learning. *Journal of Transformative Education*, *4*(4), 335–361. doi:10.1177/1541344606295318

Buchanan, J., & Young, L. (2000). Examining the relationship between material conditions, long term problematic drug use and social exclusion: A new strategy for social inclusion. In Bradshaw, J., & Sainsbury, R. (Eds.), *Experiencing Poverty* (pp. 120–143). London, UK: Ashgate Press.

Hansen, D. E. (2008). Knowledge transfer in online learning environments. *Journal of Marketing Education*, *30*(2), 93–105. doi:10.1177/0273475308317702

Johnson, R. L. (2004, Feb.). Saha, S., Arbelaez, J. J., Beach M. C., & Cooper, L. M. (2004). Racial and ethnic differences in patient perceptions of bias and cultural competence in health care. *Journal of General Internal Medicine*, *19*, 101–110. doi:10.1111/j.1525-1497.2004.30262.x

Kirkpatrick, G. (2005). 'Online 'chat' facilities as pedagogic tools: A case study'. *Active Learning in Higher Education*, *6*, 145. doi:10.1177/1469787405054239

Madoc-Jones, I., & Buchanan, J. (2004, December). Indigenous people, language and criminal justice: The experience of first language Welsh speakers in Wales. *Criminal Justice Studies*, *17*(4), 353–367.

Madoc-Jones, I., Buchanan, J., & Senior, P. (2003). Delivering the diploma in probation studies in North Wales and dyfed powys: On line learning explored. *British Journal of Community Justice*, *2*(2).

Matheson, S. G., & Lager, P. (2007). A model for developing international student exchanges. *Social Work Education*, *26*(3), 280–291. doi:10.1080/02615470601049867

Sharma, D. C. (2006). Online Technologies kill distance in learning managing participation in online journalism courses. *Journal of Creative Communications*, *1*(1), 75–81. doi:10.1177/097325860500100105

Thompson, N. (2006). *Anti-discriminatory practice* (4th ed.). Basingstoke, UK: Palgrave Macmillan.

US Department of Health. (2001). *Mental health: Culture, race, and ethnicity* (supplement). Retrieved October 5, 2009, from: http://www.ncbi.nlm.nih.gov/books/bv.fcgi?rid=hstat5.chapter.971

Wallace, R. (2003). Online learning in higher education: A review of research on interactions among teachers and students. *Education Communication and Information*, *3*(2), 241–280. doi:10.1080/14636310303143

ADDITIONAL READING

Bhui, H. S. (2008). *Race and criminal justice*. London: Sage Publications.

Chakraborti, N., & Garland, J. (2009). *Hate crime: Impact, causes and responses* (1st ed.). London: Sage Publications.

Mason, R., & Rennie, F. (2008). *The e-learning handbook: Social networking for higher education: Resources for higher education* (1st ed.). New York: Routledge.

Richardson, W. (2008). *Blogs, wikis, podcasts, and other powerful web tools for classrooms* (2nd ed.). Thousand Oaks, CA: Corwin Press.

Rudestam, K. E., & Schoenholtz-Read, D. J. (2002). *Handbook of online learning: Innovations in higher education and corporate training*. London: Sage Publications.

Thompson, N. (2006). *Anti-discriminatory practice*. 4th edition. Basingstoke, UK: Palgrave Macmillan. Webster, C. (2007). *Understanding race and crime*. 1st ed. Reading, UK: Open University Press.

KEY TERMS AND DEFINITIONS

Virtual Learning Environment: Refers to the use of a software programme to facilitate an online teaching and learning environment that is usually accessed via the internet.

E-Learning/Online Learning: Refers to structured educational study or training that is accessed via the internet.

Blended Learning: Refers to the combination of different learning and teaching approaches most commonly online learning and face to face learning.

A Social Constructivist Approach: Is one that recognizes knowledge is not fixed or given, but rather it is subject to change. A social constructivist approach to education actively involves the student and lecturer together reflecting upon experience and understanding to negotiate and construct knowledge and meaning.

Racism: Can be defined as the combination of power and prejudice, often fuelled by stereotyping, to discriminate (consciously or unconsciously) against people on the basis of their skin color.

Anti-Racism: Acknowledges racism, sees and embraces difference, and is committed to taking positive steps to counter act discrimination and promote greater equality.

Non-Racism: May acknowledge racism but a non-racist approach seeks to ignore difference by a commitment to treat everyone the same by adopting a 'color-blind' approach.

Cultural Competence: Refers to the need of those working in human services to possess the knowledge, skills and values to be able to acknowledge, understand and engage effectively with people from different cultural, ethnic, racial or religious backgrounds.

Hate Crime: Occurs when the perpetrator targets the victim because of a prejudice towards the particular group the victim represent such as; racial, ethnic, cultural, religious, sexual orientation etc.

APPENDIX

The following is a text version of the electronic online anonymous questionnaire that students were invited to complete at the end of the conference

International Race Conference Feedback

This is a short survey to get some feedback on your experience doing this race conference.
All your answers are kept anonymous.
Many thanks for your time and comments.

1. Which University do you study at?
2. Roughly how many hours would you say you have spent on the Blackboard Conference?
   Less than 2 hours
   2 - 6 hours
   6 - 10 hours
   10 - 14 hours
   14-18 hours
   18 - 22 hours
   More than 22 hours
3. What times of the day/night did you access this conference? *Tick all that apply*
   Morning
   Afternoon
   Evening
   After midnight
   Any comments: (Optional)
4. In terms of finding your way around what did you think of the Blackboard Learning Environment?
   Very poor
   Poor
   Okay
   Good
   Very Good
   Any comments: (optional)
5. What (if anything) did you particularly enjoy or appreciate in the conference? (optional)
6. What (if anything) did you dislike or wish was different in the conference? (optional)
7. How would you describe your knowledge and understanding of Racism BEFORE doing the conference, and how would you describe it AFTER completing the conference?
   a. BEFORE starting the conference my knowledge and understanding of racism:
      Very Poor - Poor - Okay - Good - Very Good
   b. AFTER completing the conference
      Very Poor - Poor - Okay - Good - Very Good
8. What (if anything) have you particularly learnt about racism from doing the conference?
9. Did it make any difference learning with staff and students from other countries?
   Yes - No - Don't Know

Please explain: (optional)

10. What do you think are the benefits of online learning? (optional)
11. What do you think are the drawbacks of online learning? (optional)
12. 'An international online conference is a good way to explore social and cultural issues' Please indicate how you feel about this statement
    Strongly Disagree Disagree Not sure Agree Strongly Agree
    Please explain: (optional)
13. Finally do you have any feedback you'd like to give about studying the issue of Racism through an International Online Conference? (optional)

Section 3
Humanitarian Technologies

Chapter 7
Humanitarian Demining Action Plan:
Humanity and Technological Challenges

Maki K. Habib
The American University in Cairo, Egypt

ABSTRACT

The presence of landmines and Explosive Remnants of War (ERW) in a place represents a major threat to civlian and affects the rebuilding process and the life of the people at that place. Hence, one of the fundamental goals of humanitarian demining is to detect and clear all forms of danger from infected areas efficiently, reliably and as safely and as rapidly as possible while keeping cost minimized. Although demining has been given top priority, currently mine's clearing operation is a dangerous, complex, time consuming, slow, labor-intensive, and costly operation. The currently available technologies are not suited to achieve the objectives of humanitarian demining. In the context of humanitarian demining it is essential to have a reliable and accurate sensor and/or an integration of heterogeneous/ homogeneous sensors with efficient and reliable data fusion and processing technique that can quickly discriminates mines from innocuous buried objects. In addition, it is necessary to overcome the constrain on the resources by developing innovative, cost effective and practical technology inspired by locality and real minefield needs to help in speeding up the demining process and enhance accuracy, productivity, operation and personnel safety, achieve higher quality of the service, and contribute to local economy. This chapter presents the facts and problems associated with landmines and their impact on health, economy, land and environment along with the difficulties in detecting and removing them. It highlights the main requirements for humanitarian demining action plan and list up solutions and priorities.Then, it presents the challenges facing technological development in different directions and concludes with the suitable actions to save human and environment from such complex problem facing humanity.

DOI: 10.4018/978-1-60960-094-5.ch007

INTRODUCTION

The concept Landmines and Explosive Remnants of War (ERW), which includes unexploded ordnance (UXO) and abandoned explosive ordnance represent a major threat to civilian. UXO refers to munitions, such as, bombs, shells, mortars, grenades and as similar. Landmines are indiscriminate weapon and they are so effective, yet so cheap, easy to make, and lay on or just under the ground surface. A mine comprises a quantity of explosive material inside casing, typically metal, plastic or wood casing, and it has a fuse mechanism to detonate the explosives (GICHD, 2007). A mine is detonated by the action of its target (a vehicle, a person, an animal, etc.), the passage of time, or controlled mean. Mines are generally categorized into two groups, Antipersonnel (AP) and Antitank (AT) or Anti-Vehicle mines (recently used term) (GICHD, 2007). AT mines are significantly larger than AP mines with a weight of several kilograms and require more pressure to detonate. All types of mines designed to be detonated by the presence, proximity or contact of a vehicle can be classified under this category. AP mines are quite small, weighing a few hundred grams at most. These mines are typically laid on the surface or buried within a few centimeters of the ground surface (normally but not always, on average 4-50mm), or buried under leaves or rocks. AP mines are designed to be exploded by the presence, proximity or contact of a person and incapacitate, injure or kill one or more persons (IMAS, 2003). AP mines are widely considered to be ethically problematic weapons with ability to kill or incapacitate their victims. AP mines commonly use the pressure of a person's foot as a triggering means (low triggering pressure), but tripwires are also frequently employed (Habib, 2007a; Habib, 2008a; Habib 2008b). In addition, mines represent a substantial barrier to economical recovery and the return to normal life while they deny access to land, food, water and other basic needs. Besides this, the medical, social, economic, and environmental consequences are immense (O'Malley, 1993; Blagden, 1993; Physicians for Human Rights,1993; US Department of State, 1994; King, 1997; ICRC, 1998, Habib, 2001, Habib, 2002b). In addition, landmines and ERW prevent the repatriation of refugees and displaced people while it hampers the delivery of humanitarian aid. The removal and destruction of all forms of dangerous battlefield debris, particularly landmines and ERW are vital prerequisites for any region to recover from the aftermath of a war. United Nation Department of Human Affairs (UNDHA) assesses that there 60-100 million mines that are scattered across the world and pose significant hazards. There are, more than 70 states were believed to be mine-affected (O'Malley, 1993; Blagden, 1993; Physicians for Human Rights, 1993; US Department of State, 1994; King, 1997; Habib, 2002b, Habib, 2008a and b; ICBL, 2009). Currently, the annual rate of mine clearance is very slow. It is believed that in the 1990s, 20 new mines were laid for every mine cleared (Winslow, 1997).

Removal of landmines carries high risks while it is slow and costly. The production costs of conventional AP mines are roughly between 3 and 30 US$ and it is slightly more for smart mines. But, the current cost rate of clearing one mine is ranging between 300-1000 US$ per mine (depending on the mine infected area and the rate of false alarms). In another way of cost consideration, depending on country, type of terrain, climate, commercial clearance, and NGOs involved, costs to clear landmines can vary from US$2 to US$39 per square meter. As an example of the size of the problem, Bosnia-Herzegovina has over 18,000 minefields mapped, with probably similar number of minefields are not yet mapped. The mapped area represents approximately 3,000 Km^2 of contamination of which around 1,000 Km^2 is good land for agricultural. In 2002, four square kilometers were cleared (French, 2006). Since 1999, at least 1,100km^2 of mined areas and a further 2,100km^2 of battle areas have been cleared in more than 90 states and other areas. Operations have resulted in the destruction of more than 2.2 million emplaced AP mines, and 250,000 Anti-Vehicle/Anti-Tank mines (ICBL, 2009).

The variety of landmines being used is enormous, including many with very small amounts of metal, such as plastic mines. There exist about many types of landmines around the world. A U.S. Army database made available to the United Nations contains profiles of more than 750 different types (Garwin & Husbands, 1997). What happens when a landmine explodes is also variable. A number of sources, such as pressure, movement, sound, magnetism, and vibration can trigger a landmine. AP mines commonly use the pressure of a person's foot as a triggering means, but tripwires are also frequently employed. Most AP mines can be classified into one of the following four categories based on their primary method of causing injury (See Figure 1 shows examples of the first two categories): blast, fragmentation, directional, and bounding devices. These mines range from very simple devices to high technology (O'Malley, 1993; US Department of State, 1994; GICHD, 2007). Some types of modern mines are designed to self-destruct, or chemically render themselves inert after a period of weeks or months. Conventional landmines around the world do not have self-destructive mechanisms and they stay active for a long time. Modern landmines are fabricated from sophisticated non-metallic materials. New, smaller, lightweight, more lethal mines are now providing the capability for rapid emplacement of self-destructing AT and AP minefields by a variety of delivery modes. These modes range from manual emplacement to launchers on vehicles and through both rotary and fixed-wing aircraft. Even more radical changes are coming in mines that are capable of sensing the direction and type of threat. These mines will also be able to be turned ON and OFF, employing their own electronic countermeasures to ensure survivability against enemy countermine operations. In addition, new trends have been recognized in having minefields with self-healing behavior. Such minefields will include dynamic and scatterable surface mines used to complicate clearance and preserve obstacles by embedding them with capability to detect breaching and simple mobility to change its location accordingly.

IMPACT OF LANDMINES

In scale, the landmine crisis is global and affecting many countries. However, the problem is ultra-hazardous to the environment and development of a number of developing nations (Berhe, 2007). The impact of landmines on mine-affected societies is devastating as mines impede the ability to fully recover from conflicts after the cessation of hostilities. AP mines kill and maim typically men, women involved in livelihood activities, such as farming, herding and the collection of firewood and water. In addition, Land-mines represent an insidious and persistent danger to children affected by war as they are likely to pick up and play with strange objects, such as toy like mines (Machel, 1996). The suffering caused by AP mines is particularly horrific. Survivors typically require medication, multiple operations, healthcare and medications, amputations, and prolonged physical rehabilitation. In addition, the victim's family members suffer, particularly if they are economically dependent on him/her. Landmines impose a heavy economic burden on mine-affected communities. Mine survivors commonly suffer permanent disability with serious physical, psychological and socio-economic implications (Habib, 2002b; ICRC, 2009). Additional economical consequence is providing artificial limbs to survivors of mine accidents that costs between 100 to $3,000 (depending on the country). Furthermore, an adult must replace his/her prosthesis once every 3-5 years to accommodate their growth patterns. However, children must have new prosthesis every 6 months (Menstuff, 2004). Furthermore, many mine-affected countries

Figure 1. Different type of blast and fragmentation AP mines (GICHD, 2007). (a) Examples of blast AP mines. (b) Examples of fragmentation AP mines.

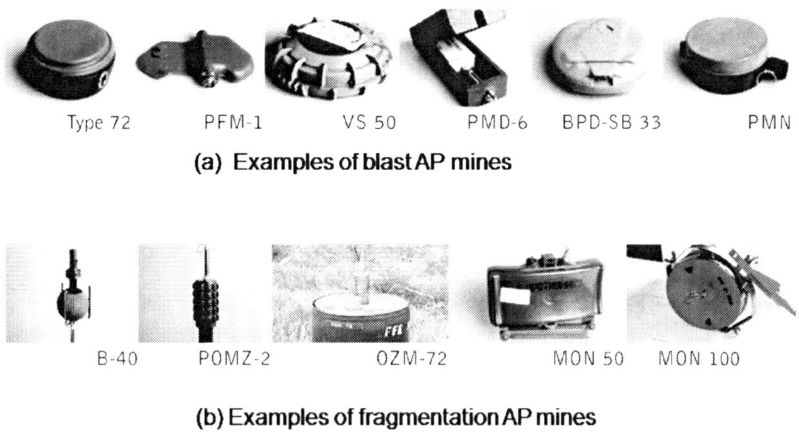

are economically poor and do not have adequate health facilities or proper roads, bridges and other infrastructure, and this prevent affected people to access services within the country. The presence of landmines in agricultural fields renders large tracts of fertile soil unusable, and at the same time, the decomposition of explosive substances in mines can cause many environmental problems because they are often water soluble, carcinogenic, toxic, and long-lasting. Furthermore, landmines prevent access to safe drinking water and this would force people to drink dirty and contaminated water. In addition, remains of animals killed by landmines turn minefields into breeding grounds for insects that transmit viruses and bacteria. Men, women, and children all suffer psychological consequences due to the presence of landmines and related injuries (Menstuff, 2004).

The International Committee of the Red Cross (ICRC) estimated in 1994 at the peak of the landmine crisis, that the casualty rate from mines exceeds 26,000 persons every year (ICRC, 1996a; ICRC, 1996b; ICRC, 1998; Brennan & Woodruff, 2003). The primary victims are unarmed civilians and among them children are particularly affected. In spite of data collection challenges and difficulties and that significant numbers of casualties go unreported (ICBL, 2008), between 1999 and the end of 2008, Landmine Monitor collected partial information on 73,576 recorded mine, ERW and improvised explosive devices (IED) casualties in 119 countries and areas, of which 17,867 were killed, 51,711 injured, and 3,998 of unknown status (ICBL, 2009).

Landmines cause particularly complex phenomena of environmental degradation across different temporal and spatial scales. Landmines produce severe environmental consequences, and it causes multifaceted and interconnected ecological and socio-politico-economic problems. Major ecological repercussions of landmine that constitute crisis include disruption of land's stability, pollution and loss of biodiversity constitute crisis (Berhe, 2007). This issue has been explored internationally by many organizations and researchers (Roberts and Williams, 1995; Winslow, 1997; Nacho´n, 2000; Berhe, 2007). Total international support for mine action for 1992–2008 was US$4.27 billion. Despite this high level of overall funding, over the past decade victim assistance has made the least progress of all the major sectors of mine action, with funding and action falling far short of what was needed (ICBL, 2009). Most efforts remained focused on medical care and physical rehabilitation, often only when supported by international organizations and funding, rather than on promoting economic self-reliance for survivors, their families, and communities (ICBL, 2009).

HUMANITARIAN DEMINING AND REQUIREMENTS

Humanitarian demining scenarios differ from military ones in many aspects. The objectives and philosophy are different. Solutions developed for the military are generally not suitable for humanitarian demining. Humanitarian demining is a critical first step for reconstruction of post-conflict countries and it requires that the entire land area to be free of mines and hence the need to detect, locates, and removes reliably and safely every single mine, and UXO from a targeted land. The terms demining and humanitarian demining are interchangeable (IMAS, 2003). Humanitarian demining activities which lead to the removal of mine and ERW hazards include: technical survey, mapping, clearance, marking, post-clearance documentation, community mine action liaison and the handover of cleared land (IMAS, 2003). It is carried out in a post-conflict context, and the important outcome of humanitarian demining is to make land safer for daily living and restoration to what it was prior to the hostilities. Demining may be carried out by various types of organizations, such as NGOs, commercial enterprises, or national mine action teams, and a small amount of military clearance. In addition, it is allowing people to use their land without fear; allowing refugees to return home, schools to be reopened, land to be reused for farming and critical infrastructure to be rebuilt (Espirit HPCN, 1997; Bruschini et al., 1999; Habib, 2002b; Goose, 2004, Habib, 2007a; Habib, 2008a and b).

The standard to which clearance must be achieved is extremely high as there is a need to have at least 99.6% (the standard required by UNDHA) successful detection and removal rate (Blagden, 1993), and a 100% to a certain depth according to International Mine Action Standards (IMAS) (Habib, 2007a; Habib 2008a and b). The amount of time it takes to clear an area is less important than the safety of the clearance personnel and the reliability and accuracy of the demining process. Safety is of utmost importance, and casualties are unacceptable. Any system to be developed should compliment this effort, not to hamper it or simply move the problem elsewhere. The risks to those carrying out the task must also be maintained at a lower level than might be acceptable in a military situation. Another consideration by humanitarian demining is the use of land for development, i.e., there is a need to reduce the environmental impact that may results from the demining operation. The currently available technologies are not suited to achieve these objectives of humanitarian demining. Until now, detection and clearance in humanitarian demining very often relies on manual methods as primary procedure. The problem resides primarily in the detection phase first, and then how to increase productivity by speeding up demining process reliably and safely.

DIFFICULTIES FACING MINE DETECTION AND CLEARANCE

Landmines are usually simple devices, readily manufactured anywhere, easy to lay and yet so difficult and dangerous to find and destroy. Landmines are remarkably durable, posing a threat years after the wars for which they were laid have ended. They are harmful because of their unknown positions and often difficult to detect.

Humanitarian demining problem is characterized by an enormous variability in the nature of explosive ordnance to be removed, and in the type of terrain and vegetation. The development of new demining technologies is difficult because of the tremendous diversity of terrains and environmental conditions in which mines are laid and because of the wide variety of landmines. There is wide range of terrains (rocky, rolling, flat, desert, beaches, hillside, muddy, river, canal bank, forest, trench, etc.) whereas mines are often laid. The environmental conditions may cover different climate (hot, humid, rainy, cold, snow, windy), the density of vegetation (heavy, medium, small, none), and type of soil (soft, sand, cultivated, hard clay, covered by snow, covered with water). In addition, residential, industrial and agriculture areas, each has its own features and needs to be considered. In addition, the mines are infesting some of the world's poorest countries, where the indigenous personnel available to undertake demining may lack technical skills and experience.

Landmines are many in terms of type and size. AP mines come in all shapes and colors are made from a variety of materials, metallic and nonmetallic. Metal detector works well with metal cased mines, but metal in modern mines has been increasingly replaced by plastic and wood as countermeasure to metal detector that making them undetectable by their metallic content. There are other methods to detect explosives and landmines. However, most of detection methods are limited by sensitivity and/or operational complexities due to type of terrain and soil composition, climatic variables, and ground clutter, such as, shrapnel and stray metal fragments that produce great number of false positive signals and slow down detection rates to unacceptable levels. AP mines can be laid anywhere and can be set off in a number of ways because the activation mechanisms available for these mines are not the same. Activation methods can be classified into three categories, pressure, electronic, and command detonation (remote control). Mines may have been in place for many years, they might be corroded, waterlogged, impregnated with mud or dirt, and can behave quite unpredictable. Some mines were buried too deep to stop more organized forces finding them with metal detectors. Deeper mines may not detonate when the ground is hard, but later rain may soften the ground to the point where even a child's footstep will set them off. Trip-wires may be caught up in overgrown bushes, grass or roots. In addition, there is no accurate estimate on the size of the contaminated land and the number of mines laid in it.

HUMANITARIAN MINE ACTION PLAN

The objective of humanitarian mine action plan is to reduce the risk from landmines to a level where people can live safely where economic, social and health development can occur free from the constraints imposed by landmine contamination, and in which the victims' needs can be properly addressed.

The process of landmine clearance (demining) comprises five phases (Habib, 2002b),

a. Locate, identify and mark any of the recognized minefields. This includes: Survey, assessment and planning, mapping, prioritization of marked minefields and resources, etc. This process can be termed as Assessments and Landmine Impact Surveys (LIS). This should be associated with mine risk education, human skill development and management, public awareness process, information management, safety and benchmark consideration, etc.
b. Prepare the marked minefields for the clearance operation by cutting vegetation and clearance, collecting metal fragments, etc. Area reduction is considered at this component too.
c. Apply suitable mine clearance techniques that suit the relevant minefield to locate and mark individual landmines within the identified area,
d. Remove the threat of the detected mines by neutralization: removal, or detonation
e. Apply quality control measures (Post clearance inspection). There is a need to verify

and assure with a high level of confidence that the cleared area is free from mine.

In parallel to the above, healthcare, rehabilitation, and medical support should be provided to affected persons. In addition, implementing continuous educational and awareness program, infrastructure building, job creation and initiating economical support should be established.

In general, it is very important to be aware that LIS often overestimate the extent of land actually affected by mines and ERW. Imprecise demarcation of suspected areas can lead to inappropriate allocation of time and resources for mine action tasks (GICHD, 2010).

HUMANITARIAN DEMINING: SOLUTIONS AND PRIORITIES

The priorities for research and development in the field of humanitarian demining require strategies that aim to start with the following needs (Habib, 2002b, Habib 2007a, Habib 2008a and b):

a. Develop reliable and accurate techniques/technologies that can enhance the performance of the demining process and allow efficient area detection and reduction of minefields. There is an urgent need to recognize and reliably locate minefields and isolate them by defining proper signs and limits to make the public aware, and to avoid further accidents,
b. Have quality-training programs that fit the needs of local environment. Such training programs need to integrate cultural, environmental and operational considerations when developed,
c. Enhance the safety of deminers by providing them with suitable training, protective clothing, tools and equipment and isolate them as possible from direct physical contact with the mines and UXOs,
d. Enhance the performance of the sensors and the deminers. To achieve this, there is a need to develop efficient techniques for sensor integration (array of homogeneous and/or heterogeneous sensors) with advance level of data fusion and signal processing algorithms that can confirm the detection in real-time and lead to the identification of mine parameters needed for the next actions.
e. Develop a portable, reliable and easy to use handheld approach to sensor movement that is still required in difficult and physically constraint environments (woods, uneven terrain, residential, etc.) although such approach is slow and hazardous for the individuals. Hence, the sensors can be integrated with vehicle-based platforms to support automatic mine clearance in open areas.
f. Use information and communication technologies with aim to enhance contact, experience exchange, research, planning and to share results and data among all parties and personnel within the demining community.
g. Mechanized vegetation cutting. However, it would be better to find a technology that can detect and mark mines without having to cut vegetation.
h. Develop simple, modular, efficient, compact and low cost mechanical machines for mine clearance that suit the target task and environment aiming to unearth mines reliably and efficiently,
i. Increase mine clearance daily performance by improving productivity, accuracy, and increase safety of demining personnel. There is a need to have a means of moving the portable mine detection device as it searches for landmines. Hence, it is important to automate/mechanize detection and removal of mines, and to improve the safety of the deminers through the use of efficient, reliable and cost effective humanitarian mine action equipment (such as robots, flexible and intelligent mechanisms, etc.), that

have minimum environmental impact. It is necessary to have a robot with efficient and modularized surface locomotion and mobility that is well adapted to unstructured environment and different type of terrain. The design should integrate proper balance between maneuverability, stability, speed, and the ability to overcome obstacles. Such robots should have decision-making capability to locate, mark or neutralize individual mine precisely, and

j. To have efficient quality control assurance methods that is reliable and accurate in ensuring no residual mines within an area declared clear of mines.

In order to approach a proper and practical solutions for the problem, there is a need for the scientists in each discipline and deminers to share their knowledge, and the result of their experience and experiments in order to design and test viable solutions for humanitarian demining without ruling out any possible technology or technique.

HUMANITARIAN DEMINING AND THE CHALLENGE OF TECHNOLOGY

Improving detection and clearance methods is a formidable technical challenge. Although demining has been given top priority, currently mine's detection and clearing operation is a labor-intensive, slow, very dangerous, expensive, and low technology operation. Current demining operation relies on careful search of mined areas with a hand held detector (the most widely used is the handheld metal detector). The other bottleneck in humanitarian demining is often the tedious exploratory probing and delicate excavation, and it that is required after a mine detector has located mine candidates. The current rate of humanitarian mine clearing is about 100 thousand per year. It is estimated that the current demining rate is about 10-20 times slower than the laying rate, i.e., for every mine cleared 10-20 mines are laid. Therefore, to stabilize the mine situation, it is necessary to increase the current capability of mine clearance by 10-20 times.

The diversity of the mine threat points out to the need for different types of sensors and equipment to detect and neutralize landmines. The requirements to develop equipment for use by deminers with different training levels, cultures, and education levels greatly add to the challenge. The solution to this problem is very difficult because, given the nature of landmines and the requirements of humanitarian demining, as any instrument must be 100% reliable for the safety of the operators and the people whom will use the land (Blagden, 1993; Habib 2002b). Hence, it becomes urgent to develop detection (individual mine, and area mine detection), identification and removal technologies and techniques to increase the efficiency of demining operations by several orders of magnitude to achieve a substantial reduction to the threat of AP mines within a reasonable timeframe and at an affordable cost.

Technology has become the solution to many long-standing problems, and while current mine detection and clearance technologies may be effective, it is far too limited to fully address the huge complex and difficult landmine problem facing the world. The challenge is in finding creative, reliable and applicable technical solutions in such highly constrained environment. Applying technology to humanitarian demining is a stimulating objective. Detecting and removing AP mines seems to be a perfect application for robots. However, this need to have a good understanding of the problem and a careful analysis must filter the goals in order to avoid deception and increase the possibility of achieving results (Nicoud, 1996). In order to approach proper and practical solutions for the problem, there is a need for the scientists in each discipline and deminers to share their knowledge and the results of their experience and experiments in order to design and test viable solutions for humanitarian demining. Technologies to be developed should take into account the facts that many of the demining operators will have had minimal formal education and that the countries where the equipment is to be used have poor technological infrastructure for equipment maintenance, operation, and deployment.

Greater resources need to be devoted to demining both to immediate clearance and to the development of innovated detection and clearance equipment and technologies. There is an urgent need to speed up the development to have compact and portable, low cost, technically feasible, fast response, safe, accurate, reliable, and easy to operate mine detector systems with flexible mobile platforms that can be reliably used to detect all types of available landmines and support fast and wide area coverage. Appropriate mine clearance technologies are those inexpensive, rugged, and reliable technical products, processes and techniques that are developed within, or should be transferred for use in mine-affected areas. These technologies should be cheap enough to be purchased within the regional economy and simple enough to be made and maintained in a small workshop. We should favor technologies that can be manufactured in mined countries; technologies that are transferable, and which provide employment and economic infrastructure where it is most urgently required.

DEMINING TECHNIQUES AND THE AVAILABLE TECHNOLOGIES

Mine clearance itself can be accomplished through different methods with varying levels of technology and accuracy, but the most laborious way is still the most reliable. Currently, almost all humanitarian mine clearance is still required to apply hand clearance method that uses 'prodding' or 'probing' within its loop to assure high reliability. Manual probing is slow, labor intensive and extremely dangerous and stressful process.

Mechanical Equipment and Tools for Mine Clearance

A good deal of research and development has gone into motorized mechanical mine clearance in which their design is influenced by the military demining requirements. The use of such machines aims to unearth mines or force them to explode under the pressure of heavy machinery and associated tools and to avoid the necessity of deminers making physical contact with the mines. A number of mechanical mine clearing machines have been constructed or adapted from military vehicles, armored vehicles, or commercially available agriculture vehicles of the same or similar type, with same or reduced size (Habib, 2001b, Habib, 2002b; Habib, 2004; Geneva Centre for Humanitarian Demining, 2006; Habib, 2008a and b). A single mechanical mine clearance machine can work faster than a thousand deminers over flat fields. They are mostly appropriate and cost effective in large and wide areas without dense vegetation or steep grades. In small paths or thick bush, such machines simply cannot maneuver. Mechanical clearance equipment is expensive and it cannot be used on roadsides, steep hills, around large trees, inside a residential area, soft terrain, heavy vegetation or rocky terrain. Mobility and maneuverability where wheeled vehicles cannot travel efficiently on anything other than flat surfaces, tracked vehicles cannot travel in areas with steep vertical walls, machines in general cannot climb undefined obstacles, and machines cannot in general deform to get through narrow entrances. In addition, mechanical clearance has its own environmental impact such as erosion and soil pollution. The logistical problems associated with transporting heavy machinery to remote areas is critical in countries with little infrastructure and resources.

In general, none of the equipment within this category has been developed specifically to fulfill humanitarian mine clearance objectives and for this, there is no form of any available mechanical mine clearance technologies that can give the high clearance ratio to help achieving humanitarian mine clearance standards effectively while minimizing the environmental impact. However, to achieve better clearance rate, these machines can be used in conjunction with dog teams and/ or manual clearance team, which double check an area for remaining mines.

A number of mechanical mine clearing machines have been tested during the past. The general trend goes from "mechanical demining" towards "mechanically assisted demining", adaptable to local circumstances. Some examples of mechanical clearance equipment include but not limited, Vegetation cutters, Flails and Light-Flails, Panther mine clearing vehicle, Armored bulldozer, Ploughs and the rake plough, the M2 Surface "V" mine plow, Earth tillers, Mine sifter, Armored wheel shovel, Mine clearing cultivator, Floating mine blade, Rollers, Mine-proof vehicles, Swedish Mine Fighter (SMF), Armored road grader, etc. (US Department of Defense, 1999; Humanitarian Mine Action Equipment Catalogue, 1999; Department of Defense, 2002; Habib, 2002a; Habib, 2004; Geneva Centre for Humanitarian Demining, 2006; Habib 2007a; Habib 2008a and b).

In addition, vegetation is a large problem facing demining (mainly in tropical countries) and often poses major difficulties to the demining efforts. The vegetation removal can take up a substantial fraction of the time and for this there is a need to properly mechanized vegetation cutting and removal. These machines should be designed to cut down on the time required for demining. In their simplest form, vegetation cutters consist of adequately modified commercial devices (e.g. agricultural tractors with hedge cutters or excavators). There is an urgent need for effective vegetation clearance technology and techniques that avoid detonating mines. Cost effective and efficient clearance techniques for clearing both landmines and vegetation have been identified as a significant need by the demining community.

Mine Detection and Sensing Technologies

The main objective of mine detection is to achieve a high probability of detection rate while maintaining low probability of false alarm. The probability of false alarm rate is directly proportional to the time and cost of demining by a large factor. Hence, it is important to develop more effective detection technology that speed up the detection process, maximize detection reliability and accuracy, reduce false alarm rate, improve the ability to positively discriminate landmines from other buried objects and metallic debris, and enhance safety and protection for deminers. In addition, there is a need to have simple, flexible and friendly user interaction that allows safe operation without the need for extensive training. Such approach needs to incorporate the strength of sensing technologies with efficient mathematical, theoretic approaches, and techniques for analyzing complex incoming signals from mine detectors to improve mine detectability. This leads to maximize the performance of the equipment through the optimization of signal processing and operational procedures. Furthermore, careful study of the limitations of any tool with regard to the location, environment, and soil composition is critically important besides preparing the required operational and maintenance skills. It is important to keep in mind that not all high-tech solutions may be workable in different soil and environmental conditions. The detection technologies are presently in varying stages of development. Each has its own strength and weaknesses. The development phase of new technologies requires a well-established set of testing facilities at the laboratory level that carried out in conditions closely follow those of the mine affected area, and at the real site. This should be followed by having extensive field trails in real scenarios to validate the new technologies under actual field conditions for the purpose to specify benefits and limitations of different methods. The work must be performed in close cooperation with end-users of the equipment and real deminers should carry out the test at a real site, in order to ensure that the developments are consistent with the practical operational procedures in the context of humanitarian demining, and that it is fulfilling user requirements. In addition, there is a need to have reliable process of global standard for assessing the availability, suitability, and affordability of technology with enabling technology represented by common information tools that enable these assessments and evaluations. The benchmarking is going to enhance the performance levels that enable the development of reliable and accurate equipment, systems and algorithms.

Methods of detecting mines vary from, simple in technology but exhaustive searching by humans using some combination of metal detectors and manual probing, to a variety of high biological and electronic technologies. The effectiveness of metal detectors can be inhibited by mines with extremely low metal content or by soils with high ferrous content and hence other detection techniques have been and are being investigated. Another technique that is widely used is the direct detection of explosive material by smell using a dog (Sieber, 1995). Trained dogs are the best known explosive detectors but they need excessive training and inherently unreliable because they are greatly impeded by windy conditions, and have only 50-60% accuracy.

New technologies are being investigated to improve the reliability and speedup the detection operation, some of these technologies are: Electromagnetic Induction Metal detectors (EMI), Infrared Imaging, Ground-Penetrating Radar (GPR), Acoustics, Acoustic Imaging, Thermal Neutron Activation (TNA), Photoacoustic Spectroscopy, Nuclear Quadrupole Resonance (NQR), X-ray Tomography, Nneutron Back-scattering, Biosensors, Commercial sniffers, etc. (Healy & Webber, 1993; Van Westen, 1993; Hewish & Ness, 1995; Sieber, 1995; McFee, 1996; Cain & Meidinger, 1996; Habib, 2001, Habib 2007b).

Currently, there is no single sensor technology that has the capability to attain good levels of detection for the available AP mines while having a low false alarm rate under various types of soil, different weather, all types of mines, natural and ground clutters, etc. If one sensor can detect a mine with a certain success rate coupled with a certain probability of generating a false alarm, could two sensors working together do a better job? The idea of developing multi sensor solutions involving two or more sensors coupled to computer based decision support systems with advanced signal processing techniques is attractive and is advocated by many as a fruitful line of development. Hence, there is a need to use complementary sensor technologies and to do an appropriate sensor data fusion. The ultimate purpose is to have a system that improves detection, validation and recognition of buried items for the purpose to reduce false alarm rates and to overcome current landmine detection limitations. A promising solution will be to apply fusion of sensory information on various sensor outputs through the use of advanced signal processing techniques, by integrating different sensor technologies reacting to different physical characteristics of buried objects. Critical to demining is the ability to distinguish fragments or stones from the target material in real time.

Sensor fusion using soft computing methods such as fuzzy logic, neural networks and rough set theory must be further explored and computationally inexpensive methods of combining sensory data must be designed. These methods should also have the capability to assess the quality of the mined area once the mines have been cleared.

Robotized Solution for Mine Detection and Clearance

The portable handheld mine detection approach to sensor movement is slow and hazardous for the individual deminers. Armored vehicles may not thoroughly protect the occupants and may be of only limited usefulness in off-road operations. Most people in the mine clearance community would be delighted if the work could be done remotely through teleoperated systems or, even better, autonomously through the use of service robots. Remote control of most equipment is quite feasible. However, the benefit of mounting a mine detector on a remotely controlled vehicle should have careful considerations that lead to decide whether the anticipated reduction in risk to the operator justifies the added cost and possible reduction in efficiency. A cost analysis should be made to determine to what extent remote control approach is a valid solution.

To increase mine clearance daily performance by improving productivity and accuracy, and to increase safety of demining operations and personnel, there is a need for an efficient, reliable and cost effective humanitarian mine action equipment with flexible and adaptable mobility, and some level of decision making capabilities. Such equipment should have selectable sets of mine detectors and work to locate and mark individual mines precisely, and at a later stage to neutralize the detected mines. Robotics solutions properly sized with suitable modularized mechanized structure and well adapted to local conditions of minefields can greatly improve the safety of personnel as well as work efficiency, productivity and flexibility. Robotics solution can range from modular components that can convert any mine clearing vehicle to a remote-controlled device, to prodding tools connected to a robotic arm, and to mobile vehicles with arrays of detection sensors and area mine-clearance devices. The targeted robot should have the capability to operate in multiple modes. It should be possible for someone with only basic training to operate the system. Robots can speedup the clearance process when used in combination with handheld mine detection tools, and they are going to be useful for quick verification and quality control. To facilitate a good robot performance in the demining process, there is a need to employ mechanized systems that are able to remove obstructions that deter manual and canine search methods without severely disturbing soil. Solving this problem presents challenges in the robotics research field and all relevant research areas.

Robotics research requires the successful integration of a number of disparate technologies that need to have a focus to develop:

a. Flexible mechanics and modular structures,
b. Mobility and behavior based control architecture,
c. Human support functionalities and interaction,
d. Homogeneous and heterogeneous sensors integration and data fusion,
e. Different aspect of fast autonomous or semi-autonomous navigation in a dynamic and unstructured environment,
f. Planning, coordination, and cooperation among multi robots,
g. Wireless connectivity and natural communication with humans,
h. Virtual reality and real time interaction to support the planning and logistics of robot service, and
i. Machine intelligence, computation intelligence and advanced signal processing algorithms and techniques.

Furthermore, the use of many robots working and coordinating their movement will improve the productivity of overall mine detection and demining process through the use of team of robots cooperating and coordinating their work in parallel to enable parallel tasks (Gage, 1995; Habib, 1998; Habib, 2007a; Habib 2008a and b).

The possible introduction of robots into demining process can be done through surface preparation and marking, speeding-up detection, and mine removal or neutralization. In addition, service robots can be used for minefield mapping too. However, the cost of applying service robot's technologies and techniques must be justified by the benefits it provides. There is no doubt that one of the major benefits would be the safety, by removing the operator from the hazardous area.

It is clear that the development of a unique and universal robot that can operate under wide and different terrain and environmental conditions to meet demining requirements is not a simple task. In the short term, it appears that the best use of robotics will be as mobile platforms with arrays of mine detection sensors and area mine clearance devices. Teleoperations are promising, but are limited too, because their remote human controllers have limited feedback and are unable to drive them effectively in real time. There are still some doubts whether such equipment will operate as effectively when the operator is at a long distance or has been removed altogether. Strangely enough, this is particularly true for urban areas normally full of rubble, while agricultural areas seem to be better, but that is not always true. A possible idea in using robots for demining is to design a series of simple and modularized robots, each one capable of performing one of the elementary operations that are required to effectively clear a minefield. An appropriate mix of such machines should be chosen for each demining task, keeping in mind that it is very unlikely that the whole process can be made fully autonomous. It is absolutely clear that in many cases, the environment to be dealt with is so hostile that no autonomous robot has any chance to be used in mid and short terms. The effort devoted to robotic solutions would be more helpful if it is directed at simple equipment improvements and low-cost robotic devices to provide some useful improvements in safety and cost-effectiveness in the short to medium term.

Several practical difficulties in using robots for mine clearance have been highlighted (Treveylan, 1997). There is little value in a system that makes life safer for the operator but which will be less effective at clearing the ground. Accordingly, a serious evaluation and analysis should be done along with having efficient design and techniques. The high cost and sophisticated technology used in robots which required highly trained personal to operate and maintain them are additional factors limiting the possibilities of using robots for humanitarian demining. In spite of this, many efforts have been recognized to develop effective robots for the purpose to offer cheap and fast solution (Nicoud & Machler, 1996; Habib, 2001b, habib 2007a, Habib 2008a and b).

Research into individual, mine-seeking robots is in the early stages. In their current status, they are not an appropriate solution for mine clearance. This is because, their use is bounded by sensing devices and techniques improvements, the difficulties facing automated solutions raised by the variety of mines and minefields, and the variety of terrains in which mine can be found. Examples of such terrains may include dessert, sides of mountains, rocky, forest, rice paddy, riverbanks, plantations, residential areas, etc. Also, robotized solutions are yet too expensive to be used for humanitarian demining operations in countries like Angola, Afghanistan, Cambodia, etc.

Many efforts have been recognized to develop effective robots for the purpose to offer cheap and fast solutions. Three main directions can be recognized (Habib 2007a; Habib, 2008a and b):

a. Teleoperated machines,
b. Multifunctional teleoperated robot,
c. Demining service robots, and
d. Unmanned Aerial Vehicles and Airships.

CONCLUDING REMARKS: HUMANITARIAN DEMINING AND THE TECHNICAL NEEDS

Demining is a time consuming process while it is subjecting deminers to high risks. With the current technology and techniques, clearing all known mined areas would take decades or centuries, even if no additional mines are laid.

Based on what has been presented and in order to have effective approach in tackling this complicated problem, with aims to accelerate the demining process, increase daily performance, improving productivity and accuracy, enhance safety of demining operations and personnel, and to achieve cost effective measures, there is a need to have innovative and efficient technology and techniques within the following areas,

a. Demining is very costly and searching an area that is free of mines is adding extra high cost and wasting resources. Hence, to avoid that, the first essential objective in the demining process should be to identify efficiently what areas are polluted by mines. Locating the contaminated land helps to separate the danger from people and makes the uncontaminated land available for use immediately. In order to accelerate the mine clearance process, and save resources, new demining methods are urgently needed to detect minefields over large and varied tracts of land in a much more cost effective, efficient, safe and reliable manner. It is important to have reliable methods that can reduce mine suspected areas. Unmanned Aerial Vehicles (UAVs) and Air-ships are integrated with remote sensing technology and sensor fusion techniques to help reducing mine suspected area and isolate the danger of mines while producing important risk assessment maps..
b. The critical element to humanitarian demining is the detection and location of every single mine. Mine detection represents the slowest and the most important step within the demining process, and the quality of mine detector affects the efficiency and safety of this process. Hence, there is a need to have practical and reliable technology that can single out mine from other objects and natural clutters, work under different climate and terrain, while it would be possible to efficiently perform without cutting vegetation.
c. It is necessary increase productivity and efficiency by having reliable and cost effective humanitarian mine action equipment with flexible mechanism and adaptable mobility, multiple operational modes, and some level of decision making capabilities. Such equipment should have selectable sets of mine detectors and work to locate and mark individual mines precisely, and at a later stage to neutralize the detected mines. Such equipment would be useful in quickly verifying that an area is clear of landmines so that manual cleaners can concentrate on those areas that are most likely to be infested. It is important to remind ourselves that there is little value in a system that makes life safer for the operator, but which will be less effective at clearing accurately and reliably the ground.

The development of such systems should be done with close cooperation and interaction with deminers in the field, and the resultant system must be inexpensive with possible use of locally available materials, and easy to use with minimal training by locals. In addition, these systems must be flexible and modular with proper logistical consideration to address a variety of clearance tasks and for case-by-case scenarios.

REFERENCES

Berhe, A. A. (2007). The Contribution of Landmines to Land Degradation. *Land Degradation & Development*, *18*, 1–15. doi:10.1002/ldr.754

Blagden, P. M. (1993). Summary of UN Demining. *Proceedings of the International Symposium on Anti-Personnel Mines*, Montreux, April 1993, CICR/ICRC, pp 117-123.

Brennan, B. M. & and B.A. Woodruff, B. A.(2003). Death and Injury from Landmines and Unexploded Ordnance in Afghanistan. *J.Am.Med. Assoc,290*(5), 2003, pp.650–653. doi:10.1001/jama.290.5.650

Bruschini, C., De Bruyn, K., Sahli, H., & Cornelis, J. (1999). EUDEM: The EU in Humanitarian DEMining. *EU report*, Brussels, 1999.

Cain, B., & Meidinger, T. (1996). The Improved Landmine Detection System. *EUREL, 1996*, 188–192.

Department of Defense, Humanitarian Demining Research and Development (R&D) Program (2002). Other Completed Mine/Vegetation Clearance Equipment. *Development Technologies Catalog 2001-2002*.

Espirit, H. P. C. N. (1997). Industrial R&D Requirements for Humanitarian Demining. (Available through <http://www.cordis.lu/espirit/ src/report43.html/).

French, B. (2006). The business of land-mine clearing. *The Economics of Peace and Security Journal*, *1*(2), 54–57.

Garwin, R. L., & Jo, L. Husbands, J. L. (1997). *Progress in Humanitarian Demining: Technical and Policy Challenges*. Amaldi Conference, Paris, 20-22 November 1997. Accessed from: http://www.fas.org/rlg/de-mining.htm

Geneva Centre for Humanitarian Demining (GICHD). (2006). *Mechanical Demining Equipment Catalogue*. Geneva, Switzerland: Geneva Centre for Humanitarian Demining.

Geneva Centre for Humanitarian Demining (GICHD). (2007). *An Overview of the Guide to Mine Action and Explosive Remnants of War*. Geneva, Switzerland: Geneva Centre for Humanitarian Demining.

Geneva Centre for Humanitarian Demining (GICHD). (2010). *Land Release: A guide for mine and ERW affected countries*. Geneva, Switzerland: Geneva Centre for Humanitarian Demining.

Goose, S. (2004). Overview of Antipersonnel Mine Stockpile Destruction. *ICBL Treaty Working Group*, June 2004.

Habib, M. K. (2001). Mine Detection and Sensing Technologies: New Development Potentials in the Context of Humanitarian Demining. *Proceedings of the IEEE International Conference of Industrial Electronics, Control and Instrumentation (IECON'2001)*, USA, 2001, pp. 1612-1621.

Habib, M. K. (2002a). Mechanical Mine Clearance Technologies and Humanitarian Demining: Applicability and Effectiveness. *Proceedings of the 5th International Symposium on Technology and Mine Problem*, California, USA, April 2002.

Habib, M. K. (2002b). Mine Clearance Techniques and Technologies for Effective Humanitarian Demining. *International Journal of Mine Action*, *6*(1), 2002.

Habib, M. K. (2004). Mechanization Technology of Mine Clearing Operations and Humanitarian Demining. *The Eleventh World Congress in Mechanism and Machine Science*. Tianjin-China, pp. 973-978.

Habib, M. K. (2007a, June). Humanitarian Demining: Reality and the Challenge of Technology - The State of the Arts. *International Journal of Advanced Robotic Systems*, *4*(2), 151–172.

Habib, M. K. (2007b). Controlled Biological and Biomimetic Systems for Landmine Detection. [Washington, DC: Elsevier Publisher.]. *Journal of Biosensors and Bioelectronics*, *23*, 1–18. doi:10.1016/j.bios.2007.05.005

Habib, M. K. (2008a). Humanitarian Demining: Difficulties, Needs and the Prospect of Technology. *IEEE International Conference on Mechatronics and Automation 'ICMA2008'*, Takamatsu-Kagawa, Japan, Paper Reference WC2-2.

Habib, M. K. (2008b). Humanitarian Demining: The Problem, Difficulties, Technologies and the Role of Robotics. In Maki K. Habib, (Ed.), *Humanitarian Demining, Innovative Solutions and the Challenges of Technology*, pp.1-56.ARS-pro literature Verlag Publishers.

Healy, A. & W., Webber, W. (1993). Sensors for the Detection of Land-based Munitions. *Naval Postgraduate School*, Monterey, California, N PS-M E-95-003, September 1993.

Hewish, M., & Ness, L. (1995). Mine-detection Technologies. *International Defense Review*, October 1995, pp. 40-46.

Humanitarian Mine Action Equipment Catalogue (1999), German Federal Foreign Office.

International Campaign to Ban Landmines (ICBL), (2008). *Landmine Monitor Report 2008: Toward a Mine-Free World.* Landmine Monitor, 2008.

International Campaign to Ban Landmines (ICBL), (2009). *Landmine Monitor Report 2009: Toward a Mine-Free World -- Special 10 Years Review*. Landmine Monitor, 2009.

International Committee of Red Cross. (1996a). Antipersonnel Mines: An Overview. Geneva, September 1996. (See also: http://www.icrc.org/).

International Committee of Red Cross. (1996b). Antipersonnel Mines- Friends or Foe? ICRC Publication, Ref. 0654, Geneva, 1996.

International Committee of Red Cross. (1998). The Silence Menace: Landmines in Bosnia and Herzegovina. ICRC Publication, Ref. 2160, Geneva, 1998.

International Committee of Red Cross. (2009). Anti-Personnel Mines: Overview of the problem, Geneva, 2009. (Available through <http://www.icrc.org/web/eng/siteeng0.nsf/htmlall/mines-fac-cartagena-021109?opendocument).

International Mine Action Standard (IMAS). (2003). *Glossary of Mine Action Terms, Definitions and Abbreviations*. IMAS 04.10, Second Edition. 01 January 2003

King, C. (1997). Mine Clearance in the Real World. *Proceedings of the International Workshop on Sustainable Humanitarian Demining,* Zagreb (SusDem'97), pp. S2.1-8.

Machel, G. *(1996). Impact of armed conflict in children: Land-mines: A deadly inheritance*. Retrieved March 12, 2010, from http://www.unicef.org/graca/mines.htm

McFee, J. E. (1996). Multisensor mine detector for peacekeeping: Improved landmine detector concept. *SPIE Technical Conference,* 2765.

Nacho'n, C. T. (2000). *Environmental Aspects of the International Crisis of Antipersonnel Landmines and the Implementation of the 1997 Mine Ban Treaty: Thematic Report. Landmine Monitor Report 2000.* NY: International Campaign to Ban Landmines.

Nicoud, J.-D. (1996a). A Demining Technology Project. *Proceedings of the International Conference on Detection of Abandoned Land Mines (MD'96)*, Edinburgh UK, October, 1996, 37-41.

O'Malley, T. J. (1993). Seek and Destroy - Clearing Mined Land. *Armada International*, *17*(1), 6–15.

Physicians for Human Rights. (1993). *Landmines, A Deadly Legacy. Human Rights Watch.* New-York.

Sieber, A. (1995). Localisation and identification of anti-personnel mines. Joint Research Centre, *European Commission, EUR 16329N*.

The National Men's Resource Center (Menstuff). (2004). Mine Fields. http://www.menstuff.org/issues /byissue/minehistory.html. Accessed on March 12th 2010.

US Department of Defense. (1999). Humanitarian Demining Development Technologies. *Catalogue*, USA, 1998.

US Department of State. (1994). Hidden Killers: The Global Landmine Crisis. *Report to Congress*, Washington D. C., Publication 10225, December 1994 (See also: http://www.state.gov/)

Van Westen, C. J. (1993). Remote Sensing and Geographic Information Systems for Geological Hazard Mitigation. *ITC-Journal*, *4*, 393–399.

William, J., & Goose, S. (1998). The international campaign to ban landmines. In Tomlin, B. W. (Ed.), *To Walk without Fear: The Global Movement to Ban Landmines*. New York: Oxford University Press.

Winslow, C. P. (1997). *Sowing the Dragon's Teeth: Landmines and the Global Legacy of War*. Boston: Beacon Press.

The Center for Defense Information. (2010). *Documentary about American landmine survivors that includes a transcript of the program*. Retrieved May 4, 2010, from http://www.cdi.org/adm/1250/index.html

The International Campaign to Ban Landmines. *(2010)*. Retrieved May 4, 2010, from http://www.icbl.org/intro.php

ADDITIONAL READING

Furuta, K., & Ishikawa, J. (2009). *Anti-personnel landmine detection for humanitarian demining*. London: Springer. doi:10.1007/978-1-84882-346-4

Geneva Centre for Humanitarian Demining (GCHD). *(2010)*. Retrieved February 1, 2010, from http://www.gichd.org/publications/

Habib, M. K. (2007). Robotics and sensors for humanitarian demining [Special issue]. *International Journal of Advanced Robotic Systems*, *4*(2).

Habib, M. K. (Ed.). (2008). *Humanitarian Demining, Innovative Solutions and the Challenges of Technology*. Vienna, Austria: ARS-pro literature Verlag Publishers.

Human Rights Watch work on landmines (2010). Retrieved May 4, 2010, from http://www.hrw.org/en/category/ topic/arms/landmines

MaGrath. R. (2000). *Landmines and unexploded ordnance.* London: Pluto Press.

Matthew, R. A., McDonal, B., & Rutherford, K. R. (Eds.). (2004). *Landmines and human security: International politics and war's hidden legacy. Suny Series in Global Politics*. New York: SUNY Press.

KEY TERMS AND DEFINITIONS

Landmines: Indiscriminate explosive traps that are victim-activated, whether the intended target is a person, a vehicle or a tank. They are so effective, yet so cheap, and easy to make and lay on or under the ground surface. Mines are generally categorized into two groups, Antipersonnel (AP) and Antitank (AT) or Anti-Vehicle mines.

Antipersonnel (AP) Mines: AP mines are quite small, weighing a few hundred grams at most. These mines are typically laid on the surface or buried within a few centimeters of the ground surface. AP mines designed to be exploded by the presence, proximity or contact of a person and that will incapacitate, injure or kill one or more persons. AP mines commonly use the pressure of a person's foot as a triggering means (low triggering pressure), but tripwires are also frequently employed.

Unexploded Ordnance or UXO: Refers to munitions, such as, bombs, shells, mortars, grenades and the like, which have failed to detonate as intended.

Explosive Remnant of Wars: Refer to unexploded ordnance and abandoned explosive ordnance (AXO). This excludes mines, booby-traps or other devices. As the case of landmines, it represents a major threat to civilian.

Minefield: An area in which explosive mines, or UXO or AXO have been placed and it has many potential hazards.

Landmines Impact: Landmines represent a substantial barrier to economical recovery and the return to normal life while they deny access to land, food, water and other basic needs. Besides this, the medical, social, economic, and environmental consequences are immense.

Humanitarian Demining: It is the critical first step for reconstruction of post-conflict countries/regions and it requires that the entire inflicted land area to be free of mines, explosive remnants of wars or other explosive devices, and hence the needs to detect, locates, and removes or destroy reliably and safely all the threats to civilian and development.

Demining: The terms demining and humanitarian demining are interchangeable.

Area Reduction: It describes the process by which one or more mine clearance tools, method, or techniques are used to gather information that locates the boundary of a suspect hazardous area, and remove all areas deemed not to be mined or free from threat.

Mine Detection and Clearance: Mine clearance can be accomplished through different technology, techniques and methods with varying levels of accuracy, such as, Mine detection and sensors, mechanization and robotization of the demining process. However, the most laborious (manual probing) way is still the most reliable.

Chapter 8
Using ICT to Foster Human Rights Defense Processes for Mexican Immigrants

María Alejandra Rocha Silva
University of Colima, Mexico

Juan Contreras-Castillo
University of Colima, Mexico

Ricardo Acosta-Díaz
University of Colima, Mexico

ABSTRACT

Frequently, Mexicans who cannot find solutions for their financial problems migrate to the United States hoping to improve their quality of life. However, they usually face abuses, mostly because they are illegal aliens, but also because they arrive to a society which is not their own. These migrants are mainly excluded from American society not only because of their race and religion, but also because they do not speak English in most cases, do not have studies higher than primary school, and are not proficient in using information and communications technologies (ICT). With this panorama in mind, the Colimenses sin Fronteras Web Portal becomes a tool to support and help them overcome the adaptation process, which might help reduce the discrimination that many of them face upon arriving to the receiving country. It also provides migrants with information about the abuses they might suffer and how to file a legal complaint.

INTRODUCTION

The era of globalization and information technology has blurred cultural and physical borders, increasing the mobility of people who migrate between countries. "Currently, citizen migration is structurally incrusted in the different economies and societies of most countries. As soon as the receiver and provider countries both depend on migration, it becomes almost impossible to contain" (Pécoud and De Gouchteneire, 2005, p. 142). This is not an exclusive phenomenon between Mexico and the United States; a world-

DOI: 10.4018/978-1-60960-094-5.ch008

wide estimation reports that around 200 million people are living in countries different from their home country.

There are important flows of citizens still colonizing the continent of Australia and more traditional European and African population movements (with an increasing number from Spain) caused, to some extent, by the socio-political reconfiguration of Eastern Europe or religious or ethnical segregation. A good example of this is the migration of women from Eastern Europe to Huelva Spain (Gualda and Ruiz, 2004). Migration from South American countries to Europe, mostly from Argentina, Ecuador and Peru is due mainly to the complex socio-economic situation of these countries. In the case of Central America, we can add natural catastrophes. Central Americans and some South Americans migrate to Mexico, but the majority only passes through, hoping to cross the border into the US.

In the early1990's, 2.2 million Latin Americans lived in countries different from their home country; however, from within the same region, this provides an overview of the high migration rates. This number has increased from 1.2 million in the 1970's. In the case of Latin Americans living in the US, their numbers were 8.4 and 1.7 million in 1990 and 1970, respectively, and half of them were Mexicans (Martínez Pizarro, 2001, p.95).

Mexicans migrate to the US because of economic reasons and its close vicinity. Mexico is a developing country bordering on the most powerful country in the world; therefore, when Mexicans cannot find solutions to their financial problems, they "look north" in figurative speech, thinking that migrating to a country with a better quality of life will improve their own. Consequently, as soon as their financial resources and social networks allow, they migrate. It is important to mention that Mexico is a large country, geographically speaking, and migrants require considerable amounts of money to travel, plus the cost of hiring a "pollero" (a common nickname used to identify persons who help migrants cross the US-Mexico border illegally). Thus, their social networks play an important role in accomplishing this complicated and dangerous crossing. This is because, many times, friends and relatives pay for the expenses of traveling and illegally crossing the border, all with the hope that the person they help cross the border will be able to send money back home to their relatives, thus improving the economic situation of those who remain in Mexico.

Unfortunately, this "American Dream" often becomes a nightmare for most migrants and their families because some of them are abused from the moment they intend to cross the border that separates one society from the other. First, they must hire a "pollero" who usually requests elevated amounts of money ranging from 500 to 4000 dollars. This money covers transportation (often through the desert) until they arrive at their destination. Commonly, migrants pay when they get to the requested address (González Velázquez in Rocha Silva, M. A. and Valencia Silva, M.M, 2009, p. 72).

The higher price for a Mexican to cross the border illegally, it's the reflection of the resurgence in the American side security measures, therefore Mexicans has to risk more by going through more dangerous places, in their pursuit for a better life.

The literature shows that between 1990 and 2002 more than 3000 people died attempting to cross the US-Mexico border. In recent years, the number of people captured by the border patrol has decreased, but the number of rescued and dead has remained equal or increased. The proportion of deceased people has increased from 15 to 35 deaths per 100,000 detentions. In comparison, Spain, who has the highest in death rate in Europe, registered only 3 deaths per 100,000 in 2000 (Alonso, 2003).

Illegal migration has not decreased; however, it has moved towards other crossing points, mainly through the desert, into more dangerous areas, where border patrol surveillance is at a minimum. At the same time, this change of crossing points makes this adventure even more dangerous and expensive.

The profile of the Mexican migrant has changed over the past few years, from people without much formal education and training from rural parts of Mexico to professionally-educated individuals with at least one year of college studies. Migration, in recent years, has significantly changed. Today, entire families migrate with the primary wage earner, which in many cases increases potential abuses against them. Customs officers, policemen, border patrol officers and even support groups have been known to take advantage of females and minors, requesting money from them when they are traveling alone.

Women face at least three different types of sexual abuse: 1) sexual abuse when women are handed over to authorities to be sexually abused as a payment to allow the group to continue the crossing. 2) sexual abuse from within the traveling group and 3) sexual abuse when they become a sexual partners of the "pollero" during the crossing to assure protection from the other illegal migrants (González Velázquez, E. in Rocha Silva, M. A. y Valencia Silva, M.M. 2009, pp.73-74).

"Children are also vulnerable to abuse as they are often forced into prostitution or have their organs harvested and sold" (González Velázquez, E. in Rocha Silva, M. A. y Valencia Silva, M.M. 2009, pp.74).

We have already discussed the dangers and abuses that migrants commonly experience during the border crossing process in order to reach their destinations. Fortunately, most of the abuse illegal migrants face is in the border crossing process while trying to find temporary or permanent housing in the US. However, even when they reside in a stable location, they are still subject to abuse because of their illegal status and their inexperience as they try to function within a society where all of the rules tend to be different from the ones they knew at home. In summary, the migrant situation could be described as follows:

- Of all foreign-born Mexicans aged 16 and over, 60 percent form part of the labor force. In numbers, this almost doubled from 2.6 million in 1990 to 4.9 million in 2000.
- In 2000, almost four percent of the civilian labor force aged 16 and over was born in Mexico, compared to just two percent in 1990.
- While a relatively small percentage of the total labor force (less than four percent of the total US labor force) is comprised of Mexican migrants, they are concentrated in much greater percentages in certain industries and occupations.
- More than one in every four foreign-born workers in the US is from Mexico.
- Almost one out of every 11 Mexican immigrants is unemployed.
- Over half of all employed Mexican immigrants work in just two areas (production and transportation, and almost 25% worked in services).
- Compared with employed Mexican immigrants, a higher proportion of all employed foreign-born workers works in management and related occupations, while native workers work mostly in professional and related occupations.
- Almost 65% of Mexican workers work in farming, fishing and forestry.
- Approximately 20% of Mexican immigrants work in the manufacturing industry.
- Compared with employed Mexican immigrants, foreign-born workers, in general, are more likely to work in the education-

al, health and social services industries (Grieco, Elizabeth y Ray, Brian, 2004)

Despite being the largest legal and illegal migrant group in the US, Mexicans are not well positioned in the labor market. They still perform the hardest, most demanding and lowest paid jobs compared with migrants from other countries. When facing this, most Mexicans tend to isolate themselves and seek other Mexican immigrants to form groups and feel more protected from abuse.

MIGRATION AND HUMAN RIGHTS

Migration has obliged international organizations to pass resolutions protecting rights. One example of legislation are resolutions 2005/47 and 60/169 of the General Assembly of the International Convention on Human Rights, which deal with "migrant protection." These resolutions provide protection for migrant workers and their relatives concerning:

- Migrant trafficking.
- Irregular detention and deportation.
- Violence in border areas.
- Migrant vulnerability.

Today, Mexico's migration is shifting from an expulsion country to a transit country, mainly for Central Americans, but also for the Asia Pacific countries, whose citizens are looking for an entry point to the US through Mexico. A good example that Mexico is evolving to a country of transit is that statistics of the National Institute of Migration statistics (INM in Spanish, http://www.inami.gob.mx) show that illegal alien detentions for Central and South American citizens looking for better employment opportunities in Mexico or transit to the US are significantly increasing. However, we must also consider the increasing internal and external mobility of Mexican farmers, students and professionals, which is largely due to lack a lack of employment opportunities in Mexico. Another example of the concern for migrant protection is the growing number of organizations and religious associations that help migrants during the border crossing process.

In fact, during the process of crossing, migrants face complex and difficult situations derived from racism, xenophobia, abuses, violence, physical injury, threats and other forms of intolerance expressed by many different forms of discrimination. This discrimination is manifest according to ethnicity, nationality, sex, age, labor market insertion and the legal status of migrants. When Jorge Bustamante took the position on the United Nations Council for Human Rights, he expressed a profound concern about the anti migrant sentiment arising in receiving countries because, in his opinion, "this could create risky situations for the Latin American and Caribbean populations, even though receiving countries still need migrant workers to work in predominantly lower-paying jobs" (Bustamante, 2005, p. 43). The intensity of anti migrant sentiment has increased greatly because of the 2008 global economic crisis. Common anti-migrant sentiments include:

- "The American culture as well as its uses and language are under assault from foreigners who arrive to live here, instead of adapting to the American way, they try to impose their cultures, languages and institutions above us" (Villegas Santana, 2008, p.61).
- "Every illegal immigrant that is detained is one less that will not go after our paycheck, fill our schools, bring more drugs for our children or rob, rape or murder another innocent American citizen" (Villegas Santana, 2008, 61).
- "Mexican immigration is a unique, disturbing and complex challenge for our cultural and national integrity and also commercially for our country's future" (Huntington, 2000).

The main supporting argument behind several attempts to pass legislation against migrant rights consider migrants to be outlaws who deserve to be punished; these types of supporting arguments frame legislation against migrant rights, such as California's Proposition187 and legislation recently passed by the State of Arizona. Fortunately, Proposition 187 and its arguments were rejected because they were blatantly unconstitutional. This law proposed that medical services and driving licenses be denied to illegal migrants in California. Arizona's legislation will undoubtedly come under judicial review.

Migrants are segregated by their race and religion but above all because they do not speak English, they have a low educational level and they are not technologically alphabetized. These three factors place them at a disadvantage with immigrants from other countries. Related to the third point, digital alphabetization is understood as communicative competency, which in its broader sense not only consists of reading and writing, but in comprehension. Furthermore, digital alphabetization includes using different language and technological skills, including basic informatics and the Internet. All of these competencies are necessary for individuals to adapt to the digital era. All of these competencies contribute to the phenomena of "information democratization," which also includes the right to a quality education and access to information. However, most importantly, they now form part of what are basic human rights. Our concept of digital alphabetization is based on the conceptual transformation of alphabetization, understood as not only reading and writing, but as transcending to basic technological alphabetization that focuses on office computer programs.

Muñoz (2001) mentions that to live responsibly in the third millennia, multicultural alphabetization is one of the basic dimensions that must comprise the integral education of individuals, groups and communities because it "strengthens our comprehension of the concept of cultural diversity and improves the communication ability of two persons from different cultures, thus increasing social interaction in culturally different individuals or groups" (Gutiérrez Martín; 2003, p. 27). Accordingly, alphabetization acquires a special sense when we are talking about Mexicans with low educational backgrounds that migrate to the most powerful country in the world, the United States, where the Internet was born. Results of this research described in the following sections show that migrants are relegated just because they are not in their own country, are more vulnerable given their financial condition and have low educational backgrounds.

MEXICAN MIGRANT PROFILE

A recent study by specialists in migration, education and technologies from the University of Colima, in Mexico, provided enough empirical data to draw a profile of the people from Colima who migrated to the US from 2000 to 2006. This information, described later, was then used to design a prototype of a Web portal inform, communicate and educate people who fit the profile. The basic idea of the design was to use a well-designed structure centered on Digital Alphabetization within the context of cultural value preservation and the search to integrate a technological platform that would be useful to persons fitting the profile.

The sample is based on The Instituto Nacional de Estadística, Geografía e Informática INEGI (In Spanish) (2000) which accounts for 542,627 inhabitants of the State of Colima (the third smallest Mexican state located on the west central coast of Mexico between the states of Jalisco and Michoacán). This information reveals that 12,581 people from Colima migrated to the US in 2006. We obtained the sample using the number of people migrating from Colima to the US during that period which was 369 migrants (Rocha, Contreras y Acosta; 2008, p.12).

The data acquired consisted of responses to a questionnaire designed specifically for this study. Its application time was less than seven minutes and consisted of 18 closed-ended questions structured in three different sections:

1. Characterization
2. Required Electronic Services from the government and
3. Technological skills and access to technology

To obtain the maximum response percentage, we selected the City of Villa de Alvarez's local festivities as the place to apply the questionnaire. The reason for selecting this time and place is that many migrants from Colima return every year to participate in the festivities during the second half of February and the first week of March. Thus, we obtained more that 50% of our sample. To complete the required 369 responses, we carried out a second stage of interviews, visiting the cities with the highest migration index, including the town of Tepames and Ixtlahuacan, capitals of their respective counties, each with the same name as their capital.

Gender

Results agree with the Current Population Survey (2000), the Cannon Zapata project (Bustamante, 1997) and the Northern Frontier Foundation (EMIF, in Spanish) (Delaunay, 1979). From these sources we observed that 74.52 of migrants were men and we also discovered a significant increase of woman migrating over time, as reported in Corona, (cited in Bustamante, Delaunay and Santibañez, 1997).

Age

Most of the respondents were between 15 and 30 years of age, which agrees with the theory that migrants are younger persons looking to improve their quality of life. The United States Census, (2001) mentions that the average age of Mexican migrants is 33. Nonetheless, around 33% were migrants older than 40. One limitation of the study was that questionnaires were responded to in Colima and this could mean that most of the respondents could afford a visit to their state of origin because they have a steady financial situation.

Educational Level

According to the INEGI 2000 Census, the average schooling of migrants is 7.3 years, equivalent to the first year of secondary school for people older than 15. However, in our study, 26 percent of the respondents had only primary school studies, but almost half of them had not finished primary school. Furthermore, 21 percent of the respondents had secondary school and 15 percent reported having a high school degree. The US census mentions that Mexican immigrants have less schooling than immigrants from other countries. According to it, 33.8% reached high school level at the age of 25 or older. The Current Population Survey of March 2000 establishes a lower percentage and also shows some differences regarding gender when referring to people who have finished high school (22.3% for men and 19.5% for women) among respondents over 16 years of age. Additionally, 66% of this group of people (16 years and older) did not finish this level. The Cannon Zapata Poll (Bustamante, 1997) also confirms this information, since the highest percentages range from incomplete primary school to complete secondary school but scarcely passing this level. This is closely related to the type of jobs Mexicans obtain while in the US, mainly in construction and services. Our study found that respondents were working in construction, services and agriculture. The concept of migrant vulnerability in the receptor community acquires its true meaning when one considers the Mexican migrants' low educational background (Bustamante, 2001).

Lack of Technological Ability

This point is indirectly related to the previous one, since migrants' low educational background may be linked to the low technological training they receive in Mexican educational programs. Two major projects to face this problem are the *Enciclomedia* (2009) and *Tu maestro en línea (2010)*. It is important to mention that these projects are not employed in all primary schools.

Results of our questionnaire indicate that 40% of respondents own their own computer, 37% have access to a computer in different places, including their place of residence (in this case we, must consider that the person has access to another person's computer who lives in the same house, but does not personally own one), their job, a cyber coffee shop, or at a friend's house. We found that 181 respondents did not have any access to a computer, and 13 did not respond. Therefore, our universe was reduced from 369 to 188 persons.

Connectivity and Technological Training

Out of the 369 sampled respondents, only 154 declared having Internet access and the distribution of their response to the question, "What do you use the Internet for?" was as follows: web surfing 100, sending e-mails 72, using messenger 54, chatting 52, and other uses 20. Some of the respondents used Internet for more than one reason so the resulting sum is higher than 154. To further inquire about their technological culture, the respondents were asked if they had an electronic mail address; only 65 responded affirmatively, but not all of them provided them.

ICT AND MIGRATION

Information and Communication Technologies (ICT) have the potential of becoming useful tools to help migrants overcome some of the migration problems they face, thus reducing their marginalization as they can obtain information about the abuses they might suffer and how to report them to the authorities. While searching for experiences regarding how the Internet has been used to support non-traditional groups, we found two important projects: "*e-migra*" (2006) and "*e-México*" (2003). The first one refers only to three countries: France, Spain and Germany, which have historically been receptors of migrants. The objective of the project is to study and promote digital culture among migrants in Europe on a large scale; the biggest asset of this project is the database with more than 130 projects that promote digital culture with a guide for migrants. Some of the hypotheses of this project are:

1. Migrants in Europe are more vulnerable to digital exclusion.
2. They have problems accessing the digital universe.
3. ICT in education can have an additional educational value when speaking about migrants. It is worth noting two situations that are completely applicable to our reality (the Mexican reality).
    a. Migrants from most countries are not well trained in the use of computer technology and the majority has no Internet access, and in the best case, they have reduced bandwidth.
    b. Several studies in Europe correlate digital exclusion, the lower educational levels and the low income levels of migrants expose them to a higher risk of digital exclusion than the native population.

However, the document points out that there are several differences with the reality Mexican migrants face in the US, and the main reason is that countries like Spain have increased the number of places for migrants to access the Internet, while in the US it is almost impossible to find a place because domestic Internet access is so common and inexpensive, and because it is so easily available in schools and in the workplace. Additionally, there is Internet access in public libraries.

The second project is a Mexican experience that was born within the e-México program. It is a section for migrants, containing a list of communities that traditionally have been providers of migrants. It has some educational options but is still primarily informative.

In Mexico, we have seen an increase in the number of places that provide low-cost Internet access and the percentages are divided as follows:

- Home 48%
- Internet Coffee Shop / Public access 34%
- Workplace 19%
- School 11%
- Friends house 4%
- Others 1%

It is important to add that some of the respondents used Internet in more than one way, which results in a sum of 117%.

Places testing e-inclusion are fundamental to help migrants maintain their links with their families and reduce their feeling of isolation, which, at a sociological level, is an extremely important issue.

The situation regarding ICT in the US and Mexico is quite different. This has motivated the creation of ONGs to train and support migrant groups obtain technological skills and gain Internet access to help them develop ICT competencies. It would be desirable to establish digital alphabetization programs for migrants in the US because their status as illegals and their socio-economic condition prevents migrants from integrating themselves to this new culture. However, this is quite difficult in the US, where public places to access the Internet are limited, contrary to Mexico where such places are common and, in some cases, even free. An example of free Internet access is the state government kiosks in Colima, Mexico, where people can access the Internet without cost.

THE COLIMENSES SIN FRONTERAS WEB PORTAL

A group of researchers and experts in migration, education and technology from the University of Colima decided to focus on this phenomenon to help reduce some problems that arise as a result of migration processes. The project centered its attention in the relationship between those who remain in their communities and those who migrate to find better opportunities. Based on the results, we developed a Web portal from migrant profiles, which were created from an analysis of similar migrant experiences.

The Colimenses sin Fronteras Web Portal (for people from Colima, Mexico living beyond Mexico's borders) is an initiative that contains several modules that help migrants remain linked with their communities of origin, offering them an opportunity to begin learning English and basic ICT, while providing them with information about health problems that they might face.

Description

The Colimenses sin Fronteras Web portal is located at: http://telematicanet.ucol.mx/colimenses/nuevo and Figure 1 shows the website map.

The interface was designed based on color psychology, combined with visual elements and important symbols of Colima, such as the Fire Volcano and the Pacific Ocean. The State of Colima is located in the west central part of Mexico and borders on the Pacific Ocean and the states of Jalisco and Michoacán. The State's geographic area comprises a minute 0.3% of Mexico's land mass, according to the national statistics and geographical institute INEGI (in Spanish). The sea and the volcano do not refer only to the directions of north and south, but also to good and evil

Figure 1. Colimenses sin Fronteras Website map

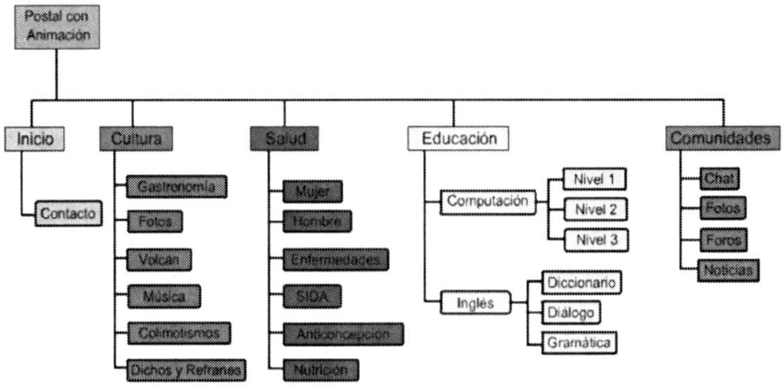

because of the high number of catastrophes related to them (hurricanes and earthquakes). They are part of the history of Colima and they represent points of reference for the population. This is the reason why the main page of the portal shows both of these features. The smoke rising from the volcano is a flash animation that displays the name of the web portal.

Figure 3 displays a welcome message with both symbols included in the upper part of the page. Within the text, objectives and contents of the four modules are explained.

Based on the results of our surveys, we developed four modules: Culture, Health, Education, and Community.

Figure 2. Main page of the Colimenses sin Fronteras portal

Culture Module

The 'Culture' module shows a menu with the following items:

- **Gastronomy:** with traditional recipes from Colima's cuisine;
- **Pictures:** from each municipality;
- **Volcano:** with a short story and pictures of the volcano;
- **Music:** with the main songs from Colima;
- **Colimotismos:** with a database containing two dictionaries about phrases and words used in Colima and;
- **Sayings:** typical of Colima.

Health Module

This module was structured to help reduce the difficulties migrants suffer when communicating their health problems. We want to add that the page contains the disclaimer that the information provided in this section is not meant to substitute physician care. Instead, information is provided to help migrants communicate their medical needs. The module is divided in two sections, each section containing information of interest according to gender with information about different body parts.

Figure 3. Starting page of the portal

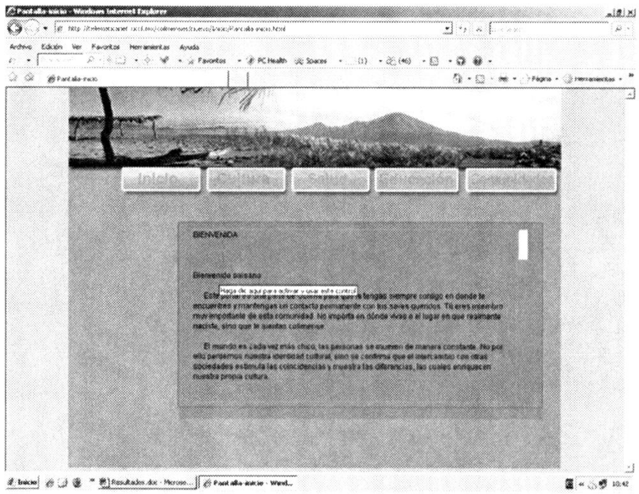

Figure 4. Culture module

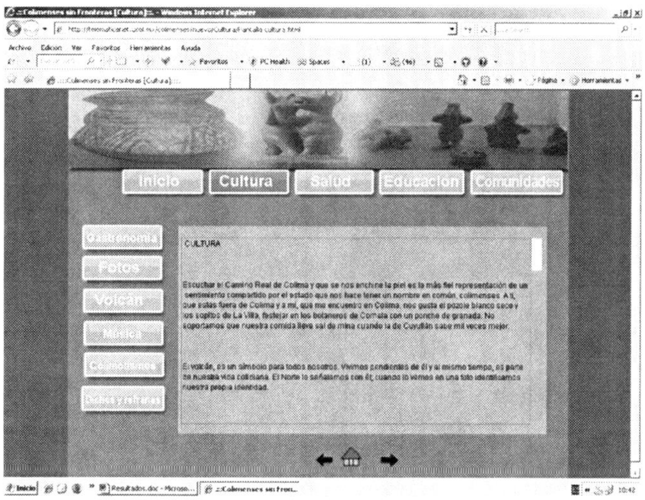

There is also general information about important health questions the migrant population faces, including AIDS, nutrition information, and how to prevent unwanted pregnancy.

Education Module

Two objectives are fundamental in this module: the first one is how to use the computer as a tool to promote digital alphabetization programs; the second one how to learn English by following the teaching methodology of the mother tongue. The proposed educational model is based on the concept of reinforcing different cultures while promoting e-integration. We do this under the premise that migrants not only need to maintain affective links with their communities of origin, but also need to integrate into the receiving community. Specifically, we maintain that if migrants do not have access to the Internet and, in some cases, even if they do have access but they lack the necessary abilities to use it, their isolation

Figure 5. Health module main page

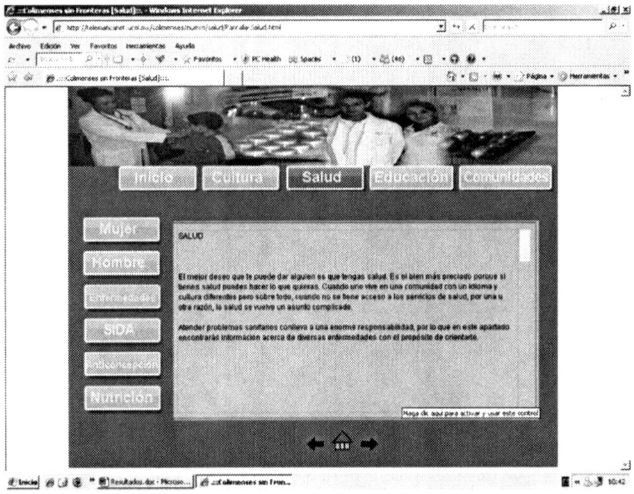

Figure 6. Male and female animation within the health module

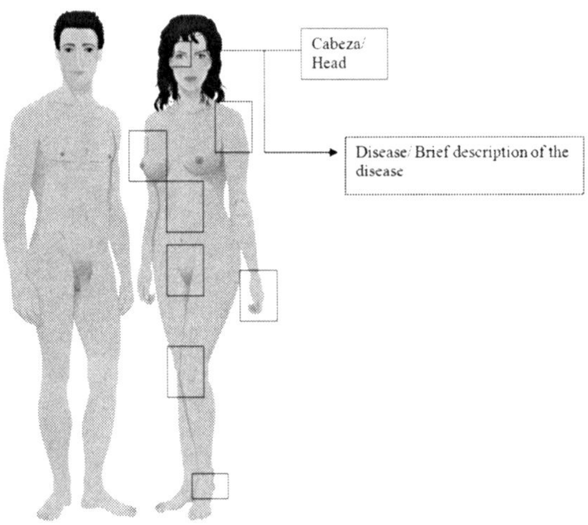

will increase because they will interact less in the receiving community, they will not be as successful in the job market, they will have lessened chances to access information and services through the Internet, where, for example, they can find cheaper flight tickets.

The primary means of communication for migrants is the telephone, which is much more costly than interacting by means of the Internet.

Based on supporting information, we propose using the portal to alphabetize migrants because they are Mexican citizens. Consequently, Mexico should be responsible for them as much as possible. Mexico must work not only to reduce emigration, but also make migrants less vulnerable to exploitation and abuse, while helping them maintain their cultural identity and links to their communities.

The educational model is Problem Based Learning (PBL) and learning by doing. This decision is based on the fact that migrants want to solve real problems in their daily lives. Representation of situations is one of the most interesting

Figure 7. Education Module

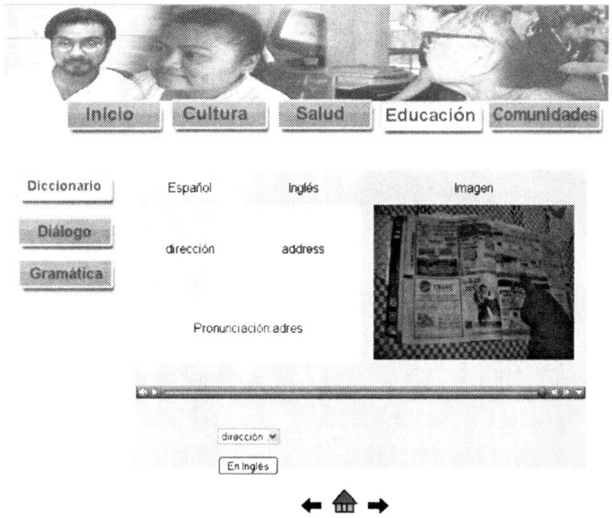

things about the portal, and in some of them, they might be able to request help from a tutor. With the latter, we are talking about distance learning; however, the *Colimenses sin Fronteras* portal can be installed and used in different centers to increase interest while at the time, providing them with the necessary equipment to use the portal's tools.

The portal administration and servers will be installed at the University of Colima main campus due to financial considerations. However, in the future, agreements with local governments and non-profit organizations may be signed to reproduce the number of servers and increase availability, as well as continually modifying the number of modules and tools developed.

In this section, we will further describe the English module and provide examples of the different sections of the Education Module.

English Module

The main difference of this program is that it employs a socio-cultural context recognized as the one in which persons can learn languages. Professionals, specialists in educational technology, language and communication experts and migrants living in the US, developed the teaching and learning content.

The learning program was developed considering how people learn languages and includes 3 levels.

Level 1. Vocabulary: this section presents single words that are shown in conjunction with images accompanied by sounds to help users mentally relate objects to familiar situations.

Level 2. Conversations: this section includes phrases, expressions and full sentences in a predetermined situations. In this section, we establish a specific context in which users face social and cultural norms, with questions and answers to help learners maintain conversations regarding topics including:

1. Restaurant
2. Supermarket
3. Gas station
4. Customs affairs
5. Migratory affairs
6. Hospital
7. Church
8. Workplace
9. Bank.

Level 3. Grammar.

The portal itself cannot accomplish anything. It is necessary to create a specific program to provide tutoring and provide content that is based on migrant data and is in accordance to migrant's technological knowledge. Families are an important part of the digital alphabetization program because they provide the motivating for migrants to learn.

Communities Module

The Communities module facilitates communication and information transfer between migrants and their respective communities. Photos of all municipalities are available for viewing and there is a special section to upload pictures sent by migrants in the US. There is also a chat to enable real-time communication. The portal also contains a section with online forums to post and discuss details about the different municipalities and all of the elements they can find in the portal. A section to upload news about the 10 different municipalities is also shown.

Digital Alphabetization Section

To fulfill the objective to digitally alphabetize migrants, we developed several modules in the area of computing.

Level 1. Objective, Digital Alphabetization. Students will learn how to identify different computer parts with the assistance of computer animations. Migrants can obtain information in Spanish about each part by clicking over it.

Level 2. Students will learn not only to use the basic software that will allow them to establish direct communication with their loved ones, but to employ basic software commonly used in the workplace, school or daily lives. Exercises are always related to daily life events in the migrant community, linked to the American society and culture, mainly in the rural environment.

Level 3. Training for Work: Specific programs developed to meet migrant requests. This portal section blends digital alphabetization and language leaning through different options. For example, to teach migrants how to use a mouse in the digital alphabetization basic module, we can reference it to a real translation while we display a picture of a cartoon mouse. Another example is related

Figure 8. Communities

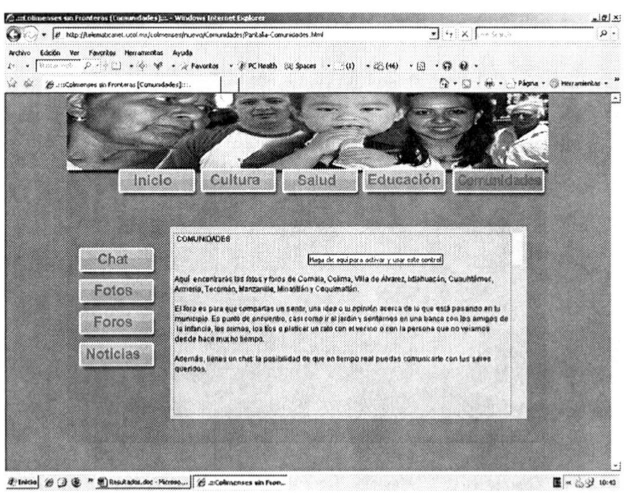

to bank transactions where we show a picture of a bank. The main objective of this is to increase vocabulary and teach the flow and development of a conversation about ATMs, how to identify each part of the device and how to use them, as well as how to access banking services through the Internet.

It is important to mention that the English teaching option cannot be accessed from Mexico. This option is only available for migrants living in the US because the objective of this activity is to help Mexican migrants in the US, and not motivate them to leave.

DIGITAL ALPHABETIZATION PROGRAM

This program arises within the "*Colimenses sin Fronteras*" web portal as an example and motivation to develop and implement similar portals in the different communities with high migration indices.

We have worked in similar projects related to community technologies, digital alphabetization and re-alphabetization. One such project is called "Enrédate," by Dra. Ana Isabel Zermeño Flores, with financing from the University of Colima and the Social Development Secretariat (Gutierrez Martin, 2003) in the county of Ixtlahuacan, the county capital located in the State of Colima with the second highest migration index, only after the county of Cuauhtemoc (Foro Estatal sobre Atención a Migrantes. Manzanillo, Colima. November 27 and 28, 2000). These projects, currently used in learning community centers (LCC), are presently located in different localities.

Learning community centers (LCC, http://www.cca.org.mx/portalcca/info_gral/como.htm) are learning spaces where participants can access educational programs through technology and interact in a permanent way with tutors, giving users the opportunity to enter the knowledge society within the framework of the e-México program (http://www.e-mexico.gob.mx/). In the near future, this experience will indicate the path to a proposal that promotes the reuse of physical spaces that already have the technological infrastructure within these communities, such as those provided by digital community centers, which are also form part of the E-México project. Some of these projects are carried out in the US, as shown in the digital map located at http://www.e-México.gob.mx/wb2/eMex/eMex_mapausa. In 2005, the government reported 7,500 active digital community centers, with 5.8 million users each year (AMIPCI, 2005).

This program was structured to approach people living in smaller communities and promote the use of technology, using social networks, implemented in the community, under the coordination of a monitor that has been previously trained in the use of technology.

The contents are mapped in a similar fashion as the Web portal and migrants are taught the different parts of the computer, its uses and basic text processing and imaging software. Furthermore, they access the internet and different communication tools through this technological platform.

FUTURE RESEARCH AREAS

The portal is currently located in a web server of the University of Colima because it was financed in a call for research projects; the electronic address of the portal is: http://telematicanet.ucol.mx/colimenses/nuevo/. Students of the School of Telematics are in charge of portal maintenance and a professor coordinates the activities. The digital alphabetization project was born within another project in which students participated as monitors. The first stage was implemented in the county of Ixltahuacan in the State of Colima, the second in Comala, and the remaining stage was carried out in the seven remaining counties of Colima. An important point to consider is female participation in this program-- all of them, mothers, sisters or wives of migrants who are already in the US.

The portal and all of its modules have been implemented and used on the Mexican side of the border; hence, an important research area is to test it on the US side, where the participants of migrant clubs will be of key importance. According to Mexican communities in the exterior (2003), the number of these clubs has increased significantly and is characterized by their support, as well as actions to benefit their communities.

Currently, we are working on the Virtual Interactive Migration Museum to integrate it within the Colimenses sin Fronteras Portal. The objective of this museum is to keep, maintain and rescue historical information of migrants who have lived in Colima and those who have left but have achieved important things.

CONCLUSION

Every year, migrants require more and more information about their rights, mostly to protect themselves against abuse, but also to know how to report violations to their human rights, a topic that is emphasized in the communities section, which includes links to the main migrant protection groups as well as the Mexican embassies and consulates around the US. Additionally, they need to overcome many educational deficiencies, including digital alphabetization and the English language. We believe that technological platforms like the *Colimenses sin Fronteras* portal can potentially help migrants obtain information, stay connected with their communities and also reduce their vulnerability to abuse.

Migration is an integral part of human evolution and we cannot live without it. Therefore, we need to implement mechanisms, policies and strategies to promote a more ordered process with respect for migrant human rights and for all of humanity's benefit.

ICT represents an important tool to maintain and promote links with close relatives who migrate to the US, and the Internet is among the most important ICT tools at migrants' disposal. The tendency of technological development is towards Internet applications and web services, as well as government programs that allow Internet access to marginalized communities within large cities, semirural or rural areas where problems related to access are focalized. Government programs such as E-México are establishing the foundations for future technological development in Mexico. The generosity of various institutions, including Telmex and the Bill and Melisa Gates Foundation has helped communities acquire educational technology. However, it is also important to integrate a digital or technological alphabetization program to provide these skills to a broader population, including women, because as this and several other studies have demonstrated, they are participating more and more in the use if the technology. Additionally, they represent the glue that binds families together. Most of the time, women are among the most willing to explore new things, which in this case, means technology. We hope this project can, in the future, provide support for improved communications among families, which is fundamental to maintaining cohesive and consolidated family units, even when some of the members live in the United States. Finally, this is not important only for migrants, but for Mexican society, in general, because an important percentage of family income in Mexico can be traced to money that is sent from migrants to family members still living in the country, which, in turn, helps improve the standard of living of the Mexican people.

REFERENCES

Alonso, G. (2003). Human rights and undocumented migration along the Mexican-U.S. border. *UCLA Law Review. University of California, Los Angeles. School of Law*, *51*(October).

Asociación Mexicana de Internet. (AMIPCI). (2005). *Hábitos de los usuarios de internet en México.* Retrieved February 28, 2010, from http://www.amipci.org. mx/temp/ResumenEjecutivoEstudioAMIPCI2005-0315079001138390784OB.pdf

Asociación Mexicana de Internet (AMIPCI). (2009). *Hábitos de los usuarios de internet en México.* Retrieved February 28, 2010, from http://amipci.org.mx/estudios/temp/ RESUMENEJECUTIVOEstudioAMIPCI2009Usuariosdeinternet-FINAL-0334725001245691260OB.pdf

Bustamante, J. (1997). *Cruzar la línea: La migración de México a los Estados Unidos*. Mexico City, Mexico: Fondo de Cultura Económica.

Bustamante, J. A. (2001). Proposition 187 and operation gatekeeper: Cases for the sociology of international immigration and human rights. *Migraciones Internacionales*, July - December, 1(1).

Current Population Survey. (2000). *Current population survey.* Retrieved February 28, 2010, From http://www.census.gov/apsd/ techdoc/cps/cpsmar00.pdf

E-México. (2009). *Portal e-México*. Retrieved February 28, 2010, from http://www.e-mexico.gob.mx/.

E-migra. (2006). *E-migra*. Retrieved February 28, 2010, from http://www.e-migra.org

Enciclomedia (2009). *Programa Enciclomedia.* Retrieved February 28, 2010, from http://www.sep.gob.mx/wb/s ep1/programa_enciclomedia

González Velázquez, E. (2009). El trayecto y problemática que viven las personas que emigran de nuestro país a los Estados Unidos. In Rocha Silva, M. A., & Valencia Silva, M. M. (Eds.), *El impacto económico y social de la migración en Colima* (pp. 67–79). Mexico City, Mexico: Gasca.

Grieco, E., & Ray, B. (2004). Mexican immigrants in the US labor force. In *Migration Information Source*. Washington, USA: Migration Policy Institute.

Gualda Caballero, E., & Ruiz Garcia, M. (2004, July-December). Migración femenina de Europa del Este y mercado de trabajo agrícola en la provincia de Huelva, España. *Migraciones Internacionales*, *2*(4), 36–65.

Gutiérrez Martín, A. (2003). *Alfabetización digital: algo más que ratones y teclas*. Barcelona, Spain: Gedisa.

Huntington, S. P. (2000). *Reconsidering Immigration Is Mexico a Special Case?* Retrieved May 1, 2010, from http://www.cis.org/articles/2000/back1100.html

Instituto Nacional de Geografía. Estadística e Informática (INEGI). (2009a). Retrieved February 28, 2010, from http://www.inegi.org.mx.

Instituto Nacional de Georgrafía. Estadística e Informática. (INEGI). (2000). Retrieved February 28, 2010, from http://www.inegi.gob.mx/est/contenidos /espanol/rutinas/ept.asp?t=medu09&c=3277

Instituto Nacional de Georgrafía. Estadística e Informática (INEGI). (2009b). Retrieved February 28, 2010, from http://mapserver.inegi.gob.mx/geografia /espanol/estados/col/ubic_geo.cfm? c=1203&e=06&CFID=591044&CFTOKEN=29002896

Martínez Pizarro, J. (2001 July - December). Reflexiones sobre la gobernabilidad de la migración internacional en América Latina. In *Migraciones Internacionales, 1*(1) (pp. 82–110). Mexico: Colegio de la Frontera Norte.

Migración internacional, Derechos Humanos y desarrollo en América. (2006). *Migración internacional, derechos humanos y desarrollo en América.* Montevideo, Uruguay: Naciones Unidas/CEPAL.

Migrants Portal, E. -México program (2009). Retrieved February 28, 2010 from http://www.e-mexico.gob.mx/wb 2/eMex/eMex_eMigrantes

Pécoud, A., & De Guchteneire, P. (2005, July – December). Migración sin fronteras: Una investigación sobre la libre circulación de personas. *Migraciones Internacionales*, *3*(2), 137–166.

Rocha Silva, M. A., Contreras-Castillo, J., & Acosta-Díaz, R. (2008). *Las nuevas tecnologías de la comunicación entre los migrantes*. Colima, Mexico: Archivo Histórico del Municipio de Colima y Secretaría de Cultura del Gobierno del Estado de Colima.

Tu maestro en línea (2010). *Tu maestro en línea.* Retrieved April 23, 2010 from http://168.255.111.18/contacto_ maestro/index.jsp

UNESCO. (2010). *UNESCO Program: Alphabetization as part of the liberty portal. Retrieved May 1, 2010, from* http://portal.unesco.org/education

US Census Bureau. (2001). *US census bureau.* Retrieved February 28, 2010, from http://www.census.gov/prod/2002pubs /01statab/stat-ab01.html

Vazquéz, C. Rodolfo. (1997). Las mediciones de la emigración de México a Estados Unidos. In Bustamante, J.A., Delalunay, D. & Santibáñez, J. (Eds). *Taller de medición de la migración internacional*. México: Colef/ORSTOM.

Villegas Santana, F. J. (2008). Conferencia: Migración Mexicana hacia Estados Unidos. In Rocha Silva, M. A., & Valencia Silva, M. (Eds.), *El impacto económico y social de la migración en Colima* (pp. 51–66). Mexico City, Mexico: Gasca.

ADDITIONAL READING

Bustamante, J. A. (1997). Retos metodológicos en la investigación de la migración de indocumentados de México a Estados Unidos. In Bustamante, J. A., Delaunay, D., & Santibáñez, J. (Eds.), *Taller de medición de la migración internacional* (pp. 231–258). Mexico City, Mexico: Colef/ORSTOM.

Castaños-Lominitz, H., Rodríguez-Sala, M. L., & Herrera Márquez, A. (2004). Fuga de talentos en México: 1970-1990, un estudio de caso. In Castaños-Lomnitz, H. (Ed.), *La migración de talentos en México* (pp. 17–48). Mexico City, Mexico: UNAM/Porrúa.

Castles, S. (2000). The impacts of emigration on countries of origin. In Yusuf, S., Wu, W., & Evenett, S. (Eds.), *Local Dynamics in an Era of Globalization*. New York: Oxford University Press for the World Bank.

Castles, S., & Millar, M. J. (2004). *La era de la migración. Movimientos internacionales de población en el mundo entero.* Mexico City, Mexico: H. Cámara de Diputados, LIX Legislatura.

Durand, J., & Douglas, S. M. (2009). *Clandestinos migración México- Estados unidos en los albores del siglo XXI.* Mexico City, Mexico: Universidad Autónoma de Zacatecas/Miguel Ángel Porrúa.

García Vázquez, N. J., Gaxiola Baqueiro, E. G. & Guajardo Díaz, A. (2007). Movimientos transfronterizos México- Estados Unidos: Los polleros como agentes de movilidad. *Confines, 5.*

González Velázquez, E. (2008). *Con todo y triques. Crónicas de migración*. Guadalajara, Mexico: Universidad de Guadalajara.

Imaz, C. (2007). *Metropolítica. Percepciones de la migración en México y Estados Unidos.* Mexico City, Mexico: Centro de Estudios de Política Comparada.

Trujeque Díaz, J. A. (2007). Minuteman Project: Segregación y activismo antimigratorio. *Andamios. Revista de Investigación Social, 3*(6), 137-172. Retrieved March 28, 2010, from http://www.uacm.edu.mx/sitios/andam ios/num6/articulo%202.pdf

Tuirán, R. (2006). La reforma migratoria pendiente. *Migraciones Internacionales, 3* (4).

KEY TERMS AND DEFINITIONS

Immigration: It is the arrival of new individuals into a habitat or population. It is a biological concept and is important in population ecology, differentiated from emigration and migration.

Emigration: It is the act of leaving one's native country or region to settle in another. It is the same as immigration but from the perspective of the country of origin.

Migration: It means that one member of the family is the one who migrates, at least in the first moment. The nuclear family disintegrates partially or reconfigures itself in the physical context, because one of its members abandons the community to go to other country (Herrera, 2001).

ICT: Information and communications technologies (ICT) are the computing and communications facilities and features that variously support teaching, learning and a range of activities.

Discrimination: It is an action that treats people unfairly because of their membership in a particular social group. Discriminatory behaviors take many forms, but they all involve some form of exclusion or rejection. Discrimination is an action that treats people unfairly because of their membership in a particular social group. Discriminatory behaviors take many forms, but they all involve some form of exclusion or rejection.

Community: It is a group people sharing an environment, there might be a number of conditions present and common, affecting the identity of the participants and their degree of cohesiveness

Digital Alphabetization: Understood as the communicative competency, in its broader perspective, that does not only include the comprehension, but also the evaluation and utilization of different languages, which allow the individual to develop technological abilities on basic computing and Internet in order to integrate himself or herself into the digital era. This, on one hand implies the democratization of information and on the other the right to education and information as a fundamental part of basic Human Rights. The concept "Digital Alphabetization" employed for the use of this research has its base on the transformation of alphabetization once understood just as reading and writing, no transcending to the digital phenomenon: "Using computing beyond its mechanical use of computational programs". "As part of that basic preparation, of what we have called multiple alphabetization, we ought to consider the intercultural education as one of the basic dimensions of the general education of individuals, groups and communities, not just in places with immigrants or minorities, as mentioned by Muñoz (2001), who also designates the systematic education of all individuals as the comprehension of the cultural diversity of our present society, within the increment of cultural interaction between people and groups of different cultures (Gutiérrez Martín, 2003)

Portal "Colimenses sin Fronteras": A website created to meet the needs and demands of Colima state residents living in the United States. This website offers all of the standard communication tools, educational resources, an up to date and accurate information and services related to both the public and private services. Providing additional institutional and life-long educational opportunities to non-traditional segments of society is an area that requires more research and development.

Digital Culture: A field of study that seeks to explore the cultural ramifications of digital technology, and how the alterations in the patterns of communication, information processing, and archiving resulting from the use of digital technology have influenced human behavior.

Chapter 9
Web-Based Multi-User Distributed and Collaborative Environment Supporting Emergency and Relief Activities

Maki K. Habib
The American University in Cairo, Egypt

ABSTRACT

This chapter presents the conceptual development, design requirements and functional descriptions of a team based multi-user, web-based and global telecooperative system as a new multidisciplinary collaborative paradigm. This system uses as a foundation the distributed and collaborative intelligent environment (DCIE) developed by the author. Distance, time, and distribution are the main attributes of the telecooperative system and its applications. The developed system aims to carry out cooperative plans and processes over geographically distributed locations that lead to make well-informed decisions, better utilization of resources, enhance safety and reduce cost. In addition, this article foresees the role of Internet, information and communication technologies (ICTs) to support and enhance emergency management and relief activities in relation to natural and human made disasters, and human assistance activities. As an example of human made disasters, landmine disaster and humanitarian demining needs are introduced, and the core collaborative functional requirements are presented to support the integration of wide range of humanitarian demining activities, cooperation and coordination between all relevant parties at global, regional and local levels.

INTRODUCTION

Disasters are tragic consequence of natural and human made hazards that negatively affect life, health, environment, economy, etc. It is a well-known fact that natural and human made disasters strike countries, developed and developing, causing enormous destruction and creating human sufferings and producing negative impacts on national economies. Cities are highly vulnerable to the effects of natural and human-made disasters. It is possible to categorize disasters into, natural, human-made, and human induced disasters. Natural disasters are inevitable and

DOI: 10.4018/978-1-60960-094-5.ch009

naturally occurring events and it is almost impossible to fully recover the damage caused by these disasters. However, it is possible to minimize the potential risk by developing disaster early warning strategies, prepare and implement developmental plans to provide resilience to such disasters and to help in rehabilitation and post disaster reduction. Common natural disasters are earthquakes, landslides, volcanic eruptions, floods, tsunamis, hurricanes and high winds, tornados, typhoons, avalanches, tropical storms, forest fires, extreme heat, extreme cold, etc. Specific disasters might occur due to location and different geographical features of a region [Alexander, 1993, IFRC, 1998, UN-ESC, 2005, HABITAT, 2009). Since 1975, there has been a fourfold increase in the number of recorded natural disasters, the highest of which was 801 disasters in 2000. Between 1996 and 2005, disasters accounted for over US$667 billion in material loss (UN-HABITAT, 2009). Examples of human made and human induced disasters are: conflicts and wars, mines and minefields, weapons of mass destructions (biological, chemical and radiation), diseases, nuclear power plants accidents, oil spills, negatively impacting climate changes and extreme weathers, ecosystems degradation and pollution, mismanagement of natural resources, etc. The social, economic, health, development and rebuilding consequences of disasters are immense.

Disaster management and risk assessment is an applied science, which seeks, by systematic observation and analysis of disasters while underlying the factors that contribute to disasters, to improve measures relating to prevention, mitigation, preparedness, emergency response, and recovery. It seeks to motivate societies at risk to become engaged in conscious disaster management. Disaster prevention, reduction and response has multi-sectors, is interdisciplinary in nature, and involves a wide variety of interrelated activities at the local, national, regional and international level. Access to information is crucial for the effective management of disasters (Helm, 1996, Smith, 1996; Stenchion, 1997)

The players in disaster management team may include governments and international organizations along with a wide range of local players and at all levels. Crisis response includes the logistics of getting medical care, food, water, awareness, shelter, and rescue teams to the scene. Regional, global, local and other resources can be provided to assist those affected. The recovery should encompass both short-term activity intended to return vital life-support systems to operation and longer-term activities designed to return infrastructure systems to pre-disaster conditions. To be prepared for unforeseen events, the integrated players must make contingency plans and coordinate their planning with other agencies and parties involved.

A major example of human made disasters is the landmines and the unexploded ordnance (UXO) hazards that have an enormous effect on human and land. By definition, UXO is any munitions, weapon delivery system, or ordnance item that contains explosives, propellants, and chemical agents. Landmines and UXO left in the ground after the cessation of hostilities are a significant threat to human life, social organization and economic development throughout the world. Mines are the cheapest, deadliest, and ubiquitous weapons. The fundamental goal of humanitarian demining is to make the land safer for daily living and restoration to what it was prior to the hostilities. Hence, it requires that all forms of dangerous battlefield debris, particularly landmines in an area are located, uncovered and removed or destroyed from infected areas efficiently, reliably, and as safely and as rapidly as possible. In addition, it is important to have almost 100% clearing performance in order to allow people to use their land without fear. This involves a great effort, time, and large risks, which results in high clearance cost per surface unit. To achieve proper demining outcome, many varied prerequisites have to be observed, such as, soil and type of topology, environment, type of mines, as well as level of contamination (O'Malley, 1993, Blagden, 1993; Hewish, & Ness, 1995, King, 1997; US-Department of Defence, 1998; Bruschini et al., 1999; Humanitarian Mine Action Equipment Catalogue, 1999; Habib, 2001a; Habib, 2001b, Habib, 2007; Habib, 2008).

The critical effect of mines is to deny access to land and its resources, causing deprivation and social problems among the affected populations. In addition, the medical, social, economic, and environmental consequences are immense. United Nations Department of Humanitarian Affairs (UN-DHA) assesses more than 70 million landmines around the world that pose significant hazards in more than 68 countries around the world. These landmines are killing or maiming a person every 20 seconds. The victims could be children playing in a field or roadway, farmers tilling soil, young or old persons gathering firewood or a displaced person approaching his or her former home. The international Committee of the Red Cross estimates that the casualty rate from mines currently exceeds 26,000 people every year. It is estimated that 800 people are killed and 1,200 maimed each month by landmines. The primary victims are unarmed civilians; children are particularly affected. The direct cost of annual medical treatment and rehabilitation exceeds US$ 750 million. This figure is very small compared to the projected cost of clearing the existing mines. The current cost rate of clearing one mine is 300-1000 US$ per mine (depending on the mine infected area). There exist many types of mines around the world; including more than 650 types of antipersonnel (AP) landmines (O'Malley, 1993; Gros, B., & Bruschini, C., 1996; Nicoud, 1997; Bruschini et al., 1999, Habib, 2001a, Habib 2007; Habib, 2008).

It is essential for experts of many disciplines within the fields of disaster management and humanitarian assistance to come together to identify common problems and share successful strategies. This will help to develop synergy within and between communities working in the fields of disaster management and humanitarian assistance in general and within humanitarian demining community in specific while orienting technological innovation towards the operational needs. Communication, coordination, and collaboration are becoming basic needs in the information society that demand to develop technologies to support mobility and able to bridge gaps in time and space in working and living environment.

New information and communication technology tools and applications are needed to offer promise for assisting emergency management, relief and human assistance, professional access, integrate-acquired-process and disseminate required information in real time, efficient utilization of distributed resources and expertise, minimize the impact of hazards, create awareness, facilitate training, and manage the information needed to save lives and property from natural disasters. Integrating multiple systems worldwide to enable relevant people to access and get timely information that help them to make decisions, optimize use of resources, and enhance performance. Merging virtuality and reality in unique and innovative ways will take new meanings to support collaboration mediated by technology that facilitate the realization of joint work beyond time and space constraints, and its implication is extensive.

COOPERATION AND COORDINATION: THE NEED

Cooperation is an essential part of modern work, business, research, and other professional and social work that aims to improve quality, speedup technology transfer and expertise. It provides an effective means for sharing information and other resources, exchanging ideas and techniques, and enhancing problem-solving capabilities. Cooperation is a combination of technology, people, and organizations that facilitate the communication and coordination between those cooperating to establish a shared understanding that enables working together in the pursuit of a common goal, and to achieve benefits for cooperating members. Cooperation implies synergistic relation while reducing costs through improved resource sharing to perform task, enabling parallel performance of operations, speeding-up operation, and improve operational efficiency and quality. In a business/organizational context, people communicate and collaborate to share information and ideas in order to accomplish jointly purposeful work. From such a perspective, communication and collaboration are critically important for effective and efficient work when people are working within a shared context and guided by a common sense of goals. Cooperation can take place over varying dimensions of time and space. Cooperation among team members can vary depending on the degree of their interdependence. Cooperative work can be achieved by a large scale of processes distributed across time, participants, and functional perspectives.

Effective coordination is critical to the success of such cooperative processes since the distributed activities are interdependent due to shared resources, input-output relationships and so on. Maximizing the outcome from cooperation, awareness of others' action, thoughts, and feeling among collaborated members is important. In addition, cooperation is a fundamental part of effective decision-making and problem solving in a complex environment. Support for coordination can be viewed as being divided into three layers, each built on top of the ones beneath it. At the bottom is the communication layer that allows participants in the decision process to share information (this involves networking infrastructures); collaboration layer: allowing participants to collaboratively update some shared set of decisions, and coordination layer: ensuring the collaborative actions of the individuals working on a shared set of decisions are coordinated in order to achieve the desired result efficiently.

Organizations that communicate and make information sharing processes easier, more effective, and more efficient are more responsive, productive, profitable, and flexible; people and resources are well leveraged and the organization is robustly connected with its environment (Klein, 1997). Efficiency is also a fundamental goal for communication and collaboration, especially in today's increasingly decentralized world. People working in specialized roles must efficiently interact with others and with the relevant and available information resources in order to minimize coordination costs while maximizing organizational effectiveness, as an example in research collaborative environments, researchers multiply their output through synergistic collaboration while reducing hardware costs through improved equipment sharing. In addition, they improve operational efficiency and quality by distributing knowledge functions, information processing, and enabling parallel performance of operations.

Traditional information sharing processes are becoming more expensive and more difficult: travel to facilitate in-person collaboration is becoming more expensive; acquisitions and mergers are creating organizations with more geographically distributed workforces; labor expenses continue to climb, making user productivity more critical; competitive pressures are forcing companies to provide better and faster support and services or face risk of losing. Moreover, as society becomes increasingly global and interconnected through the use of information and communication technologies, it is essential to use collaboration as a means to foster common understanding, enhance performance in different fields, improve productivity and have a better use of distributed resources (physical and expertise).

TELECOOPERATION AS A NEW COLLABORATIVE PARADIGM

Collaboration deals with all the various efforts aimed at improving joint work mediated by technology locally or remotely. The goal here is an effective and agreeable interaction that comes close to habitual and natural ways of communication and cooperation. Collaboration technologies that allow team members to share information, make presentations, communicate, interact, conduct, or receive training and experiments, monitoring processes and systems, developing new design, jointly control real facilities in real time, etc. in different ways are important features and capabilities that make team members more efficient toward achieving their goals. Adding presence capabilities to collaboration will make distributed collaboration (synchronous and asynchronous) over long distances possible. The availability of presence enabled by real time collaboration technology can enhance existing workflows by providing information on the availability of all working parties.

The quality of information and communication technologies, infrastructures and techniques of networking are forcing a rethinking of the physical structures and spaces needed for living, working, and learning as the need to look for alternative forms of physical movement of people, equipment, records and information, etc. is evolving. It addresses the issue of universal connectivity and global access to information; easy control and use of globally linked physical equipment/devices, systems, and processes, facilitating real time interaction, space sharing in mixed realities and awareness, supporting mobility and portability of people, and enhance access to experts, and other dispersed and valuable resources. The ubiquitous paradigm has the potential of drastically changing the way in which users interact with computers by providing ubiquitous access to services and applications through a large number of cooperating devices.

Telecooperation is an outstanding example for the power of enabling technologies as well as it is the latest appearing manifestation of those guiding visions and paradigms that have governed application development. Telecooperation stands for the fusion of computer science, telecommunication, and multimedia facilities to carry out a cooperative process among organizations, individuals, systems, or combination of them by having better access to resources and share timely information over geographically distributed locations that lead to make well-informed decisions. Distance, time, and distribution are the main attributes of telecooperation applications. Telecooperation comprises procedural and collaborative modes of work and its focus lies on cooperation in the broader sense. The effective use of telecooperation will affect almost every aspect of society, industry, economy, and other services. In addition, telecooperation entails new skills for individuals, new organizational structure, and open the prospect for new applications and services. Telecooperation concerns with a series of issues ranging from particular application domains, to tools for communication and cooperation. Application Domains may include global office, innovative services, medical services (medical records, medical information, diagnosis, consultation), technical and professional services (businesses and technical consulting, accounting, architectural design, legal advice, travel service, etc.), entertainment products (motion pictures, videos, sound recordings), education (classroom, libraries, seminars, training, books), information services (databases (general and technical), on-line newspapers, magazine), product licenses, advertising, etc.

As the amount of information and communication increases dramatically, new working environments must provide efficient mechanisms to maximize the benefits of these developments. Due to their ability to overcome the restriction of time and space, telecooperation systems promise to enable the virtualization of workplace, project and research teams, manufacturing, business units, etc. Organizations, professionals, and individuals that successfully apply telecooperation methods can enhance customer and supplier communications, improve quality of services, sharing unique and expensive resources, minimize risk associated with hazardous and dangerous environments, access timely information and expertise that support diagnosis and decision making, and reduce overall operational and maintenance cost.

The emergence of global, decentralized, and distributed organizations and businesses along with the growing of computer network and their on-line users substantially increase the need for integrated solutions of workflow management systems, telecooperation and communication tools, aiming to support overall business processes and cooperative work between peoples within and outside their organization in an efficient way. Working together in co-located or geographically dispersed individuals, groups, and organizations needs particular technologies. In addition, methods for analysis and design have to be developed that can cope with the characteristics situation of management, cooperative and interactive work. However, convincing solutions that enable organizations, individuals and groups to adopt it are still weak due to a variety of reasons ranging from social and organizational problems to purely technical issues.

Telecooperation affects many areas such as, sociology and economics, tele-working, tele-learning (education and training), tele-healthcare, tele-design and development, tele-services, utilizing distributed resources and expertise beyond time and physical constraints, network technologies, software architecture, distributed systems etc. Telecooperation is the road towards ubiquitous intelligence pervasion in both physical and virtual realities. The vision of ubiquitous intelligence will greatly reform our living and working spaces beside our way of life to create a smart world filled with a variety of embedded intelligence, smart real and virtual components ranging from software to hardware, from everyday appliances to sophisticated systems, from stationary places to moveable vehicles, etc.

TELECOOPERATION AND THE ROLE OF THE OPEN NETWORK

The most suitable environment that can realize telecooperation among dispersed teams is the open network, Internet. The Internet as a universal network and the World Wide Web (WWW or the Web) constitute a ubiquitous network. This allows to electronically connecting places that are thousands of miles apart regardless of any physical constraints. The Internet represents the suitable infrastructure technologies that facilitate virtual presence via electronic communications between entities, human populations, systems, and physical facilities (Leifer, 1996; Wolf & Foritzheim, 1997; Dalton, & Taylor, 2000). The Internet and the Web can make remote and distributed resources accessible to a broad range of users for different purposes regardless of geographical location. The Internet enable to deliver seamless access to live data, which empowers users to enhance performance and safety, improve productivity, reduces costs, and increase profitability. The Web creates the social framework required for field placement of sophisticated technology, and offers suitable infrastructure that integrate computer-based services from multiple platforms and provide a flexible and effective tool for examining and monitoring remote physical environments (Habib, 2000; Fong, 2001). The adoption of Web based systems to provide security, monitoring services, and the richness of media for communications offered through the Internet and the high level on interactivity provided by the Web can make resources accessible to a broad range of users regardless of geographic location, enhance efficiency, and maximize cooperation outcome.

The Web provides a low cost, portable, interoperable, and widely available platform with its multi modal rich interface to applications. It enables any device, equipment, and software system to become part of a seamless intelligent environments linking variety of activities and make resources accessible to a broad range of users while enabling multi users to access such environment, interact with each other, and access distributed physical facilities independent on physical and time constraints. The WWW model provides flexible and powerful mechanisms to allow links and access from one document to other text, audio, image and movie documents residing anywhere on the Web (King, 1997). The availability of Web browsers for the most common platforms and the presence of the Web throughout the world ensure very wide dissemination of information and access of other resources. The prediction for the Internet was to support, online multimedia collaboration, sophisticated information retrieval, and remote control of expensive or unique scientific instruments via tele-experimentation. Nevertheless, the Internet of today gives access to a wealth of distributed hypertext documents. Extensive efforts to realize the evolution of such predictions are going on with positive research results and new applications. In addition, Internet will lead to share ideas with a very large community, improve quality of services, utilize efficiently resources, minimize risk associated with hazardous and dangerous environments, access timely information and expertise that support diagnosis and decision making, and to reduce overall operational and maintenance cost.

The advent of the Internet as a major communication channel has triggered a great deal of interest in real-time communication over packet-switched networks. While the Internet is not originally designed to handle real-time traffic, the spontaneous growth of the WWW over the past several years has resulted in having special focus on remotely controlled mechanical devices that can be accessed via the WWW. This technology enables users to maintain a physical presence remotely in any one of many remote sites.

Affordable and ubiquitous Internet connections are rapidly linking together organizations, laboratories, and offices across nations and continents. The term ubiquity assumes and calls for a considerable increase in the degree of ubiquitous access to computing, communication, and media, leveraging off both novel mobile terminal devices and an adequate global infrastructure. Furthermore, Internet browsing programs are fundamentally changing the ways we use computers to display and exchange data. These two factors are making Internet the ideal testing ground for sophisticated new applications, such as video-conferencing and scientific experiment management. Internet browsers owe their success to innovative data visualization and communication technologies. However, these technologies cannot yet support the development of Internet programs for direct control applications, such as haptic interfaces and telerobotics. These protocols introduce random time delays between packets that make a control system unstable if it were to close the control loop over the Internet. Such a control system cannot even be designed, since there are few analysis or design tools with limited capabilities for modeling the effects of packet switched communication on a digital control system.

Internet based systems with tools supporting human communication and interaction with the real world, will inevitably lead to many useful applications in various sectors of the society, professionally and socially. The Internet enters the mainstream of daily life, such as, education, healthcare, finance, work, and leisure activities, and will affect all aspect of human life. Dispersed groups and populations that were once separated by distance and time are now experiencing these changes as part of a global community. Improvements in speed, accessibility, and usefulness coupled with the lowest prices ever for computers and Internet access will likely make for the most explosive growth yet. In addition, business is expecting to make the most of that growth.

Entrepreneurs are able to start new businesses easily with smaller up-front investment requirements by accessing the Internet's worldwide network of customers. Engineers, product developers, and managers thousands of miles apart can collaborate to design and manufacture new products more efficiently; Businesses can work more efficiently with their suppliers and customers; Consumers have greater choice and can shop in their homes for a wide variety of products from manufacturers and retailers all over the world. They will be able to view these products on their computers or televisions, access information about the products, and order and pay for their choices, all from their living rooms. Teachers and Students across the world are discovering vast treasure troves of data via the Web. They can have immediate access to the world's information from their classrooms. Possibilities are opened for distance educators to overcome time and distance to reach students. Doctors are utilizing tele-medicine to administrate off-site diagnoses to patients in remote parts of the globe from their offices. Citizens of many nations are finding additional outlets for personal, professional, and political expression.

The Internet is changing classic businesses and economic paradigms. It revolutionize commerce by accelerating the growth of trade and by dramatically lowering transaction costs and facilitating new types of commercial transactions and new arrangements of buyers and sellers that would make commerce easier.

The network makes it possible to have complex relationships with individuals and groups of people all over the globe. Virtual corporations, hospitals, factories, offices, communities, universities, teams, governments are coming into existence on a regular basis due to the support and the capabilities of the Internet. This leads to think about the convergence of television, the Internet, cable TV and the telephone and try to predict how and when these communication mediums will intersect. Hence, it is difficult to predict exactly what the Internet will develop into, but it is safe to say that the world is heading to an age of rapid dissemination of products, services, and information facilitated by technology and shared by businesses and individual alike.

ICTS FOR EMERGENCY AND RELIEF ACTIVITIES

Sudden natural disasters are often believed to cause not only widespread death but also massive social disruption and outbreaks of epidemic disease and famine that leave survivors entirely dependent on outside relief. Disasters kill at least one million people each decade and leave millions homeless. However, it is not possible to avoid completely the natural disasters, but the sufferings can be minimized by creating proper awareness of the likely disasters and its impact by developing a suitable warning system, disaster preparedness, and management of disasters through application of various technical tools. Early successful warning systems are a direct result of the ability to collect, interpret, and disseminate reliable and timely information to populations at risk (Helm, 1996, Smith, 1996; Stenchion, 1997). All those who are concerned with managing disasters necessarily have the need to access timely and accurate information before, during, and after the incident takes place. In the aftermath of natural disasters, and particularly in the context of complex emergencies, civil sectors need to be functioning before stability can be restored and made lasting in the affected area. Perspectives ranged from acute and chronic medical care to mental health, disease prevention, health promotion, and other public health applications including disease surveillance and early warning.

To be prepared for unforeseen events, they must make contingency plans and coordinate their planning with other agencies and parties involved. Information, and information management and manipulation, is vital to effective and successful disaster response and recovery. In addition, emergency users need to access, retrieve, and process a wide range of distributed information resources from the disaster area, regional offices, headquarters, etc. In many countries of the world, new applications of telecommunications and information technologies (ICTs) are proving to be valuable in crisis management and in improving the health status of rural, isolated, or otherwise underserved populations. These technologies can be used to provide a range of services to support tracking and management of patients, provide consultative support to clinicians in remote areas, deliver training for health providers, and to permit exchange of medical information whenever and wherever is needed.

Technologies are increasingly becoming available to help in the effort for disaster reductions and the role of telecommunication is critical in this regards. Timely and effective warnings of natural and related hazards coupled with local capability to take avoidance or mitigating actions are fundamental requirements for disaster reduction. Increasingly, earth observation, hazard analysis, and telecommunications technologies are used to support this process. Maps and spatial information are important components of the overall information in case of any disaster event.

Communications are critical elements in the effective management of disaster response and recovery efforts. Remote sensing and geographic information system (GIS) technologies are excellent and well developed tools that can be integrated in the mapping of the spatial distribution of disaster related data within a relatively short period. GIS has the ability to collect, process, store, retrieve, transform and display spatial data efficiently from a real world for a particular set of purposes. It may be observed that advancement in information technology in the form of Internet, GIS, Remote Sensing, satellite communication, etc. can help a great deal in planning and implementing hazards reduction measures. Satellite communications, remote sensing images from space, and location data from global positioning system (GPS) will become critical to the success of emergency organizations. These types of information can be integrated into the Internet. In emergencies where contaminants have been dispersed over a wide area; the gathering and analysis of samples can be crucial. The Internet can be used to communicate sample analysis results from multiple locations and integrate them into databases. Wireless technology that is rapidly evolving plays an increasing role in the lives of people all over the world. Examples include cellular phones, pages, GPS systems, and cordless phone sets. Wireless, satellites, and global telephone technologies become helpful in supporting wireless Internet and seamless communication.

The convergence of computer and telecommunications technologies into the information technology, with all its components and activities, is distinctive in its extension and complexity- and while it is undergoing rapid and fundamental changes, its growth and development has led to their wide diffusion and application. ICTs are the broad description for the technologies, systems, services, and tools that enable information storage, retrieval, communication, and dissemination. The digital revolution has a fundamental impact and it is rapidly transforming social, economic, cultural, and political interactions all over the world. It has drastically changed the way we think, the way we live, and the environment in which we live. Undertakes a wide range of activities aimed at improving our understanding of how ICTs contribute to sustainable economic growth and social well-being and their role in the shift towards knowledge-based societies

The new ICTs that have emerged over the last two decades lend themselves to greater possibilities of integration of different communication systems. The interoperability of various communication systems including Internet, mobile phones, fax, e-mail, radio, and television is increasingly becoming functional. As a result, the possibilities for application of communication technologies in mitigation and prevention of disasters are also increasing. The Internet brings together people with common interests irrespective of their geographical location and the distance separating them. In its role as an effective tool for communication, it can be invaluable for disaster management. Furthermore, there is a move towards globalization of disaster networks to provide speedy assistance to every disaster victim, irrespective of the national boundary and geographical location.

The last decade has seen the emergence of new ICTs. On one hand, this opens doors for innovative application of communication technologies in different phases of disaster management. On the other, this poses new challenges for disaster management community. The effective application of new communication technologies would require that these new technologies be integrated with the more conventional technologies. The Internet could support many command, control, and surveillance tasks in emergencies. One of its advantages is that a great deal of real-time information can be made available to a large audience. Access to these data can be either restricted or open to any Internet user. However, the Internet has some drawback, such as hardware software requirements, computer literacy requirements, traffic jams, dependence on power and communication network, and risk to information integrity and. security.

It is essential for experts of the many disciplines within the fields of disaster management and humanitarian assistance to come together to identify common problems and share successful strategies. The sharing of experiences in the use of telecommunications and information technologies for various applications in humanitarian and health initiatives, and critically evaluating the ways in which those technologies can improve the collection, management, exchange, and dissemination of data and information for strategic decision-making and operational coordination of humanitarian health initiatives as well as for teaching and training. The need for tele-centers to fulfill the medical and health awareness, education, training, information and communication needs of rural residents, relief workers and refugees supported by voice, fax and internet connectivity. In addition, such centers can stimulate development and growth for local and displaced community, and developing ICTs skills. In addition, there is a need to establish monitoring systems to forecast and monitor the impact of natural and human made disasters, particularly in developing countries. Access to information by all citizens is the prerequisite for effective disaster prevention, mitigation, and preparedness.

THE ROLE OF ICTS IN SUPPORTING HUMANITARIAN DEMINING NEEDS

Humanitarian demining is a critical first step for reconstruction of post-conflict countries and its goal is to remove lingering remnants of war, allowing refugees to return home, schools to be reopened, land to be used for farming and critical infrastructure to be rebuilt. The important outcome of humanitarian demining is that the community is confident that cleared land is safe. Therefore, there is urgent need to enhance safety, maximize performance and enhance efficiency, reliability, quality, and to reduce overall cost are always constitute the main factors that motivate people to create alternatives and new ideas in various fields.

The elimination of landmines remains an extremely complex mission. Humanitarian demining is a global problem that needs to be addressed and managed through a proper strategy among all interested parties in order to have better solutions and actions, to exchange experiences and expertise and to accelerate the phase of demining process. The fundamental function of demining activities is data management and integrating demining community with their resources to facilitate technology development, long and short-term planning, and to have effective coordination of mine action activities. Accordingly, there is a great need to identify requirements and opportunities in the use of information and communication technologies to develop synergy in the demining community while orienting technological innovation towards the operational needs. In addition, this will help to increase the efficiency of humanitarian demining operations and help to set priorities. For this purpose it is essential to develop a global strategy covering planning and execution phases while considering local priorities, local resources, and while integrating resources, results and experience locally, regionally and internationally. In addition, it is important to have a mechanism to review and adjust any part of a developed plan based on local needs, feedback through implementation at the field, and sharing experiences and lessons learned with others.

To facilitate the work in the field of humanitarian demining, there is a need for

a. Provide mine awareness, training programs, and information centers. This aims to enhance the capability of populations to live in a mined environment through mine-awareness and public information programs.
b. Establish coordinated, sustainable, and indigenous demining capabilities and support the training of personnel from mine-affected countries while it is important to consider the unique demining challenges faced in each country, and enables them to become trainers of deminers and managers of demining programs.
c. Facilitate smooth execution of humanitarian demining operations through cooperation and coordination globally, regionally and locally.
d. Adopt a systematic process to help in assessing the availability, suitability, and affordability of technology. In addition, there is a need for international cooperation in technology assessment, stimulating information exchange, development of international technology demonstration, and certifying mechanism.
e. Develop monitoring mechanisms that are able to track demining progress at each location, follow up plan implementation, and maximize technical and other assistance.
f. Buildup strong international coordination and cooperation network among members, such as, nations, international organizations, governmental and non-governmental organizations, private commercial operations, etc., that demonstrate leadership in humanitarian demining activities in terms of time, energy, material, and money.
g. Establish interdisciplinary forum aiming to enhance mutual exchange of information among all parties engaged in humanitarian demining.
h. Focus to have quick, safe, and easy demining tools is to exploit existing technology to put better equipment into deminers' hands.
i. Develop a realistic evaluation environment is required to help facilitate new technologies and offers quantum improvements
j. Link research laboratories and researchers with real activities at mine areas and experienced field practitioners to enable the exchange of views and experience that support the development of effective technologies for demining.
k. It is important to consider local resources and circumstances while developing new technologies for humanitarian demining. Innovations that are very expensive, risky, and hard to fit into existing work practices or that do not address high priority problems are not viable.
l. Develop testing methodologies and the design of standards for sensors for mine detection and system assessment.
m. Assist mines afflicted nations to establish a comprehensive, long-term, and sustainable humanitarian demining programs that meets local needs and requirements.
n. Promote economic and political stability by returning mined areas to productive use, utilizing available labor resources during and after the demining, program for qualifying people for new jobs, education, health and medical services, etc. In addition, it is necessary to promote national and regional stability.

o. Enhance the synergism between all research centers, academic institutions, and industry.
p. Establish international network of test, evaluation, and certification centers to validate mine detection and clearing technologies.
q. Enhance quality control assurance, awareness, medical support, and other services to help rehabilitate mine-affected areas.
r. Leverage work, experience and enhance data sharing in overlapped domains to improve demining planning and performance and to improve the level of cooperation within the demining community at large.
s. Having different modules to support and improve minefield mapping, management, and landmine database for landmine identification and assessment.
t. Allow major donors to share information seamlessly among the members of the global demining community including authorized demining centers and demining policy makers, donor states, teams supporting humanitarian demining, non-governmental organizations (NGO) and contractor staff.
u. Promote exchange of information, collaboration, and establishment of mechanisms for supporting the appropriate research, training, and applications at local, regional, and global levels.
v. Establish mines safe world through effective coordination and sustainable support from both public and private sectors.
w. Establish international information standards for collecting and representing demining and relevant information, and developing tools that enable different systems to exchange of different data structures and/to having unified information exchange platform.
x. In addition, it is necessary to have systems that support different modalities, support heterogeneous, and hybrid security requirement among different systems, facilitating access and share accurate, reliable, and timely landmine information, experience and lessons learned.

ICTs are increasingly becoming an important tool for supporting the development and validation of new demining technologies and action plans, and as a management tool for national, regional, and global humanitarian demining activities. Accurate information management system is critical for successful mine action programs. The system needs to be usable in the field and compatible with modern information and communication technologies.

TELECOOPERATIVE ENVIRONMENT FOR HUMANITARIAN DEMINING

Increasingly, large-scale processes distributed across time, space, participants and functional perspectives are being achieved through cooperative work. Effective coordination and awareness are critical to the success of such cooperative processes since the distributed activities are typically highly interdependent, e.g., due to shared resources, input-output relationships, workspace sharing, and so on. Reviewing existing and emergent technology will help to find-out how people, technology and other assets can be networked to detect, classify and neutralize or destroy landmines, and unexploded ordnance (UXO), from humanitarian demining perspectives.

While considering the difficulties and needs related to humanitarian demining, it is important to develop synergy with the involved communities by orienting technological innovation towards the operational needs. Activities will assist collaborative team members to concentrate their R&D and experimentation efforts by providing information on existing work, experiences, user needs, and opportunities and by encouraging collaborative partnerships. This will lead in developing cooperative relations in many areas that may include,

a. Deminers needs and safety,
b. Operational problems,
c. Better use of current technology while developing effective mine detection and

clearing technology through research and development activities,

d. Define future research and development requirements.
e. Mission statement and policy making
f. Train local demining personnel and support of the local populations.
g. Medical support and other services within mines affected area.
h. Support local industry in relation to demining tools and equipment.
i. Landmine safety, awareness and reporting
j. Minefield surveys, markings and verification
k. Develop a comprehensive training program for mine detection,
l. Landmine clearance technique,
m. Create relevant landmine databases in the host countries that are transferable to local, regional, and international groups as necessary.
n. Share new technology and information, and integrate efforts from the minefields to research institutions, governmental bodies, and nongovernmental organizations
o. Validate mine detection technologies, testing and evaluation
p. Enhance quality control assurance for demined area.
q. Benchmark for testing,
r. Increase the efficiency of humanitarian demining operations, and
s. Help set priorities.

Accordingly, the author is introducing the development of a global telecooperative system using team based distributed and collaborative intelligent environment that merges virtuality with physical reality while using all distributed resources efficiently and beyond time and physical constraints for humanitarian demining and other related activities. The development of a cooperative platform, which embodies the principles and policies for a coordinated effort, aims to create and build a generic team model for telecooperation among all actors related to humanitarian demining. This can be realized by having component based open system architecture to facilitate the addition of new features and tools, integrate team members and resources. This system can be used as an environment to integrate and support managing, planning, monitoring, and designing solutions concerning global demining and disaster/hazard activities, policies and finance, cooperation on technology, research and development, medical and social needs, awareness, and training. Such system enables the integration of demining teams that are geographically distributed, demining related organizations, research centers, and technical professionals to avoid duplicating efforts and to enhance overall demining performance through proper use of the distributed resources beyond the limit of time and physical constraints. The system is to be supported by different type of ICTs for localization, mobility, interoperability, interaction awareness, conferencing, multimedia, etc. to provide objective and easily collectable data, and to facilitate decision-making process. This environment facilitates interaction, sharing information about mines available at each regions, strengthen cooperation and coordination between experts in all fields related to demining, exchange views and experience, difficulties and problem solving, technological updates research, environmental problems, medical problems and assistance, etc.

The new environment can be used for the purpose of multimedia supported training for new deminers, for public education/demonstration, and provide multimedia-based information that enhances mine awareness among public. This can achieve promotion for the wide spread usage of technologies know how, adequate maps that are also needed in a number of circumstances, threat and danger, etc. The new system has to support remote testing benchmark tools for equipment evaluation and detection performance using an independent and large set of data. Media technology and seamless user interface that support interaction through distributed and collaborative intelligent environment specified for each team and enabling inter-teams cooperation. The system includes continuous tracking and mapping update tools that manage and maps mine areas with its operational mine clearance status and associated information and difficulties. In addition, it has the ability to monitor, track, and communicate quality operational data. This system brings together all the information about the status of mine clearance at local, regional, and global levels. It is necessary to have a central or distributed document repository where developers, team members, and organizations can submit and locate important documents with little effort. It should have the capability to index new document for future keyword/content searching to allow quick access to the content of documents. Information sharing of whom is doing what in the field to allow for greater collaboration among project planners.

Landmine database management enables to find detailed information on specific activities in the field of mines and demining, groups and teams, organizational profiles, researchers profile, demining-teams location, detection technologies, incidents, difficulties, experiences, testing results, contact names, project abstracts, funding, causalities, economical impact, environmental, and health impacts. It is necessary for such database to be featured by multi-user database management system that supports relational object oriented, free text and multi media data with high-speed information retrieval. In addition, it should facilitate main events and calendar of meetings, seminars, workshops, conferences, topics, dates, agendas, venues, etc. Furthermore, the database should be able to deal with incidents and problems analysis and utilization of lessons learned, integrating single location links of members database of all types of mines with other accompanying information, sensor technologies, and information about demining mechanization and applications of robot technology, technologies for safely clearing and naturalizing mines. The system should make necessary demining information available to scientist, researchers, experts, and the public bounded by some level of access control to share what is already known and to agree upon standards for storing and accessing data and for the classification of test results.

DISTRIBUTED AND COLLABORATIVE INTELLIGENT ENVIRONMENT (DCIE)

Collaboration in shared, virtual environments saves money by eliminating hardware duplication; improve operational efficiency of wide range of systems and processes, while it increases their effectiveness by bringing people and technologies together beyond time and physical constraint. Merging virtuality and reality in unique and innovative ways will take new meanings and its implication is extensive. Users of the system are immersing themselves into 3D shared multi layers distributed and collaborative intelligent environment, which creates illusion of physical presence at the remote site supported by sensory devices that can develop a better understanding and appreciation of reality. This will extend distributed team members capabilities at different locations to inhabit and shape a common synthetic environment. Team members can move between the layers of the distributed and collaborative intelligent environment seamlessly according to the task need. The appearance and capabilities of graphically represented task environment as space, equipment, and people are displayed and tracked at the logged-in time to the system. Shared synthetic environments are a representational container that can empower a broad range of uses.

Cooperation is enabled using the distributed and collaborative intelligent environment as a metaphor. This environment enables,

a. Separation in space and time without the need for physical presence,

b. Communication through computer network supported by various ICTs and allowing arbitrary distance between team members,
c. Sharing interfaces is an effective way to allow geographically dispersed experts to participate and work together in custom operations
d. Simplicity to reorganize working group and working space, and interact over a network to discuss issues and solve problems in shared virtual scenes.
e. Enable multiple participants to view simultaneously dynamic and geometrically defined environments. Each viewer can exclude extraneous detail or include additional information in the scene as desired. Enable to save and replay gathered information in a stand-alone mode.
f. Synchronize the local geometric model with remote simulators and sensory systems
g. Interaction and Awareness among multi-user accessing the same working environment,
h. The integration of shared applications concept,
i. The integration and real time interaction with distributed physical devices, equipment, and other resources.
j. Design flexibility to achieve higher equipment utilization than can be achieved with single-operator systems.
k. Visualizing and share decision making process.
l. Extending techniques of collaborative design and visualization to accomplish collaborative control in real to distributed equipment,
m. Collaborators that are spread over great distances to solve efficiently complex engineering problems.
n. Supporting integrative heterogeneous security approaches, scalability, interoperability, and mobility.
o. Combine the specialized skills and tools of varying disciplines to create highly effective teams.
p. Linking kinematics and other high-level model data to enable high performance virtual reality based visualization by distributed user groups.
q. Experts collaborate, test ideas, generate test programs, catch potential mistakes, and then give the go ahead to execute the desired set of actions/operation.

The features of the developed distributed and collaborative intelligent environment are as follows:

a. The environment is created through commercial 3D supporting graphical software,
b. The environment needs not to have a real physical counterpart. The environment is divided into sections,
c. Each section can be arranged based on the user needs. The sections should be task related instead of usual user related ones,
d. The intelligent environment should abstract from the characteristics of the real physical environment,
e. It should be intuitively understandable, and
f. The graphical representation of the intelligent environment needs not to be dimensionally accurate.

The distributed and collaborative intelligent environment should offer the possibility to

a. Establish communication channels between users (Audio/Video Conference)
b. Display information about other users (their availability, activities considering their privacy),
c. Visualization tools that support synchronized views,
d. Supporting editing and sharing sessions,
e. Join and leave discussion sessions (invited and general),
f. Create, edit, copy, delete, etc., documents,
g. Public and invited shared whiteboards,

h. Sharing applications,
i. Service should be offered to the users in transparent way, i.e., no need for user to know and involve with any underlying mechanisms,
j. Integrate sensory information from real processes.

The distributed and collaborative intelligent environment utilizes these capabilities as well as network control and intelligent system modeling, programming, control, and interface technologies to allow teams to collaboratively operate, evaluate, and maintain robots, machine tools, and other Mechatronical systems.

Challenges include delivering audio, video, and data streams with variable data rates, real time control, and applications with bounded delay, low packet loss as required by application, and service guarantees. A difficult element of the shared data environments is the particular type of data being shared (e.g., large visualization data sets, complex 2D and 3D virtual worlds, and simple 2D drawings) and the maintenance of shared state and multiple views of the data (synchronized and unsynchronized). Security is a big problem. Individual and group authentication is required so you know whom you are communicating with and that the data you access is from the organization you think it comes from. Another aspect of security is privacy, particularly person information privacy and access controls.

WEB-BASED INTELLIGENT MULTI-USER INTERFACE

It is important to have media technology and seamless user interface to facilitate interaction through a distributed and collaborative intelligent environment specified for each team, and enabling inter-teams communication. Multicast communication is needed to facilitate point-to-point, point to many and many to many communication support. The integrated user interface should support multi operators that share control through an integrated collaborative relation to achieve shared goal. Displays provide information that enable human decision-making and mode change, and it has to support dialogue and serve the remote process. The design of the user interface should be natural and friendly in use to enable efficient human-machine interaction, support multiple operators' collaboration to enhance control, minimize use of human resources, and minimize response time. What kind of information is needed, how that information should be displayed, and what kind of information must be monitored when an operator interacts with the remote task environment are critical questions that need to be considered and answered carefully and understand its impact on the design of user interface.

Teleoperated systems have emphasized the use of multi-modal user interface that provides human operator with a variety of control modes (individual actuator, coordinated motion, etc.) and displays (text, visual, haptic, etc.). The targeted user interface is an active and dynamic Web based, platform independent, and support geographically dispersed teams and distributed environments. The interface should be designed to maximize usability by a target set of users. Yet, this can be difficult for Web based teleoperation because the interface needs to support users having diverse skills, knowledge, and experience. In particular, Web interfaces need to be designed so those novices will feel comfortable yet must not unduly constrain experts. If the interface fails to do either, users will quickly lose interest and avoid using the system.

To support a wide range of users, a Web based interface must present data clearly and concisely. Complex data should be hidden from the user with the possibility to be accessed when needed. Parameters and operations, which are best handled by the machine, should be removed from operator control. User interface should inform the user about the reliability of the data it displays so that the user can make informed decisions. Because of varying transmission delay, the information displayed may not match the actual state of the remote system. Thus, the interface has to be designed so that commands will not affected by delay. Specifically, motion commands should be sent to the robot relative to the state displayed by the interface.

The need to have clear, compelling, and user-friendly interface that featured by:

a. Modern Interface design approaches the problem through maximizing the usability by a target set of users.
b. Web interfaces need to be designed so that novices will feel comfortable, yet must not constrain experts.
c. Low level and complex data should be hidden from the user. Instead, it is necessary to have tools that enable users to represent and access such data at high-level of abstraction and in an understandable way.
d. Spatial, temporal and conceptual consistency of user interfaces
e. Web-based teleoperation interface must present data clearly and concisely.
f. The interface must inform the user about the reliability of the data it displays (so that the user can make informed decisions).
g. The interface has to be designed so that commands will not be affected by delay. Specifically, motion commands should be sent to the robot relative to the state displayed by the interface.
h. The interface must be robust enough to deal with communication failure and warn the user if the connection to the remote site is lost.
i. Enable shared use of multiple services by multiple users using multiple devices. In addition, it should support dynamic adaptation among several dimensions: devices, users, services, etc.
j. Allow the user to generate commands and to receive feedback.
k. Display the status of the system status through the user interface.
l. Bridge and merge the physical and virtual worlds.
m. Should enable 3-dimensional space of interaction and support space reconfiguration.
n. It should enable tracking of individual and entities accessing and leaving the environment dynamically.
o. physical, visual, and auditory design of ubiquitous Interfaces.

A variety of tools can ease the difficult task of developing and implementing multi-user user interfaces.

ACTORS ASSOCIATED WITH HUMANITARIAN DEMINING ACTIVITIES

The actors for the scenario of team based teleco-operative system augmented with mixed realities supporting and integrating humanitarian demining activities may include but not limited to,

a. Deminers and demining experts
b. Researchers/Research institutions and organizations
c. Academia and Universities
d. Companies and industries
e. Governmental institutions, policy makers, local authorities, Humanitarian departments or agencies for guidance and information on counter-mine measures
f. Local and global demining centers
g. Non-governmental organization (NGO)
h. United Nation (UN)
i. Aid agencies
j. UN demining database
k. Independent aid agencies
l. International committee of red cross/crescent
m. Public
n. Others

These actors can be configured into different teams. The teams can be at different hierarchy locally, regionally and internationally. A team may constitute a group of the same professional or a mixture of different professionals. Teams may have inter-teams communication by enabling configuration that facilitates more than one team to interact and cooperate according to local, regional and global needs. The question is how to make all groups that are making-up the teams assist those who are tackling the issue of humanitarian demining across the globe.

CONCLUSION

This paper has introduced and discussed in the design development of a team based multi-user, web-based and global telecooperative system as a new multidisciplinary collaborative paradigm. The developed telecooperative system uses the 3D distributed and collaborative intelligent environment to support merging physical and virtual realities and facilitates the realization of wide range telecooperative applications with focus given to the integration of disaster management and humanitarian assistance activities. The design current design supports portability, interoperability, flexibility, interactivity, access control, and integrative friendly web based interactive user interface. The paper has presented the detail design requirements and functional descriptions that are necessary to integrate humanitarian demining activities and relevant communities with aims to utilize efficiently all distributed resources and embedded them with the developed distributed and collaborative intelligent environment on the open network. Such environments can help demining community to cooperate beyond the limit of time and physical constraints. This work provides the framework in having a strategy and implements a methodology and evaluation criteria to develop effective and better demining action plans globally, regionally and locally with aim to speed up demining process, helps relieve human suffering, and develop indigenous demining capability. It is necessary to have different solutions to help improve the quality of demining activities, their management, and their relative policies. The realization of such project needs to formulate cooperative relations among all parties involved in humanitarian demining. Currently, the author is proceeding with a prototype implementation of the presented system. Finally, the international community must act to foster and further research in the area of disaster management and humanitarian assistance enhanced by the role of ICTs, and coordinate their efforts in order the proper technologies, procedures, and standards that could enhance the effectiveness, cost efficiency, reliability, and safety.

REFERENCES

Alexander, D. (1993). *Natural Disasters*. London: University College London Press ltd.

Banger, S. K. (2000). *Remote Sensing and Geographical Information System for natural disaster management*. Accessed March 27, 2010 from http://www.gisdevelopment.net/application/natural_hazards/overview/nho0011.htm

Blagden, P. M. (1993). Summary of UN demining. *Proceedings of the Symposium on Antipersonnel Mines,* Monteux CICR/ICRC. April 1993, 117-123.

Bruschini, C., De Bruyn, K., Sahli, H., & Cornelis, J. (1999). *EU-Report-EUDEM: The EU in HUmanitarian DEMining*. Brussels, Belgium: European Commision.

Dalton, B., & Taylor, K. (2000). Distributed Robotics Over the Internet. *IEEE Robotics & Automation Magazine*, *7*(2), 22–27. doi:10.1109/100.848264

Fong, T. (2001). Collaborative Control: A Robot-Centric Model for Vehicle Teleoperation. *Doctoral dissertation. Techical report CMU-RI-TR-01-34*. Robotics Institute, CMU.

Gros, B., & Bruschini, C. (1996). Sensor technologies for the detection of AP mines: a survey of current research and system developments. *In Proceedings of the 6th International Symposium on Measurement and Control in Robotics (ISMCR'96)*. Brussels, Belgium, May 1997, 564-569

Habib, M. K. (2001a). New Potential and Development of Mine Detection and Sensing Technologies in the Context of Humanitarian Demining. *Proceedings of the 5th International Conference on Mechatronics Technology (ICMT'2001)*, 222-230.

Habib, M. K. (2001b). Service Robots in Humanitarian Landmine Clearance. *Proceedings of the International Symposium on Robots (ISR'2001)*, Seoul, Korea, 1333-1338.

Habib, M. K. (2007). Humanitarian Demining: Reality and the Challenge of Technology - The State of the Arts. *International Journal of Advanced Robotic Systems*, *4*(2), 151–172.

Habib, M. K. (2008a). Humanitarian Demining: Difficulties, Needs and the Prospect of Technology. IEEE International Conference on Mechatronics and Automation 'ICMA2008', Takamatsu Kagawa, Japan, Paper Reference WC2-2.

Helm, P. (1996). Integrated Risk Management for Natural and Technological Disasters. *Tephra.*, *15*(1), 4–13.

Hewish, M., & Ness, L. (1995). Mine-detection Technologies. *International Defense Review.*, *28*(10), 40–46.

(1998). *Humanitarian Demining Development Technologies Catalog*. USA: Department of Defence.

Humanitarian Mine Action Equipment Catalogue (1999) *German Federal Foreign Office*

IFRC (International Federation of Red Cross and Red Crescent Societies). (1995). World Disasters Report 1995, *from*http://www.ifrc.org/pubs/wdr/95.

Julian, B. R. (1990). Are earthquakes chaotic? *Nature*, *345*, 481–482. doi:10.1038/345481a0

King, C. (1997). Mine Clearance in the Real World. *In Proceedings of the International Workshop on Sustainable Humanitarian Demining (SusDem'97)*. Zagreb, S2.1-8.

Klein, M. (1997). Coordination Science: Challenges And Directions. In Conen, W. & Neumann, G. (eds.) *Coordination Technology for Collaborative Applications*. Lecture Notes in Computer Science, LNCS, Springer Publishing.

Leifer, L. J., Toye, G., & Van der Loos, M. (1996). Integrating the socio-technical framework of human service through the Internet World Wide Web. *Robotics and Autonomous Systems*, *18*(1-2), 117–126. doi:10.1016/0921-8890(96)00019-X

Nicoud, J. D. (1997). Vehicles and Robots for Humanitarian Demining. *The Industrial Robot*, *24*(2), 164–168. doi:10.1108/01439919710165725

O'Malley, T. J. (1993). Seek and Destroy - Clearing Mined Land. *Armada International*, *17*(1), 6–15.

Smith, K. (1996). *Environmental Hazards: Assessing Risk and Reducing Disaster* (2nd ed.). London/U.S.A./Canada: Routledge.

Stenchion, P. (1997). Development and disaster management. *The Australian Journal of Emergency Management*, *12*(3), 40–44.

UN-Economic and Social Council (UN-ECS) (2005). Third Annual Report of the Information and Communication Technologies Task Force. *E/2005/71*. New York United Nations Human Settlements Program (UN-HABITAT) (2009). *Global Report on Human Settlements: Planning Sustainable Cities*, Technical Report.

Wolf, H., & Foritzheim, K. (1997). Web Video: A Tool for WWW Based Teleoperation. *Proceedings of the IEEE International Symposium on Industrial Electronics (ISIE'97)*. July 1997, SS268-SS273

ADDITIONAL READING SECTION

Habib, M. K. (2000). Collaborative teleoperation: Design requirements and development Issues. *IEEE-IECON 2000*, Nagoya-Japan, October 2000, 19-27.

Habib, M. K. (Ed.). (2008). *Humanitarian demining, innovative solutions and the challenges of technology.* Vienna, Austria: ARS-pro literature Verlag Publishers.

Hyndman, D. (2008). *Natural hazards and disasters*. Belmont, CA: Brooks Cole Publisher.

Kock, N. (2007). *E-collaboration in modern organizations: initiating and managing distributed projects*. Hershey, PA: Information Science Reference publisher.

Kruse, E., & Dai, F. (2001). Virtual training using web and interactive visualisation. *In IFAC International Conference: Telematics Applications in Automation and Robotics*, 517-522.

Talwar, S. J. A. K. (2009). *Human made disaster management*. Delhi, India: Commonwealth Publishers.

KEY TERMS AND DEFINITIONS

Disaster: is a tragic consequence of natural, human made and human induced hazards that negatively affect life, health, environment, economy, etc.

Disaster Management and Risk Assessment: It is an applied science, which seeks, by systematic observation and analysis of disasters while underlying the factors that contribute to disasters, to improve measures relating to prevention, mitigation, preparedness, emergency response, and recovery. It seeks to motivate societies at risk to become engaged in conscious disaster management.

Cooperation: is an essential part of modern work, business, research and other professional and social work that aims to improve quality, speedup technology transfer and expertise. In modern life, it implies synergistically the integration of technology, people, and organizations that facilitate the communication and coordination between those cooperating to establish share understanding that enables working together in the pursuit of a shared goal, and to achieve benefits for cooperating members.

Ubiquitous: Exist or seeming to be being everywhere (all places) at the same time. This term often used in ubiquitous computing to indicate that something seems to be found anywhere and everywhere.

Information and Communication Technologies (ICTs): is a broad and widely recognized term that fosters sustainable development and concerned with information and communication technologies and other aspects of managing and processing information of wide range applications especially that in large organizations. Technically, it covers the use and development of wide range of computers, mobile devices, communication protocols, networking, information processing, software applications, intelligence and decisions, etc. The growth and development of ICTs has led to their wide diffusion and applications, thus increasing their economic and social impact with strong role in the shift toward knowledge-based societies.

Telecooperation: It stands for the fusion of computer science, telecommunication, and multimedia facilities to carry out a cooperative process among organizations, individuals, systems, or combination of them by having better access to resources and share timely information over local and geographically distributed locations that lead to make well-informed decisions. Telecooperation is an outstanding example for the power of enabling technologies.

Web-Based Intelligent User Interface: it refers to media technology and seamless user interface through the open network to facilitate interaction through distributed and collaborative intelligent environment specified for each team, and enabling inter-teams communication. Intelligent interface enables people from wide range disciplines and research areas where they can meet, debate and collaborate.

Section 4

Moral Principles and Responsible Use of Technology

Chapter 10
"Ethics" and "Social Responsibility" in Financial Crises as Parameters of the Financial Environment

Alex Stefan
Bloomfield College, USA

Doru Ursutiu
University Transilvania of Brasov, Romania

Cornel Samoila
University Transilvania of Brasov, Romania

ABSTRACT

The world financial crisis and the economic recession have introduced major ethical issues to our international banking institutions. Employees perform unethical acts primarily to maintain job security and to ensure a competitive advantage in the job market. Increased hacking has also caused damage to client confidence, resulting in a customer base decline and increased litigation. These problems are further complicated when bank steering committees decide to outsource IT and help desk departments to IBM[1]. The legal implications of bank hacking issues are included in the Computer Misuse Act, the Evidence Act, and different state and federal laws that define the contractual relationship between the bank and its clients. To ensure financial systems are secure, banks must vigorously apply policies and controls. Unscrupulous persons both from within and outside of investment banks seek to gain profit by employing, at best, very questionable practices. Banks must implement a business continuity plan to integrate previous records and financial statements, as well as back up transactions and provide proper receipts. This is a case study that can be used for students of Computer Security or Computer Science.

DOI: 10.4018/978-1-60960-094-5.ch010

INTRODUCTION

The communication environment of the worldwide banking network and its interactions raises many ethical questions that were not frequently posed before the present world financial crisis. Suddenly, the world has discovered that bank interactions without special rules devoted to stricter standards of behavior and regulation might significantly contribute to a dangerous financial environment.

Many times considered a guide to social interaction, ethical principles have been revealed to be far more than just abstract ideas. The present financial crisis has demonstrated that the observance of ethical principles creates a framework where interaction with others can take place in an equitable, honest and productive manner.

We know that business values, in general, tend to be driven by amoral behaviors, in great part due to competition, efficiency or simply profit, not to mention greed. For many years previous to the 2008 crisis, the world simply accepted unethical and even immoral behaviors because they led the world to believe in a false wealth, a lie which led many individuals to "lose" a significant part of their personal wealth. A short list of justifications commonly accepted by a corrupt corporate environment includes:

- *"I had no choice"* ….i.e. denying responsibility;
- *"No one is hurt"* .....i.e. denying injury;
- *"Blaming violated parties"*...i.e. denying victims;
- *"They are worse than we are"*...i.e. social weighting;
- *"It was for a good cause"*...i.e. appeal to higher loyalties;
- *"I have the right to be unethical because of poverty (or wealth)"*...i.e. self justification.

For many years, codes of ethics were considered binding and were respected by the private sector. In fact, the Internet environment has contributed to new and different ways to recruit, mislead and coerce the public, in addition to more traditional methods. Only after the actual crisis did information technology become a prime concern for researchers seeking a connection between Internet technology and unethical behavior, which in a larger context, is deviant behavior in that it is not socially acceptable to steal to take what is not one's own. Without a doubt, behaviors by investment banks and the stock market contributed overwhelmingly to the impoverishment of millions and hardship to countless others.

- ***Computer crimes***:
  - Computer crime has become an instrument for unethical actions and frauds because of unauthorized access to a data, theft, defamation, intimidation, bullying, etc.;
  - Computer technology has become a tool used to violate the law and deny the rights to information access and privacy, leading to anti-social practices like cyber bullying, sexual solicitation, information piracy, etc.
  - The interconnectedness of the Web has increased the incidence of electronic fraud, including credit and debit card theft, identity theft and the unauthorized transfer of funds, as well as legal practices such as automatic buying and selling in the stock market.
  - The Internet has contributed to a growing number of security problems related to e-mails, the theft of personal or financial information and the loss or manipulation of records.
  - The Internet has also expanded the scope of crimes involving obscenity/pornography, child solicitation for sex, violence against minorities, terrorism, etc.

- ***Computer risks:***
  - Aircraft control might be affected by hackers;
  - Nuclear power plant control can be compromised by insecure computer network environments;
  - Military computer networks have insecure spaces that compromise national security as hackers might steal user names and passwords, which can lead to the leaking of confidential or proprietary information;
  - The privacy and security of E-commerce can be compromise by unauthorized persons.

Again, deviant behaviors are behaviors that do not conform to societal norms. These behaviors are often very harmful either physical, emotionally, psychologically or economically. Excessive greed that leads persons or organizations to obtain wealth by harming others is one of the most common deviant behaviors in society. What is certain is that after the crisis, important changes must be made in legislation and enforcement to impose professional and ethical guidelines. We must accept that ethics, as a branch of the social sciences, will in the future not be the only science that deals with concepts such as right and wrong, good and bad, fair and unfair, just and unjust, legal and illegal, moral and immoral or proper and improper, regarding human actions. Instead ethics will move towards the public interest, generally stressing actions concerning crime prevention and antisocial behavior. The concept of *client confidentiality* may become a key value in ethical practice.

Ethics will become a central point of IT development and deployment, and because humans are not born with values and IT is ethically neutral, we need to develop IT that provides learning, observation, and social interaction -- without neglecting constraints.

CONSIDERATIONS ABOUT ETHICS IN THE FINANCIAL ENVIRONMENT

Ethics comes from the Greek word "*ethos,*" which means moral character and morality and comes from the Latin word "*moralis,*" meaning custom or manner. In current speech, morality and ethics are often used as though they have the same meaning. Often, they are also used as if they have no relationship to one another. As is stated above, most people consider ethics or morality similar to the notions of good or bad. Many times the definition of immoral is connected with sex. In short, we want to stress that for the great majority of the population, these two notions are connected so that morality or ethics of the persons in charge of financial institutions should know the difference. However, the interpretation of what is good or bad, right or wrong, is often in the eye of the beholder. Sometimes, the beholder can be blinded by greed or personal benefit at the cost of others. This type of exploitation is socially deviant as wealth is transferred in an unfair way from one party to another. As a matter of fact, there has been a marked increase in the net worth of many of the actors who caused the economic meltdown, many of which used electronic means to make their financial schemes work.

To better understand the current financial crisis, we must accept the fact that if humans can be smart or foolish, financial institutions can be ethical or unethical. To be exact, they may be more ethical in their declarations that actually hide unethical actions. It means that we must understand that people from the business sector may be otherwise moral but have different opinions about fairness. Sternberg mentions that: "*...making ethical decisions while conducting business requires not only a good understanding of the business' nature but also understanding of ethical framework...*" This statement implies that some words are closely connected with ethics: morals and morality, values, virtues, ethics, ethical principles, which might be explained using the well-known definitions of Lisa Newton:

- *Morales and morality:* rules and prima facie duties that govern our behavior towards other persons;
- *Values:* the state of affairs that people desire and we wish to increase;
- *Virtues:* conditions of people that are desirable both for the people themselves and for the good functioning of the society;
- *Ethics:* properly speaking, the study of moral duties, values and virtues to find: their theoretical links and relationship and how they work together (or do not) in practice;
- *Ethical principles:* very general conceptual schemes that sum up a range of morals, values and virtues from which moral imperatives can be derived.

All of the above definitions stress that the ethical framework plays an important role in the financial framework. This role is to protect critical stakeholder interest. But this interest is not the same as general interest. The field of application of a financial ethical framework is very large: stakeholders, investors, employees, community and the environment. Reading the above list, we can understand that it is impossible to have the same ethical rules applied with equal influence in all of these domains. The breadth of application domains will increase possible conflicts generated by the differing ethical responsibilities that drive internally mentioned groups and external ones to consider new approaches and new considerations when the design and applications parameters will be analyzed.

The lack of understanding of the roots of the financial crisis are revealed by public opinions "*...they should be authentic leaders, focused on serving their clients and all the institution's constituents, rather than charismatic leaders seeking money, fame and power for themselves...*".

Because we are now in a period of analysis and critique, we understand that all of the above statements are only a succession of nice words. Reality shows that putting the financial future of people in the hands of persons with questionable ethics guarantees the recurrence of crises.

Business ethics is not the same as personal ethics. It is compulsory to implement a new system of financial ethics that does not depend on "*...authentic leaders...*" or "*...charismatic leaders...,*" where the difference is not due to the personal morality or qualities of the individual. Instead, ethics become compulsory limits are well defined by the ethics imposed by an embedded system. Many people involved in business or persons who are beneficiaries of business results concur that integrity is the foundation on which everything else depends. But integrity must act in an environment which does not allow skidding, not so much due to a person's morality, but due to the automatic compulsory limits of the system.

We must accept that due to GLOBALIZATION, there is an important change in the relationship between business, society, government, the environmental and corporate governance. The financial crisis has its origin in corporate irresponsibility and the socially deviant behavior of excessive greed. Despite the fact that moral people work in all of the aforementioned areas, risk management systems, a lack of transparency in decision making, and business interdependence have a domino effect that crosses ethical boundaries.

It might be a great mistake to consider the origin of the financial crisis as simply the wrong interpretation of ethics embedded in the business culture. All of the persons involved in creating the crisis were perhaps moral persons with a high degree of ethics and a solid knowledge about morality, because without these qualities they usually cannot obtain high positions within the financial system. It means that APART FROM GENERAL ETHICS we need other rules that are specially tailored for businesses. We know that these special rules exist and have the name BUSINESS ETHICS. The present crisis, however, allows us to state that these rules are not enough and we need others that do not depend on entirely on personal decisions. The decision of people in financial institutions must be embedded in a strong system that effectively covers people's decision making and does not allow individual greed or mismanagement. These new rules must evaluate risk continuously and have the power to stop unethical practices as they appear.

Financial ethics is a subset of general ethics, but if we compare general human behavior with financial behavior, we notice that financial behavior is binomial and is either "*...right...*" or "*...wrong...,*" similar to personal ethics. According to game theory, this difference is generated by the fact that individuals are egoistic and pursue their own self interests. Researchers in the field of financial ethics have discovered the existence of a paradoxical situation generated by the following aspects:

- Ever-increasing needs;
- Individual weakness;
- The lack of interest of persons to monitor permanently the financial environment;
- Selfishness of individuals.

These paradoxical situations, when a financial crisis appears, become obvious. People now want a new approach regarding ethical violations and strict adherence to existing codes of ethics.

New ethical principles must provide us the answer to two fundamental questions: "*...what should I do...*" and "*...how should I act...*" Immanuel Kant wrote the following: "*...every person is obligated to act only in ways that respect human dignity and moral rights of all persons...*" If we consider the problem of acting as above, we neglect an important aspect: "*...what should people be...*"? This remark moves the above two questions towards only one: "*...what kind of people, based on virtue ethics, should be in charge with our money...*"? Will these persons apply and respect ethics and moral principles so that no other financial crises are ever caused again due to excessive greed and speculation?

Honesty, generosity, integrity, self-control, prudence and courage are examples of virtues. All of us agree with these examples of virtue, but who is the person who can declare that all businessmen possess the virtues necessary do prevent another financial crisis from occurring due to dubious ethical practices? This question is extremely important in light of how some of these very people have profited from the crisis. Instead of losing personal wealth, they actually gained wealth at the cost of millions of people who today suffer the consequences of their excessive greed and socially irresponsible actions.

The people may be divided on whether any specific action is right or wrong, being influenced by the own short- or long-term interest. This is the simplest explanation for why some of these persons assumed excessive risks in the financial market—risks that were unwarranted and often unethical. It indicates that decision making under ethical considerations is not proprietary, belonging to only a few persons. We consider that the decisions must be made by responsible and ethical persons, but a separate system of risk evaluation must be in function.

Why is it possible for a person with ethical principles and good mortality to make unethical decisions? A few reasons are listed below:

- *Ethics in business is what I feel regarding what is right or wrong;*
- *My religious beliefs lead me to make good ethical decisions;*
- *Respecting legal requirements make me ethical.*

From the abovementioned answers, many of us probably agree most with those referring to legal requirements, because in many cases the law often integrates ethical standards. The financial crisis occurred within an extensive legal framework, but wrong or unethical actions were made by exploiting existing gaps in this framework. People with the power to make decisions sometimes believe that all actions not explicitly prohibited by the legal or regulatory system are allowed *de facto*. This mentality, in large part, led to the crisis. Fraud is one of unethical actions that was extensive and resulted in legal infringements. The involved persons mentioned above do not violate laws. They only exploit unmentioned restrictions.

The conclusion of all the aforementioned considerations is that ethical standards in the context of lawful behavior must be sustained not only by persons, but by a new system of checks and balances.

A more simple comparison can be made with industry. For many years, the quality of products depended on qualified workers and by the individual company's ethical behavior which obliged them to deliver only quality products. When human qualifications became insufficient, ethics were taken out of the manufacturing process by automation, robotics like computer nummeric command language. As a result, worker intervention was limited to design and requirements. The quality of the final product was above worker ethics or degree of specialization because these technologies removed the variable of quality from human involvement.

Similarly, the existing ethical standards practiced by financial institutions, including business ethics, virtue, integrity, etc., must be complemented by instruments that do not depend on the dubious human character that, in acting in self interest, sometimes resorts to lies, deception, fraud, and theft. The following case study intends to place these control instruments in the field of information technology (IT). As we have mentioned in the introduction (computer crimes, computer risk) this field is not exempt from ethical considerations. However, the lack of ethics is manifested externally towards the investor, and is not overseen internally, as greed often overrides other considerations. We propose a new system which acts inside the financial system as a brake against abuse. It is important to note that in the scenario to be discussed, we consider all types of dishonest behaviors as deviant behaviors because dishonesty, in itself, is not considered an acceptable behavior, and it can harm more people worldwide than what are usually considered deviant or unacceptable behaviors.

SCENARIO

The world financial crisis and economic recession have impacted bank revenues. Our hypothetical online bank is a large size international bank. We have faced increasing hacking problems, loss of customers and litigation issues. Additionally, our IT and help desk departments need to downsize and begin to layoff staff. Our bank has limited security resources to monitor and protect valuable customer information. Due to serious financial and legal issues, the IT executive committee determined that the bank cannot afford to hire full time in-house security IT staff to work on hacking issues. Therefore, outsourcing our IT department and our help desk is our only viable solution. The impact of outsourcing would introduce major ethical issues for bank employees.

The IT and help desk department is composed of 300 employees worldwide with a headcount burden of $50M in operating annual expenses. These IT data centers and help desks are located in the US, Canada, Hong Kong, Japan and UK. The goal of the IT executive committee is to reduce the size of worldwide IT and help desk operations to a centralized management team of 10 people and outsource comprehensive IT and help desk functions to IBM. The term of the contract will be 10 years. Terms will be negotiable every 2 years to allow adjustments to market conditions.

The IT executive committee understands the legal, financial and ethical dimensions and impacts to the bank's operation and its employees.

We are inviting IBM to sign non-disclosure and confidential agreements.

The bank has several major requirements:

1. It requests IBM develop a comprehensive transition plan to protect Bank's information assets from hacking including internal employees of our Bank as well as IBM.
2. It asks IBM to provide a comprehensive business and technical proposal to satisfy Bank's IT and help desk requirements. These will include a description of the IBM outsourcing business plan and network architecture drawings.
3. It must identify any financial, legal and ethical risks of the outsourcing plan.

LEGAL ISSUES IN ONLINE BANKING

The legal issues of online banking arise from current events, activities, and actualities, all of which will have to be treated in our agreement with the contractor IBM. One legal issue involves determining local jurisdiction of IBM and the bank, where data and/or telecommunication equipment are accessible and not affecting the business. In addition to facility property, there are other tangible properties to include applicability of contractual obligations and probative value of digital documents. This leads to direct discretionary ownership of electronic evidence, and should convey a dialogue of handling electronic money.

Other legal issues include careful consideration of statutory laws and common laws for the bank and its clients. If here were fraudulent activities and hacking events, the bank must retain its image as a credible bank. Even if hacking occurs in which no fraud was committed, the bank must provide the notification of the attack and prevent possible future attacks. In 2007, Commerce Bank suffered a serious intrusion in which 3,000 customer profiles were obtained, and Commerce Bank had to inform all clients whose profiles were illegally obtained. Though not many were used in fraudulent activity, the persons affected received protection in the form of credit monitoring for 24 months (Kirk, 2007).

Other legal issues include the obligation of the bank to provide protection of privacy for the IBM contractors. When IBM employees are working on bank property, such as day-trading facilities, what transparencies and walls need to be in place for the IBM employees to function and perform IT tasks and help desk support while bank employees perform their functions? At the same time, this issue spans across to the virtual realm as well, including visibility of web marketing of lending services and electronic bills of lading. Other vital information such as internet stock offerings and electronic determination of titles also should be considered sensitive information that needs to be protected. Some electronic banking functions are already outsourced, and this becomes a legal issue if IBM needs to access information systems which IBM has no allegiance to.

Another legal issue that may arise would be about the dismissals of the current IT staff supporting the system. (Boatright, 2007) The current employees may have contractual agreements with the bank that need to be addressed when considering dismissing those employees. Naturally, the existing employees would be transitioned out as IBM employees and become more familiar with the bank's system. For term contracted employees, those employees would not receive renewed contracts. The contracted employees must be legally respected according to the requirements of their contract. One solution would be to offer substantial severance packages in exchange for them to nullify their contracts.

Recommendations for IBM Agreement

Our experts have highlighted a variety of solutions that our online bank should undertake to counteract threats to the online banking system.

The most effective way to prevent these hacking issues begins with typical protections, including webpage access controls, content filters, malicious code and virus protection, and firewalls. In addition to the typical protection settings, IBM can provide further comprehensive security strategies including other hardened processes such as analysis and hardening tools, for identity authentication, and virtual private networks with encryption, PKI and digital certificates. A stringent network administration can be setup with clients using trusted operating systems. IBM-Business Service Management (BSM) tools ill be used to assure operational success through improved service visibility.

Aligning IT with the objectives of the business requires a transformative change in how IT operates. IBM's BSM – Tivoli Service Management solutions provides the tools to help with that transformation by providing an essential business context from the top to the bottom of the online banking business service delivery chain.

However, while the technical aspects of the server-side security may be beefed up constantly to the latest cryptographic standards, more often than not, security threats to the system are human in nature. Human weakness in the area of ethics poses the greatest security threat as greed and theft are strong motivators. Therefore, employees have also been identified as the worst security threats of an organization. While technological sophistication is still the keyword, certain key principles can be followed to ensure the security of a computer system including human resource management, whereby sensitive and critical procedures are not under the purview of one person and duties are segregated and access controlled within the organization itself (BankInfoSecurity, 2005). Our Bank should put in place specific internal administrative policies in this regard and also specific policies regarding IT contracts.

Furthermore, the education of customers and employees is an integral part of ensuring that the combination of the online vault is safely beyond cyber-safecrackers or unscrupulous administrative personnel who might take advantage of security flaws. Customers must be educated to set up client side firewalls as well as secure their respective username/passwords. At the same time, customers should also be dissuaded from using a generic pass phrase for every single one of their accounts given that the compromise of one system may inevitably compromise all other accounts. Furthermore, any potential security holes should be readily identified and patched with the assistance of the IBM contractors.

Financial Issues

Our hypothetical bank is already plagued with financial and cyber-liability issues and guidelines must be developed to ensure certain laws are in place. Some financial issues must address how daily work would continue while the IT system is being integrated with IBM service solutions. For auditors, the transition must be as seamless as possible as to not disturb cash-flows, as business banking is time dependant (monetary differences in world currency exchanges). Auditors need to have special tools to make sure the bank is using ethical practices that are not involved in questionable practices. The case of investment banks and their legal fraud could have been largely prevented with proper IT tools to properly audit them.

Policy Statements

Online bank departments are to ensure that financial systems have adequate internal controls over the processing of financial information. These controls are explained and described through general policies and controls.

General Policies and Controls

Online bank departments are to ensure that the functions of initiation, authorization, and recording of transactions and custody of assets are separated to ensure no employee or group of employees has exclusive control over a financial transaction or group of transactions. A lack of oversight and the concentration of decision making by relatively few actors were key contributors in the worldwide financial crisis. Online bank departments are to ensure that all inputs to the system are complete and accurate and all transactions are valid and properly authorized. In this way, the system can keep an electronic trail of the persons involved in transactions and the people who authorized those transactions. If IBM decides to outsource portions of the work, these considerations must also be transposed and applied to the sub-contracted company.

The solution to these financial issues is for online bank departments to ensure that accuracy of data is maintained during processing and computations are carried out accurately. The many different departments are to establish procedures to ensure errors are detected, corrected and corrections are re-entered into the system to minimize mistakes. The system also needs to differentiate between what is an error and what might be systematic abuse. Again, in investment banks, systematic abuse caused the massive sale of documents of dubious value; sometimes with the knowledge of personnel and sometimes without their knowledge. Also, the departments are to ensure the system facilitates tracing input documents and transactions through to the creation of output reports and vice versa. This would include verifying that reports are not copied or reproduced by external parties. This would require that the system output provide the information needed to confirm all authorized transactions are processed promptly and accurately; and adequate consideration is given to the comptroller general's requirements respecting accounting records and financial statements, which again brings us to the problem of how to insure that the comptroller is actually doing is/her job correctly and how to document it. That is why all departments are to ensure that all aspects of the financial system are adequately documented. Finally, backups and transmission receipts must be created as redundancy for failures and as cyber-liability insurance. Again, a digital trail that can constantly and easily be monitored is of great importance to detect the first signs of abuse. Most of the ethical issues wind up here where the human element comes into play.

BANK FUNDAMENTAL ETHICAL PROBLEMS AND CHALLENGES

The bank has been confronted with internal and external ethical issues which are described as follows:

Internal Ethical Issues

Because of the economic recession, the bank intends to outsource the IT department and help desk to IBM. Certain layoffs of IT and help desk staff is inevitable. The bank steering committee is very concerned about ethical issues of the affected employees who may lose their jobs. A 2008 survey found that because of the global recession, many workers would perform unethical acts to maintain job security and secure advantages within the company. It also indicated that employees would abuse their virtual rights to steal sensitive and confidential company information to ensure a competitive advantage if they were to be laid-off. (Cyber-Ark, 2008) The bank steering committee requires a comprehensive plan to mitigate this unethical risk.

Other internal ethical issues include the embarrassment of the reputation of the company if vitally sensitive information were to be penetrated and stolen. This would include personnel information that may be retrieved from a human resources server. Other sensitive information includes business contract deals and future plans for the company. Without any mitigation, it would be a matter of time before the system would be critically breached, which would be more serious and costly than just having the bank robbed.

External Ethical Issue

Because of security system tests, IBM employees will require to access bank IT systems, and may potentially access sensitive information and documents. The Bank is challenged by avoiding attacks by IBM employees who use the bank's information systems. It can be just as dangerous for the bank if hacking occurs as the result of information leaking out as IBM employees phish through the banking infrastructure system. In the case of the World Bank being hacked from the outside by their IT contractors, the responsible individuals along with the contractor gained complete access to the bank's systems and vital information was examined. Passwords and identities were stolen and abused to achieve further infiltration into the bank's systems. Obviously, the contractor was terminated, but it took nearly four months before the last contractor left their facilities so internal IT employees could "transfer knowledge" (FoxNews, 2008). The ethical issue becomes evident when the bank becomes too reliant on the contractor because of the difficulty of canceling services with them and transitioning to a new contractor.

The bank steering committee needs to implement a comprehensive standard rules and process following the general rules and processes.

Recommendations and Solutions to Internal and External Ethical Issues

To address both internal and external ethical issues, internal (Bank) and external (IBM) employees require training on the following ethical declarations.

Social Engineering

The first ethical declaration is reinforcing the basic moral requirements in a work environment. Mostly, this must make employees maintain honesty and integrity as two of their principal professional values. Both the bank and IBM employees will use that to support their principles of professional operation to uphold professional creditability as well as confidentiality amongst each other and the bank clients. It must be stated that both bank and IBM employees must be respectful of clients, colleagues and competitors to run an ethical operation in the bank, and adhere to respect and fair treatment for all. This is similar to social engineering in the workplace so that principles of governance and social responsibility can be nurtured. At the same time, management must be in charge of responsible corporate governance to reinforce these declarations. All indiscretions must be treated as equally as possible, regardless if they were the responsibility of the bank or IBM employees. Management responsibility is to the shareholders. The bank's management must apply practices to manage risks. This is to realize an effective operation and independent audit of business controls.

Social responsibility signifies an ongoing commitment to sustainable economic growth. This is realized through cooperation with employees, their families, local communities and society in order to improve standards of living. It may sometimes not be evident to employees why the bank requires them to be socially molded. The bank must offer reasonable justifications and processes for why these virtues are important to the business. To prevent internal employees from performing unethical acts, the bank must motivate the employees to focus on the client. The employees must be viewed as a trusted financial partner from the client's perspective to regain assurance and confidence in the bank; confidence that has been considerably undermined by their past practices that contributed to the financial meltdown we are currently recovering from.

Protecting Trade Secrets

Management needs to ensure trade secrets are not disclosed and for involved company employees to keep secrets. Insider trading by unscrupulous bank employees in conjunction with trading firms created a lot of market uncertainty and instability. Trade secrets may include contract deals and plans for the future which may be visible once internally. Bank information is a valuable business asset that provides significant advantages over competitors. Our bank must ensure that IBM sign non-disclosure and confidential agreements to protect the bank's information assets from IBM internal hacking.

Dealing with Conflicts of Interest

Conflict of interest occurs when personal interests come into conflict with an obligation to serve the interests of others. This situation may happen to both IBM and bank employees when some tangible interest or personal interest motivates them to leak sensitive information outside the bank. Therefore, teaching and guiding employees to handle conflicts of interest is an important ethical issue. Employees must refrain from all activities that conflict with the interests of the bank or its clients and make decisions without prejudice and bias. Many traders, for example, commonly invest their clients' resources in stocks that benefit the trader, not the investor. They must reject gifts and unethical benefits and may not imply or express that they expect these. In the case that employees are unsure of any scenario or situation, the bank must provide conflict management guidelines to help the employee resolve the issue. Some issues mentioned in "Ethics and the Conduct of Business textbook" (Boatright, 2007) addresses the listed below.

Non-Competition Agreement

Through this agreement, there is a legally restrained relationship between employees and the bank. If employees leave the bank and obtain confidential information, this agreement obligates them to protect the bank from sharing the confidential data. However, the terms of agreement must be legal and be in line with the judicial requirements. (The time period specified, the geographic area, the kind of work related to the secret)

Confidentially Agreement

This agreement requires the confidentiality obligations of IBM employees and internal employees to protect the bank from stealing, sharing or leaking confidential information by IBM employees or internal employees who would leave the bank or want to trade the sensitive information for tangible wealth or career opportunity. The agreement defines legal restrictions of the employees (IBM and Bank).

Actual and Potential Conflict of Interest

From the bank's ethical perspective, there is a potential conflict because both employees of IBM and the bank can possibly act against the bank by stealing and trading confidential information.

Personal and Impersonal Conflict of Interest

There may be an impersonal conflict of interest when an IBM employee works for the bank as a contractor. The impersonal conflict of interest arises when this IBM employee is obligated to act in the interests of two different organizations, namely the bank and IBM. Administrators from either the bank or IBM may provide conflicting directions to the IBM contractor.

Individual and Organizational Conflict

In certain instances, the policy of the bank and IBM may be different on personal work/life leave (emergency leave) issues. The IBM Life Leave policy was established to help employees manage situations that range from attending a school function, to moving house, to providing emergency care for an elderly relative.

Study Leave: provides the opportunity for employees to sit for exams and attend management-approved courses.

Guidelines

Employees are required to have mandatory certification and subsequent recertification to keep them updated on current compliance cases and to read and understand corporate business conduct guidelines using real or near real examples. Conflict management guidelines would be included in the business conduct guidelines (that involve the basic situations and operational principles in daily work). Following general policies and controls will outline basic operations of daily work. To ensure that the guidelines are working, the bank must develop and design an exam to check if the employees have understood the guidelines. Management can utilize the results to organize the employees to study their habits. Examples and scenarios can be created through case studies for those who do not clearly understand or feel uncomfortable about the guidelines. Representatives comprised of subject matter experts will be provided for clarification and interpretation of the guidelines.

In addition to reinforcing the understanding of employee guidelines, the bank must continue to consolidate the study of guidelines among the employees, especially with both bank and IBM employees together to foster an ethically positive culture. At the same time, they must report anyone who violates business conduct guidelines or is in conflict of interest.

Open Door Policy

The bank and IBM must develop a communications channel or establish an open door policy. This is to encourage employees to anonymously report other employees who break guidelines. Both the bank and IBM must offer a complete privacy process to protect the employees who freely talk about unethical actions in the workplace so that the employees will feel safe reporting infractions, which will lead to improving dynamics.

This could be done by setting up an internal channel or access to provide employees the ability and accessibility to report any unethical behaviors that violate business conduct. A vertical management structure could also be established so that a direct manager or superior at work can be addressed if an employee has an issue with his/her direct supervisor. This strategy would include the head of the Independent Compliance Department of the bank and IBM. A secretary of the ethics committee could be created to handle the complaint management system to address how to properly deal with ethical issues or ethical offences that must be defined in the code of ethics and/or business conduct guidelines. To allow transparency for employees who do not actively report ethical issues, but feel that some things should be changed, an interactive assessment system comprised of self-assessments and evaluation among employees can gather information and forecast some potential ethical problems before they arise.

CONCLUSION

With all the problems of the economic downturn, banks have to be confident that the legal, financial, and ethical issues of outsourcing its IT and help desk departments outside companies can be resolved. By focusing on bank employees and rebuilding moral standards, investment banks can regain client confidence and begin to recover their reputations. The relationship with outside companies is contractual, but must be treated more as a partnership since most of the technical portions of the banking process will require the IT company's expertise and control.

Final Comments

Fields in which the new IT ethical system must act are in our opinion:

- An integrated macro-financial policy;
- Market confidence restoration;
- Prevention of immoral behaviors;
- Enhancement of depositor and investor protection;
- Sustaining cooperation at a global market level.

What will remove the new IT ethical system?

- "*...dog eat dog...*" competitiveness without limits;
- "*...every man for himself...*" applied at corporate level;
- Business success measures does not mean eliminating competition;
- Better controls will not kill creativity and threaten autonomy.

The IT ethical system in financing will become a compulsory solution one day. "*The information*" has traditionally has had great importance because it guaranteed successful management. Ironically, today "*the information*" has become the problem. We are overwhelmed by the amounts of information produced or required to optimally function. Because of this, the global coordination and global transparency regarding information exchange will add consistency and reestablish trust.

The financial framework might be compared to a game in which there is no clear distinction between the players and audience. All of us are players in the global economy. There is a need to carefully assess of the direct effects of stock exchange speculation and resulting financial difficulties and subsequent actions by the people.

One should not confuse and account balance with a score card. The most disastrous financial policies are those in which the primary players concentrate on "*...victory...,*" without considering the effects of the spectators sitting alongside the playing field (e.g. Enron, Parmalat, Citicorp, etc.)

Milton Friedman states that managers have the responsibility of answering exclusively to their shareholders and that their only moral obligation is to maximize profits. This view was true when Friedman obtained the Nobel Prize because the goal of business was not to promote public well being. We agree with this point of view, however, when business interests come into conflict with moral values, the public good must be given priority. Why? Because in the short term, illegality, incorrectness, and fraud might be advantageous, but in long term these practices result in extreme financial losses.

Max Weber made a clear distinction between two types of ethics: "*...ethics of belief...*" (morals) and "*...ethics of responsibility...*" (business). The interpretation of the above two views of ethics was the separation between business and morals. In fact, Weber sustains that the osmosis between the above views of ethics represents a paradox for businessmen: *He/She must choose between responsibility and morality.* The solution to this paradox is efficiently applying IT to enforce a well-conceived ethical framework for a banking and investment system that defrauded and created great hardships for hundreds of millions of people throughout the world.

REFERENCES

BankInfoSecurity. (2009) *Reports Shows Internal IT Attacks Rising*. (Rev. ed.). Retrieved February 1, 2010, from: http://www.bankinfosecurity.com/articles.php?art_id=56&search_keyword=internal+it+attacks+rising&search_method=exact

Boatright, J. R. (2007). *Ethics and the Conduct of Business* (5th ed.). Upper Saddle River, NJ: Pearson Prentice Hall.

Cyber-Ark. (2008). *The Global Recession and its Effect on Work Ethics.* Retrieved February 1 2010, from: http://cyber-ark.com

Kirk, J. (2007). Commerce bank thwarts a major database hack. *Computerworld, 41*(42), 16–16.

ADDITIONAL READING SECTION

Behar, R. (2008). *World Bank under siege in 'Unprecedented Crisis' (Rev. ed.) FoxNews. com, International News.* Retrieved February 1, 2010, from: http://www.foxnews.com/story/0,2933,435681,00.html

Bush, V., B. & Bush, A. (2000). Ethics and Marketing on the Internet: Practitioners' perceptions of societal, industry and company concerns. *Journal of Business Ethics, 23*(3), 237–248. doi:10.1023/A:1006202107464

Castells, M. (2000). Information technology and global capitalism. In Giddens, A., & Hutton, W. (Eds.), *On the Edge: Living with Global Capitalism* (pp. 52–74). London, UK: Vintage.

Chown, J. F. (1994). *A History of Money from AD 800.* London, UK: Rutledge.

Cooper, D. E. (1996). *World Philosophies: A historical introduction.* Oxford, UK: Blackwell.

French, W., & Mühlfriedel, B. (1998). Discourse instead of recourse: Is it worth using a guided approach to negotiation when working across cultures? In Lange, H., Löhr, A., & Steinmann, H. (Eds.), *Working Across Cultures: Ethical Perspectives in Intercultural Management. Issues in Business Ethics* (*Vol. 9*, pp. 263–285). London, UK: Kluwer.

Habermas, J. (1992). *Moral Consciousness and Communicative Action. Translated by Lehnhardt, C., & Weber Nicholsen, S. Moralbewusstsein und kommunikatives Handeln, Suhrkamp Verlag, Frankfurt am Main.* Cambridge, MA: MIT Press.

Hamelink, C. (2000). *The Ethics of Cyberspace.* London: Sage.

Ind, N. (2000). *Living the Brand.* London: Kogan Page.

Kalakota, R., & Whinston, A. B. (2001). *Electronic Commerce: A Manager's Guide.* (2nd ed.). Harlow, UK: Addison Wesley.

Kelly, E., & Rowland, H. (2001). Ethical and online privacy issues in electronic commerce. *Business Horizons, 43*(3), 3–12. doi:10.1016/S0007-6813(00)89195-8

Kettner, M. (1996). Discourse Ethics and Health Care Ethics Committees. In Byrd, B. S., Hrushka, J., & Joerden, J. G. (Eds.), *Jahrbuch für Ethik und Recht/Annual Review of Law and Ethics* (pp. 249–272). Berlin: Duncker & Humblot.

Lorenzen, P. (1968/1987). *Constructive Philosophy* (Rev. Ed.). Amherst, Mass.: The University of Massachusetts Press.

Mannion, J. (2002). *Essentials of Philosophy: The Basic Concepts of the World's Greatest Thinkers.* Colorado Springs, Colo.: Fall River Press.

Mason, R. (1986). *Four ethical issues of the Information Age.* Minnenapolis, MN: MIS Quarterly.

Melamed, L. (1990). *The Need for Risk Management.* Chicago: Addison Wesley.

Preuss, L. (1999). Ethical Theory in German Business Ethics Research. *Journal of Business Ethics, 18*(4), 407–419. doi:10.1023/A:1005983124799

Rogerson, S. (1999). *E.Society – Panacea or Apocalypse? The Rights and Wrongs of the Information Age.* Leicester, UK: Innaugral Lecture, Professor of Computing and Social Responsibility, De Montfort University.

Spence, L. J. (1998). Comparative European Business Ethics: A Comparison of the Ethics of the Recruitment. *Interview in Germany, the Netherlands and the UK: Using Erving Goffman's Frame Analysis.* Unpublished PhD Thesis, Brunel University, UK.

Spence, L. J. (2002). Like building a new motorway: Establishing the rules for Ethical Email Use at a UK Higher Education Institution. *Business Ethics (Oxford, England)*, *11*(1), 40–51. doi:10.1111/1467-8608.00257

Steinmann, H., & Löhr, A. (1994). *Grundlagen der Unternehmensethik* (2nd ed.). Stuttgart, Germany: Poeschel.

KEY TERMS AND DEFINITIONS

Access Control: The process of limiting access to an information system to authorized users, applications, processes, or other systems.

Backup: Storage of system files to facilitate recovery in the event of data loss.

Certificate: Document that binds a public key to an identity. It is digitally signed, includes the Certification Authority's identity, and the user's identity and public key.

Data Encryption Standard (DES): An algorithm for encrypting and protecting unclassified data. Details of the standard were released in Federal Information Processing Standard (FIPS) Publication 46 by the National Institute of Standards and Technology (NIST).

Encipher: A process for transforming plaintext to ciphertext using a cryptographic algorithm.

Firewall: A system developed for a private network to combat unauthorized access to or from the network.

Gateway: An interface that enables compatible transmissions between different networks by altering protocols, codes, security measures, and transmission speeds.

ENDNOTE

1 The IBM case described in this chapter is hypothetical.

Chapter 11
Law, Deviation and Paradigmatic Change: Copyright and its Metaphors

Stefan Larsson
Lund University, Sweden

Håkan Hydén
Lund University, Sweden

ABSTRACT

Drawing on debates in Sweden about Internet freedom, particularly those connected to copyright and file-sharing, and on the European legislative trend of amending copyright, this chapter analyses metaphors and conceptions in terms of a societal paradigmatic shift and the collision of mentalities. Kuhnian paradigms are wedded with the mentalities of the French Annales school of historic research. The chapter argues that the "building blocks" of these mentalities and paradigms can be studied in metaphors, in public debates or in legislation, which may reveal the conceptions they emanate from. This chapter touches upon ethical, moral and legal issues related to the digitisation of society. The relevancy of this chapter in relation to the theme of the book is found in the problematisation of "deviancy". One has to ask from what perspective or paradigm the judgment of the behaviour takes place, and in what historical context it is made.

INTRODUCTION

Somewhat more than one hundred years ago, labour strikes were still illegal in most European countries. Labour unions had no right to represent their members and negotiate with employers. Collective agreements were not formally accepted in Sweden until 1928. These legal instruments had quite a dramatic history before they became the leading mechanisms (especially in Sweden) for regulating the labour market. Indeed, less than 80 years ago, workers were killed in Sweden when taking part in a demonstration for labour rights (Ådalen 1931). Those supporting collective labour rights were united in their opposition to the prevailing logic of production within the guild system in the handicraft and agricultural sectors. Despite the fact that the large-scale conditions of industrial production had long been present in their sectors, it took some time before these

DOI: 10.4018/978-1-60960-094-5.ch011

instruments were accepted. Today, however, they are widely cherished in the industrialised world. Now, as we transition from an industrial society to more of an information society based on digital technology, we have reason to bring with us the experiences from earlier and similar periods of transition in industrial history.

The relevancy of this chapter in relation to the theme of the book is that it examines the question of what perspective or paradigm one is judging "deviant behaviour" from, and in what historical context it takes place. Part of what is considered deviant behaviour online is, rather, a consequence of the new system's expectations and conditions around the social norms and behaviours that the digital context offers. An understanding of this is required to effectively regulate any behaviour connected to this emerging context. If, for example, a legislator chooses the wrong battles on this issue, there are clear negative consequences for the overall respect for and legitimacy of laws and the legal system.

Ulrich Beck claims that sociology needs to change if it is to understand and explain the changing needs of a transitional society (Beck 1995, p. 231). Social science cannot rest too rigorously upon the "truths" related to the structures of the industrial age. Take copyright law as an example—it is developed in industrial society as a means of stimulating creativity and ensuring a return in profit for investments in intellectual products such as literature, music, film and other media. The "risk society", in Beck's terminology, is seen here as a label for the transitional society, since that society is still in conflict over the new practices, which are not yet legally codified. The focal point is shifted from a purely hierarchical, top-down structure towards an increasingly local influence facilitated by networking. The transition towards a new society is initiated by an unregulated bottom via an emerging core technology and its initial drivers. We can only learn how to cope with these changes in society and law by comparing them with corresponding shifts in the past. It is in times like these that labels used for describing key conceptions can be questioned and renamed. The labels used to describe phenomena in the digitised milieu online, which are often metaphorical in nature, are quite naturally borrowed from the analogue context that created them. We point out a few metaphors or conceptions that have been the subject of particularly heated debate.

This chapter touches upon ethical, moral and legal issues of the digitisation of society. A few of the illustrations used are connected to the debate in Sweden around Internet regulation, such as the copy-based formulations of the Swedish Copyright Act, the rise of a Pirate Party successful enough to win two seats in the European Parliament in 2009, and the rise of a blogosphere with political ambitions strong enough to affect the implementation of surveillance laws and other legislation. To a certain extent, we will use file-sharing as an example of deviation from copyright regulation (making it be regarded as illegal). The example is interesting from a historical perspective since Svensson and Larsson's study (2009) shows that among Swedish 15- to 25-year-olds, the file-sharing of copyrighted content is not perceived to be a deviation from social norms, despite being a deviation from legal norms. Furthermore, the debate around file-sharing and privacy has also been going on in relation to European legal trends connected to the creation and implementation of directives expanding copyright legislation (INFOSOC), its enforcement (IPRED), and the internal market (Telecom Reforms Package). These legal initiatives amend copyright or affect its enforcement, meaning that the metaphors embedded in copyright and the conceptions behind it are of interest if one seeks to understand the overarching paradigmatic battle or incompatible mentalities.

THEORETICAL BACKGROUND

Mentalities and Paradigms

Historically, research stemming from the famous Annales school has often used the term "mentality" to describe different mindsets in different cultures or historical times.[1] The journal *Annales d'histoire économique et social* was founded in 1929, marking the starting point of the Annales school. It also came to mark a turning point in French historical research. Lucien Febvre and Marc Bloch criticised contemporary historical science for focusing too much on details and events, becoming exceedingly specific, and losing its grip on the bigger, explanatory contexts and its connection to other social scientific research (Burke 1992). The focus then shifted towards unspoken or unconscious assumptions and the structure of beliefs over longer periods of time, the so-called *long durée*.[2] We can talk here about civilisations, which Wallerstein (1974) argues have a history of development that consists of periodic cycles "such as switching between growth and stagnation, and the alternation between hegemonic power and rivalry, and related long-term trends such as increasing commoditisation and commercialisation, and increasing polarisation between the privileged and non-favoured."

Every society has its own set of core "mentalities" or belief systems that define it (Burke 1986). Using the language of the Annales school, we can talk about *la moyenne durée*, a kind of middle-to-long period of mentality dominance. We will introduce the paradigm concept in order to describe this phenomenon and what characterises a transition from one type of society to another over time.

The long historical lines of mentalities, as with any ideas of consistency, are slightly problematic as an explanatory instrument for societal change. When—and why—would there be a break in something that has persisted throughout the ages? The explanation and description of change when it comes to mentalities lies close to Kuhn's description of paradigmatic shifts. The anthropologist Robin Horton, elaborating the ideas of Evans-Pritchard and Popper, has sketched a general picture of change in modes of thought, emphasising the importance of awareness of alternatives to a given intellectual system or, as he now puts it, the relative importance of competition between theories in different societies (Horton 1982). Burke approaches the problem of explaining change with the "history of mentalities" and the passage from one system to another by stating that there is an obvious case for taking up the paradigmatic shifts of Thomas Kuhn:

> *"If the great stumbling block for the history of mentalities is, as suggested earlier, 'the reasons for and the modalities of the passage from one system to another', then there is an obvious case for taking up Thomas Kuhn's notion of an intellectual tradition or 'paradigm' which may absorb or resist change for long periods thanks to relatively minor 'adjustments', but will finally crack and allow a 'Gestalt switch' or intellectual 'revolution' (Burke 1986: 446).*

The concept of a paradigm was developed in relation to changes in science. Classic examples are Isaac Newton (1643–1727) and his theory of gravitational force, and Noble Prize winner Albert Einstein (1879–1955) and his theory of relativity. The theory of paradigms, however, has mainly been developed over the last 40 years. The starting point is Thomas Kuhn's well-known book about *The Structure of Scientific Revolutions*, first published in 1962. When Kuhn refers to scientific revolutions he primarily uses empirical examples from the natural sciences. A classic example is how physics was established during the seventeenth century when Galileo Galilei rejected the hitherto dominant paradigm of Aristotelian physics and created a new conceptual system that made it possible to construct new objects for scientific knowledge. Galilei found certain anomalies in the way the motion of material things was explained within Aristotelian physics. This is often the starting point for paradigmatic change—when anomalies occur there is a risk of a crisis of legitimacy within science. The crisis might then cause some kind of revolution in the way things are perceived, which initiates searches for other explanations in relation to relevant phenomena: "The weight of anomalies leads to a cumulative switch to other exemplars and, ultimately, to logical incompatibility between disciplinary matrices, differences in prediction, differences in vocabulary, and to an argument over competing world views and competing ways of doing science" (Fine 2002, p. 2061). It starts with the emergence of a kind of pre-paradigmatic stage of scientific development, later followed by a stage of multi-paradigmatic science. Different scientific explanations compete in relation to being the bearer of scientific truth vis-à-vis the actual problems. After some time, one of the contenders among the multiple paradigms will be regarded as more adequate than the others and thereby "wins". We then reach a stage that Kuhn calls normal science.

Different aspects dominate each of these different phases of scientific development. One of Kuhn's most important contributions in relation to understanding science is that he raised awareness of the fact that science is not only a question of cognition and theoretical aspects (see Brante 1980); what is equally important to science is the structure of the group, which collectively holds a paradigm (Kuhn 1970, pp. 176–181). A paradigm presupposes an integrated community of practitioners that shares a consistent body of belief such that a consensus emerges with regard to the phenomena one investigates, the methods one uses, and so on (Eckberg and Hill 1979, p. 928). Science is not composed of a specific type of cognitive framework alone, but is also related to psychological and sociological factors. Thus, according to Kuhn, science can be characterised as systematised and institutionalised cognitive systems. It can also be characterised as structured and institutionalised social relationships—that it is something practiced in certain institutions in society, representing the sociological dimension. Finally, science can be regarded as connected to structured and institutionalised subjects; in other words, not everyone can claim to be a scientist, only those who are accepted as scientists by belonging to certain institutions (Brante 1980, pp. 24–31).

These different dimensions alternately play a dominant role during the different stages of development within science, as mentioned above. Thus, the theoretical part dominates the normal science stage, the psychological dominates when new perspectives are developed in the pre-paradigmatic phase, and the sociological/institutional dimension dominates in the multi-paradigmatic stage, when different paradigms compete to be the one and only "truth". In these phases, metaphors and conceptions play an important rhetorical role in the development of social sciences, see more below.

As mentioned above, the theories about paradigms have been developed in relation to the natural sciences. Since nature can be regarded as fairly stable, the development of science is related to cognitive progress in understanding the way physical and biological systems are functioning. Many have tried to relate paradigm theories to sociology and social sciences.[3] Martin has argued that there is a difference between the natural and social sciences in relation to paradigms (Martin 1972, p. 54). Whereas Kuhn's natural science paradigms relate to segments of disciplines, paradigms in sociology seem to be discipline-wide or even, as in the case of historical materialism, behaviorism or action theory. While sociology lacks a clear-cut puzzle-solving tradition (Eckberg and Hill 1979, p. 925), still more important seems to be the difference in the role of science in relation to nature, compared to society. While science in relation to nature has mostly been about accumulating knowledge to help mankind exploit nature for production, science in relation to society is more a question of mirroring society. Natural science works ex ante—developing knowledge in order to make it possible for mankind to use nature. Social science, on the other hand, works ex post—trying to understand and relate to changes in society.

Paradigms in the natural sciences develop due to innovations within science, such as when Newton developed the theory of gravity after having heard an apple hit the ground while he sat contemplating, or when Einstein invented the general relativity theory to explain certain planetary motions that Newton's theory on classical mechanics could not. The basics of what is called classical mechanics goes back to the research efforts of the early modern period, performed by, among others, Galileo Galilei, Johannes Kepler and Isaac Newton. This part of classical mechanics had been the basis for engineering and construction techniques. While the motions of the planets have not changed over the course of humankind, natural science has over time developed more precise theories for understanding these motions. In social science paradigms, societies develop continuously; they are never the same. There are specific motions for specific types of societies, and there are motions that have to be understood in terms of transitions from one type of society to another.

The shift from an agricultural society to an industrial society is one such example. Here we face a societal shift that affects the conditions for science in such a radical way that we can talk about a paradigmatic shift with consequences for all social sciences. This fundamental paradigmatic change on a societal level is captured by several social scientists using law as an indicator for describing the development of society. One of the first representatives of this school of thought was the legal historian Henry Maine, who described the evolution from status to contract (Maine 1861, 1959). The main representatives of this scientific approach include Émile Durkheim and his study, *De La Division du Travail Social* (Durkheim 1933), in which Durkheim uses the transition from criminal to civil law over time as an indicator of a society transitioning from what he calls mechanical to organic solidarity. Even Max Weber's analysis of various types of authority in which the legal authority represents a modern society can be mentioned in this context (see Gerth and Wright Mills 1991).

According to this understanding of social science paradigms, one can talk about paradigmatic changes in society as a whole, which are reflected in the conditions of the daily lives of people and in the scientific interpretation of society in different respects. Changes in society are the driving force for the paradigmatic changes of social science. Since paradigmatic changes follow a certain logic (as described above), we can count on a time lag before society and science move in lockstep. This affects not only science in a narrow sense but the understanding of society as a whole—what is regarded as right or wrong, true or false, good or bad, etc. As a consequence, we can expect that what is regarded as normal behaviour or "deviant" behaviour will undergo shifts over time. Even though many old principles of what is right and wrong in relation to human behaviour in Western societies today have their roots in canonical law and Christianity, paradigmatic changes in society create new conditions that alter opinions of what is good or bad. The same applies for principles of economic activity—new techniques may give rise to new circumstances, which make the old "rules of the game" outmoded and eventually obsolete, as in the historical case mentioned at the outset of this chapter. History is full of examples of people who, in these situations, have been punished for being forerunners and, thereby, norm-breakers.

In the following parts of the chapter, we will argue that we find ourselves at just such a critical stage in history—a point where society is undergoing dramatic changes without the benefit of having science and present-day mentalities catch up with, and articulate, these new conditions. This is not meant as a critique, since we know that these necessary mental adjustments take time, and that the time lag depends on broad societal acceptance before the new reality can take root and develop in ordinary peoples' minds. We will just attempt to create awareness about these processes and, thereby, a level of humility in relation to what is happening in the digital world.

Metaphors and Conceptions

If mentalities describe core structures of beliefs or assumptions, they can be investigated in the context of the conceptions or "mental grids" they are constructed of, while looking for the symbols and metaphors that represent and reproduce them (Burke 1986, see Allwood 1986, p. 132-136). This can be complemented by Lakoff and Johnson's (1980) research on metaphors and metaphorical concepts. They strengthened the idea that human thought processes are mainly metaphorical—that the "human conceptual system is metaphorically structured and defined." (They equalled "metaphor" with "metaphorical concept.") (Lakoff and Johnson 2003, p. 6). Their work inspired various disciplines to develop in this direction. There are many ways to study metaphors and several schools to follow (see for instance Cameron and Low 1999, pp. 29-30). The purpose of this chapter is however not to in detail outline how to study metaphors but to connect metaphors and conceptions to mentalities and paradigm shifts. The presentation leans on the mentioned Lakoff and Johnson rather than other schools of metaphorical research, although an important distinction is here made between metaphors and conceptions.

The choice of what metaphors are used, Lakoff and Johnson argue, are sometimes connected to power (2003, p. 159 f). By this, they point out the changeability of language as depending on those who have the ability to control it. To state this is to state that a picture used in language to describe a phenomenon not necessarily is the most "true" way to describe it, that there are alternatives, and these alternatives can be limiting and controlling the conceptualisation of a phenomena in different ways. It can hence be "imposed" on us in our need to conceptualise a phenomenon, consciously or unconsciously, and to be able to control this process is to exercise power.

The conceptual system is not something that we normally are aware of. This is also the reason for our search for metaphors connected to law, in order to draw out the conceptions hiding behind them, in order to scrutinize parts of the principal foundation copyright law and its connected debates originates from (see Larsson 2009; Larsson and Hydén 2008). Metaphors carry with them a heritage of the context they are derived from. They are not always easily translated from one context to another without some kind of distortion. Metaphors controls the way people think, and describe the way in which we understand life, our world and our place in it (Morgan 1999). The problem, however, is that metaphors can be both informative and deceptive. They can be borrowed from a context where they function well, only to be used in another context where they deceive and distort. The metaphors reveal the conceptions behind them, the mental structures that form, for instance, debates on legal solutions and shapes. By looking at the linguistic labels (the metaphors) one can determine how phenomena are conceptualised in a given context. We look for metaphors in order to display "conceptions" rather than "concepts". The difference between "conception" and "concept" is here similar to how Eberhard Herrmanns exemplify the difference (2008):

"Our conceptions of gold, for instance, are the different understandings we get when we hear the word 'gold' whereas the concept of gold consists in the scientific determination of what gold is" (Herrmann 2008, p. 63. Emphasis added).

It is the understanding or perception of phenomena, rather than some type of definition of object, that we focus our attention to when it comes to conceptions. It is the thought structure on the one hand and its label or metaphor on the other, although it has to be said that this distinction occasionally is neither easy nor meaningful to make.

To be more specific and simultaneously focus in on the subject of copyright legislation, one can use the Jessica Litman's "sleight of hand" example (2006, pp. 77–88). Litman argues for a "metaphorical" progression (which could be described as a change of underlying conceptions here) behind American copyright legislation during the twentieth century: from the initially less expansive conception of what rights authors and creators should have, to a more reciprocal, quid pro quo model between creators and the public, where dangers from "over-protection ranged from modest to trivial" (2006, p. 79). In the 1970s, copyright law began to be perceived as a construct that was full of holes and a threat to the interests of copyright owners. Litman argues that the bargaining "conception" has gradually been replaced in favour of a model drawn from a economic analysis of law, which characterises copyright as a system of incentives. She further argues that the success of this model lies in its simplicity, as it posits a direct relationship between the extent of copyright protection and the amount of authorship produced and distributed: "any increase in the scope or subject matter or duration of copyright will cause an increase in authorship; any reduction will cause a reduction" (2006, p. 80).

"When you conceptualize the law as a balance between copyright owners and the public, you set up a particular dichotomy—some would argue, a false dichotomy—that constrains the choices you are likely to make. If copyright law is a bargain between authors and the public, then we might ask what the public is getting from the bargain. If copyright is about a balance between owner's control of the exploitation of their works and the robust health of the public domain, one might ask whether the system strikes the appropriate balance." (Litman 2006, p. 79).

The point here is that depending on how copyright is conceptualised, the debates, the arguments and the regulatory efforts will be constrained within the logic walls of the leading conception. When the leading conception of copyright changed from a balance of mutual interest between creators and the public to a system focused mainly on the rights of creators, the remedy to this (newfound) lack of control would be more enforcement, more protection and more criminalisation of actions regarding unlawful distribution of content. This contemporary repressive legal trend (see Larsson, in press) rests upon this leading conception of copyright.

In short, the conceptions behind, for instance, law and metaphors create the building blocks for mentalities and paradigms. These "belief systems" could be looked upon as bundles of "schemata" or mental "grids" (or *grilles* as Foucault called them) that generally support one another, but can sometimes be in contradiction. Burke supports the idea of studying metaphors in terms of outlining the mentalities within a society or culture: "the notions of 'schema' and system may themselves be clarified if we look more closely at language, and especially metaphor and symbol" (Burke 1986, p 447). This is especially the case in terms of outlining the dissimilarities: "...if we are trying to describe the differences between mentalities, it seems a good idea to look at recurrent metaphors, especially those which seem to structure thought" (Burke 1986, p. 447, see also Allwood 1986). The fact that important metaphors are "in battle" and different imperative conceptualisations of reality seek to gain advantage over the other can be interpreted as signs of a bigger paradigmatic struggle or societal shift.

THE STRUCTURES OF SOCIETAL DEVELOPMENT

Societies can be said to develop over time, in waves or in a cyclical manner (Hydén 2002; Ewerman and Hydén 1997). They follow the cycle any system follows: society is born, it grows up, matures, and after a time it dies and begins a process of decay. One society emerges as a reaction to the existing society, meaning that when a society has reached its peak, a new society is already under development.

We have no way of knowing what will constitute society in the future. We can, however, use the cyclical model of societal development to predict what form society may have. Societal development shows one common feature over time. Before explaining this, we will say something about the driving forces behind development. From a long-term perspective we live in a market epoch. The fundamental conceptions and metaphors of our time are, to a large extent, formed by the mentalities belonging to the logic of a market. In the wording of the Annales school, the market represents *la longue durée* or, using Immanuel Wallerstein and the World system theory, a civilisation. In the market epoch, technology has been the prime driver of change. However, not all technological innovation has had an effect on a societal level; instead, a "core" technology is necessary for driving change, such as the steam engine at the beginning of the eighteenth century, which was used to develop new engines that could, in turn, lay the foundation for further technological advancements. The same applies to the computer today, which also represents a core technological change. A significant factor in relation to a core technology is that it is potent enough to stimulate the fantasy and imagination of people, such that their application of the new technology promotes the development of new modes of fulfilling existing human needs.

These factors also influence legal trends and developments. The development of law "follows" the cyclical development of society. We know, for example, that the feudal system in rural areas, the guild system in towns and cities (with its statutes and regulations of who was entitled to obtain a certificate as master craftsman and carry on craftsmanship), and the mercantile system with its strong regulation of trade, were deregulated during the eighteenth century and gradually replaced by a policy of non-restrictive practices and free trade policies. In the nineteenth century, new kinds of regulatory principles emerged. The *Code Napoleon* in France and *Burgerliches Gesetzbuch* in Germany heavily influenced the regulation of the market economy regarding property, contract and economic security rights. The public law system grew significantly during the twentieth century, particularly during the First and Second World Wars, when a great amount of public administrative laws were introduced. Finally, a new type of legislation flourished when we reached the peak of industrial society: the intervening regulation. During this time, from about 1970 to the present day, society has been covered by an enormous legal superstructure—just as when the handicraft and agricultural society was at its peak at the beginning of the eighteenth century.[4] Therefore, a process of deregulation is not only expected; to a large extent it has already taken place.

The first phase of the new information society faces a period of self-regulation and competing, pluralistic efforts to set the standards for the new society. The characteristics of the shift from the old (industrial) society to the new (information) society are always related to a change from large-scale to small-scale. It is a question of going back to basics (or perhaps forward to basics), namely fulfilling old human needs in a new way with new technology. With this shift we have what we can call a society in transition, which affects all of the societal dimensions mentioned above. Technology makes it possible to produce goods and services in a much more efficient way. Social conditions will change, with growing social tensions in society followed by greater differences in wealth among different sectors of society. Those who have receive more; those who do not have receive less. This is a nearly inevitable consequence of the clash between the old and new society, which creates winners and losers. For a long period of time, however, hegemonic power continues to be related to the structures and strata that belong to the old society. The existing society (reality) always has the preferential power of interpretation with regard to what is right and what is wrong. Therefore, it is not until the new society has managed to articulate its own societal solutions that one can expect a tendency to shift from the old way of living and fulfilling human needs to the new one. For a considerable period of time, the emerging society will be without these articulations and will therefore be an unknown phenomenon. This is apparent in contemporary science in their labelling (their metaphors) of present-day Western society as being post-modern or post-industrial—labels that remain focused on technological aspects and not reflecting the transition to the new society. These articulations are what newcomers in the "fractal political" scene are trying to aggregate, such as the French La Quadrature du net, the Swedish Juliagruppen, and thinkers and academics in the blogosphere.

LAW IN A SOCIETY OF TRANSITION

The Swedish Case

Sweden is an interesting case since it has a developed information technology (IT) infrastructure and a high degree of Internet usage. This is tied to the political vision of Sweden as a "leading IT nation", particularly as an "information society for everyone" (Prop 1999/2000:86, p. 1; Larsson 2008, p. 30f). This is significant for Swedish IT politics in general (Sundqvist 2001; Prop 1999/2000:86, p. 130; Larsson 2005, p. 39). Of a selected group of countries in which Internet usage is high, Sweden ranks among the highest, with 80 per cent of the population online. The file-sharing of copyright-protected material is, naturally, connected to the systematic conditions of a society: its infrastructure, degree of Internet access, and usage.

File-sharing and copyright has been a widely debated issue in Swedish politics for years. In 2008, the Centre Party, the third-largest party in Swedish Parliament at that time, suggested a thorough revision of the Copyright Act. Several of the youth segments of Sweden's political parties support free file-sharing for private use, as does the Left Party, one of the smaller parties in Parliament. Also in 2008, The Pirate Bay—claimed by Wired magazine as "the world's most notorious BitTorrent tracking site"—received global attention when the four individuals behind the company were prosecuted in Stockholm (Wired Blog Network 1 February 2008)[5].

The issue of file-sharing and media content was addressed at a hearing in Swedish Parliament in April 2008. The setting itself can be questioned from the perspective of a society in transition: only legal representatives were allowed to present their cases and no advocates of file-sharing were invited to the hearing. Although the point of the hearing was to discuss the issues and how they should be handled, with no one representing the file-sharing community, it was an unbalanced approach that undermines any attempt to understand the dilemmas of modern copyright.

On 17 April 2009, four men were sentenced to one-year prison terms and fined €2.84 million (SEK30 million) for assisting in the violation of copyright law through The Pirate Bay website. Three of these men had started a so-called BitTorrent tracker site in 2003 that, over the following years, grew into one of the most used and likely the most famous file-sharing site in the world. The Minister for Culture expressed support for the conviction, which she was reported to the Constitutional Committee for (Konstitutionsutskottet, which scrutinises the government and its ministers). The Pirate Bay case has been appealed by both sides. The date for the trial is not set, but has been postponed once; it is likely to start in summer 2010.

The Pirate Bay case, along with other unpopular legal reforms regarding surveillance laws and an EU-initiated expansion of the enforcement of copyright, has most likely fuelled interest in launching a Pirate Party focused on Internet-related issues. In the June 2009 national vote, the Pirate Party won two seats in the European Parliament. The legal reforms of interest here highlight the conflicts of interest that are at play in Sweden today, reflecting a society in transition.

Svensson and Larsson's 2009 study, in which 1,000 respondents between 15 and 25 years of age expressed very little negative social pressure with regard to illegal file-sharing, showed that this social norm barely exists. Moreover, the extreme popularity of the Pirate Bay BitTorrent tracker site shows that there is something dysfunctional with copyright law in the digital domain. The response to this dysfunction so far has been an expansion of efforts to monitor and enforce existing copyright laws in the EU, mainly through the INFOSOC and IPRED directives.

In 2008, a law was passed in Sweden regarding surveillance and signals intelligence. The law and, more importantly, the debate around the law, marked a key point: it is during this debate that Internet-related outbursts from politicians and the media became a critical force in the legislative process in Sweden. Bloggers and loosely-knit networks of intellectuals, "citizen journalists", academics, programmers and others joined forces, under the common theme of privacy and integrity, to voice their opinions against the law. The expression "blog quake" was used to describe the events. The law was called the "FRA law", after the authority responsible for carrying out the surveillance task, Försvarets Radioanstalt. This authority was previously only allowed to focus its surveillance activities on radio traffic, but this was expanded to include Internet traffic at "cooperation points" (Internet service providers). The law came into force on 1 December 2009 (see Kullenberg 2009, Ds 2005:30, and Prop. 2006/07:63).[6] The exceptionally stormy debate over increased government surveillance and signals intelligence is a good example of the blogosphere and Internet activists becoming an important entity with regard to knowledge-gathering and democratic journalism, as well as establishing a political voice on issues such as free communication, privacy and file-sharing. The "FRA law" seems to have had triggered an outburst of widespread discontent regarding how the politics around the Internet had been run. It also seems to have triggered a new type of political organisation and online activism that is here to stay. Protestors have highlighted the problems that this type of mass surveillance can bring (see Kullenberg 2009, p. 39).

This can be seen in the subsequent legislative processes regarding Internet and copyright enforcement (IPRED, see below) and the European internal market (Telecommunications Reform Package), as well as, to some extent, the directive on data retention that has yet to be implemented in Sweden.

The Telecommunications Reform Package and Copyright Amendments

The European Telecommunications Reform Package was heavily debated during 2009. Although it was presented to the European Parliament on 13 November 2007, the first vote on the legislation only occurred on 6 May 2009. The Reform Package is a cluster of directives (COM (2007) 697)[7] that is significantly focused on the role of Internet service providers. One battle over the legislation has revolved around whether or not it should be possible to regulate the ability of ISPs to disconnect Internet users based on suspected copyright violations before they are proven guilty in court. This was recently debated in France in the context of its HADOPI law.[8] Indeed, it was the French representatives in the European Parliament that sought to withdraw Amendment 138, which would ensure that a court trial preceded any potential disconnection. Another issue raised was whether ISPs should be able to determine which web pages users were allowed to visit. The battle within this debate was focused on the strength of the clauses to be included in the Reforms Package regarding the protection of individuals' rights.

Directive 2001/29/EC of the European Parliament and the European Council of 22 May 2001 regarding the harmonisation of certain aspects of copyright and related rights in an information society (the INFOSOC directive) included narrow exemptions to the exclusive rights of the rights holder as well as protection for "technological measures", often referred to as Digital Rights Management, or DRM (Article 6). The effect of this directive was that more actions were criminalised, and that copyright regulations around Europe were generally expanded and became stronger. The directive has been criticised for focusing on the aggregators' rights rather than those of the creators (Hugenholtz 2000). Indeed, the INFOSOC directive caused some debate in Sweden, but nothing like the 2008 and onwards debate on the FRA law. Moreover, the implementation of the directive was somewhat delayed. The changes in the Swedish Copyright Act came into force on 1 July 2005 (SFS 2005:360; SOU 2003:35; Prop 2004/05:110; Larsson 2005, p. 28–29).

When the IPR Enforcement Directive (IPRED)[9] was approved by the European Parliament (9 March 2004), it caused a stir among civic organisations in the United States and Europe.[10] The directive deals with the enforcement of intellectual property and industrial rights, and its most debated aspect was the fact that the directive gives copyright holders the right (via a court decision) to retrieve the identity information behind an IP address once they have "presented reasonably available evidence sufficient to support its claims" (Article 6.1). The "competent judicial authorities" could then order the provision of such information. The implementation of IPRED in Sweden meant that most of the provisions in the IPRED directive were implemented by 1 April 2009. Its implementation was intensely debated in Sweden in 2008, and especially throughout 2009—particularly on the rights that copyright owners' representatives (such as the International Federation of the Phonographic Industry (IFPI) and Svenska Antipiratpyrån, an association of producers and distributors of film and video in Sweden) have to apply to the courts for the release of identity information from ISPs. This has even led to an increase in the use of online anonymity services (see Larsson and Svensson, in press), ISPs stating that they discard the information that IPRED targets as soon as possible, and even initiatives within online communities to create new, encrypted file-sharing services.[11]

METAPHORS REVEALING CONCEPTIONS

The European legal trend builds on conceptions that have worked well in the industrialised and "analogue" paradigm, but less well in an Internet-connected societal paradigm. The debate and the protests show how these conceptions and metaphors are being challenged by attempts to replace them with other metaphors that better relate to conceptions of the new context of a digitised society. The rhetorical power of the old metaphors gains by having the preferential power of interpretation as mentioned above, what we also can call the "darling conceptions" of our time (see Larsson 2009; Larsson and Hydén 2008).

The Swedish Copyright Act, as likely most copyright acts, is a complex set of rules that is a patchwork of amendments from an early draft. It is not all these technicalities of the actual law that people argue and debate or think of when they think of copyright, but rather a few principles or conceptions that they mean the law should be based upon or not. These conceptions are often expressed through, or labelled by, various metaphors that do not exactly describe what they are used for, but to a lesser or higher degree are functional for the phenomena they are intended to represent. Some of these conceptions and metaphors can be found *in law* or preparatory works (an important legal source in Sweden), some can be found *outside law*, in arguments and debates aiming at the legal conceptions, or *in between*, for instance in the extensive interpretation of the process of the court case against the men behind the Pirate Bay. This section aims to discuss a number of these conceptions and related metaphors.

The arguments that support copyright and its enforcement often build on the conception, that characterises copyright as a system of incentives, as Litman has showed (2006). The argument then leads to the fact that if copyright fails in its enforcement, there will be no incentives for new cultural expressions to be born. Since much of the debate and legislative efforts centre around copyright in a digitised society, we will put forward here a few examples of metaphors that are problematic—some of them embedded in law, others that are a part of the debate around it.

The Swedish Copyright Act divides the rights of the creator into two parts: the economic right and the non-profit (or ideal) right. The economic right itself has two parts, namely the right to produce copies of the work, and the right to make it publicly accessible. Economic right is limited in some ways, however. One example, which is of interest in the context of moving from an analogue to a digital era, is the right to produce a few copies for private use, as expressed in Section 12 of the Copyright Act, which outlines the right to "produce one or a few samples of public work" for private use.

The Exemption for "Private Use" in Copyright Law

The exemption for "private use" builds on the concept that there is a viable dichotomy between private and public use. Generally, in Swedish legal tradition, the private sphere has been left unregulated. Copyright legislation has followed this logic, as demonstrated in Section 12 of the Copyright Act. With the digitisation, and organisation of networks, this private-public dichotomy has become a regulated conception that functions increasingly less well as a regulatory method (at least in the field of copyright). Behavioural and societal norms change in accordance with how the conditions in society change. User-generated application emerges, many industries transition from producer lead to consumer lead, and copyright is unavoidably affected by the introduction and distribution of information technology in society. This development takes place in contrast to the basic economic principles and thereby has to struggle against the long term mentalities of the market economy

"One or a Few Samples" and "Copy"

The word "copy" elicits the act of replicating an original, which can be described as an action better situated in an analogue setting. The idea that each copy is valuable and should be protected comes from the idea that copying involves a cost. The Swedish term for copyright is more tied to "the originator's right" (Upphovsrätt) and is non-specific with regard to its content, more than it is some type of right of the individual who has created something. Traditionally, the reproduction of copyrighted content was not an every-day act. Now, when you can't do anything online without reproducing copyrighted content, the conception that the exact numbers of copies should be controlled and protected is less well adopted to the modern conditions of society (see Lessig 2008, p. 269, compare Yar 2008, p. 611). What conception you argue for here likely depend on what mentality you base your arguments upon. "A few samples" is problematic in a digitised context from two perspectives: it makes little difference from a production cost perspective if you create three or three thousand copies, and "private use" is not private in the same sense as it used to be.

"Theft"

When the idea of property rights are established in an analogue reality and then transferred to a digital one, certain problems will occur. An obvious one, which reflects the two sides of the debate over the handling of media content, is the "copyism" of Internet communications on the one hand and "theft" on the other. From a traditional perspective, the illegal file-sharing of copyrighted content has been called theft. The metaphor is problematic in the sense that a key element of stealing is that the individual who has been robbed physically loses the stolen object; this of course is not the case with file-sharing, since files are copied. The Swedish Penal Code expresses this as: "A person who unlawfully takes what belongs to another with intent to acquire it, shall, if the appropriation involves loss, be sentenced for theft to imprisonment for at most two years" (Penal Code, Chapter 8, Section 1, translation in Ds 1999:36). More specifically, the problem in arguing that file-sharing is theft lies in the phrase "if the appropriation involves loss". There is no loss when content gets copied, and the loss is radically different from losing a physical product, such as a bicycle. The loss in this case is cast as the individual likely losing a *potential buyer* of the product. The "theft" argument, therefore, is an example of how one idea or conception tied to a traditional analogue context is transferred to a newer, digital context and creates problems in the transfer. Ultimately, something is simply "lost in translation" (See Larsson 2009 p. 38, Yar 2008, p. 612-613).

"Piracy"

"Piracy" is problematic in a similar sense. Even though file-sharing advocates have adopted this term and have used the Jolly Roger symbol as a logo to identify their resistance, genuine acts of piracy such as hijacking ships under violent and cruel circumstances have nothing in common with the act of copying media content and sharing it freely. A problem with a metaphor like this is that, as Lakoff and Johnson write, "the acceptance of the metaphor forces us to focus only on those aspects of our experience that it highlights, leads us to view the entailments of the metaphor as being true" (Lakoff and Johnson 1980, p. 157). This means that whatever negative value originates from the original use of a concept can remain inextricably linked to the concept and contaminate the new actions that the concept is now used to describe metaphorically. The use of the word "piracy" to describe file-sharing is a way of describing a complex new activity from the perspective of the traditional paradigm, while adding a characterisation to make it sound ruthless and "bad".

As such, this term will be functional and meaningful for the brief period of time when file-sharing represents something rebellious or otherwise deviant from a widespread and accepted value system (including one supported by laws). By the time the flows of Internet is the defining paradigm, file-sharing is not likely to be seen as rebellious or deviant, and therefore will not fit well with the "piracy" metaphor.

ISPs as Customs Officers or Caretakers of "Mere Conduits"?

The leading principle in the EU on the liability of Internet service providers has been that of "mere conduit" (Article 12, Directive on Electronic Commerce).[12] Critics believe that legal proposals such as the Telecommunications Reform Package attempt to make the ISPs liable for the data that is being run though their systems—thereby creating a monitored Internet (see Horten 2008). The debate draws on different metaphors such as the postal system and the mailman, revealing the different conceptualisations of what it is an ISP does and, hence, what an ISP should have liability for.

The abovementioned cluster of legislation seeking to harmonise national laws on copyright within the European Union are all part of a trend of increasing control over the flow of information on the Internet. More data is being generated and retained in order to support copyright owners in their fight against illegal file-sharing of protected content. At the same time, the copyright holders' representatives have been given easier access to identification data via regulation that hands greater responsibility to Internet service providers for content that is being trafficked through their infrastructure. This is one of the reasons why the debate around net neutrality has increased.[13] Europe, with France in the forefront, has shown tendencies of further increasing ISPs' regulatory responsibilities (Larsson 2010, in press).

PARADIGM SHIFTS AND THE LEGITIMACY OF LAW

Problems of Transition

To borrow from the abovementioned work of Lakoff and Johnson on metaphors, but in the wider context of this chapter: metaphors are unavoidably attached to discourse, and although they may have a very specific meaning in the discourse this meaning can change, and their use can be altered. This implies that metaphors can represent conceptions, that can be tied to an arranging order—an administrative pattern—which in and of itself stems from the analogue context of media distribution (for instance). These conceptions are likely to stand in the way when the administration is in need of change due to an evolving context. In short, digitisation has changed the context for media distribution, and the conceptions behind some parts of the way copyright is regulated today are standing in the way of the necessary changes to copyright legislation.

Many of the conceptions and metaphors scrutinised above regards the boundaries and ways of thinking about property in the digitised milieu. It is around these that a type of battle is being fought, a battle of who is to "impose" metaphors and conceptions on others, whose conceptions will lead and take precedence over the others'. This has much in common with the "rhetoric" in educational actions made by copyright interest organisations targeting children in school between eight and thirteen years that has been studied by Majid Yar (2008). Yar has looked at the ways in which

"...the boundaries of criminal and deviant behaviour are rhetorically redefined. It suggested that current attempts to moralize intellectual property rights and criminalize their violation make recourse to a range of repertoires of justification that attempt to naturalize a capitalistic conception of private property" (Yar 2008, p 619).

What about this "conception of private property", when it is translated to a digitised environment? Many individuals practicing illegal file-sharing do not believe that what they are doing is morally wrong or an illegal infringement on someone else's property rights. The natural and spontaneous feeling of ownership is related to the "use value" of a specific thing, not its "exchange value" on the market. Intellectual property rights are an abstract construction which, in part, has no reference to the moral world of ordinary people. It is motivated by market reasons, introduced from above, and forced upon the relevant actors. When the capitalistic economy—based on the concept of exchange value—emerged in the nineteenth century, modern society attempted to find solutions in the transition from the old to the new society, where use value and exchange value could co-exist (Christensen 1994). But when the exchange value of the product eventually took over as the dominant paradigm, the legal and regulatory framework lost its legitimacy. The new practices were guided norms other than those the laws were built on; in these situations, the law will lose, especially if the norms have "history as a tailwind". Legal regulation has to be supported by existing norms in society. While these norms can sometimes be changed by law, the law must be an expression of a desirable state in society. Otherwise, the regulation will be too costly to implement and uphold, and will be unstable over time.

Cognitive Jurisprudence and the Predicaments of Digitised Property

Information technology changes the context of regulation vis-à-vis the concept of property. The Austrian sociologist of law and one-time federal chancellor of the Austrian empire, Karl Renner

(1870–1950), has described how the legal context of property has been the same since the time of Roman law, despite the fact that the socio-economic consequences have changed significantly since then (Renner 1949). This legal context remained unchanged by staying connected to different complementary legal instruments, related to the contract and credit systems, and the concepts of the legal person and state regulation. From the perspective of information technology, property as a legal institution has become complicated in relation to the question of how property is transferred from one owner to another. In the classical legal understanding, this is constituted by the Latin word *tradera*, which means that the thing or a representation of it on paper is literally handed over to the new owner. In a situation where an increasing amount of transactions and changes in ownership take place in an electronic form, the antiquated notion of *tradera* no longer fits. This becomes even more apparent when more and more goods take the form of software. The challenge here is whether intellectual property rights will be developed in a way that aligns with the new regulatory requirements or if the legal concept of property has reached the end of its useful life in this context.

Within the emerging discipline of cognitive jurisprudence (which builds on cognitive neuroscience), Oliver Goodenough and Gregory Decker have asked the question: Why do good people steal intellectual property (Goodenough and Decker 2006)? The authors built on the findings created by the link between the physiology of our nervous systems and how we think and translate thought into action (Goodenough and Decker 2006, p. 2). The idea is that our decision-making is formed by a combination of the genetic organisation of our brain and the influence of our physical, social and cultural environments—which all come together to inform memories and habits, and develop capacities such as conscious thought, logic and the ability to create and shape external institutions such as the law (Goodenough and Decker 2006). Emotion is also a key component in an effective legal regime (Maroney 2006). Property rules need to be powerfully rooted in our emotional reactions in order to make us recognise and respect property. It is also connected to some primitive cognitive reactions in the human brain: the brain has a structure that helps humans assign the characteristics of property to those things that we recognise as possessions. Furthermore, there appears to be a deep emotional component to our property rules, since they apply to our physical possessions. Intellectual property law, by contrast, faces more serious challenges in promoting voluntary compliance. The problem is thus not doctrinal, but emotional (Goodenough and Decker 2006, p. 13). Instead of emotions, the parts of the brain that assign property rights to creative expression and invention are activated. A different set of pathways may have evolved to reward creativity—pathways related to respect and prestige on the one hand and to keeping secrets on the other. The authors provide the following conclusion:

"The act of spreading intellectual works sends a deeply understood message to the recipients that runs counter to the concept of property. It is a message that demands respect but not money. In this context, file-sharing makes perfect sense and is not a crime; in fact, what better way to show respect for the creator than to pass the subject matter of the acclaim on to others?" (Goodenough and Decker 2006, pp. 16-17).

If these hypotheses are correct, there is no use of "more of the same, only harder", as reflected in the abovementioned description of legal developments in the field (Goodenough and Decker 2006, p. 18). This will not produce noticeably better results in terms of compliance. As such, it seems as though the new practices produced by technological developments will be regarded as normal and legitimate, albeit illegal. The concept of deviance pre-supposes the existence of a yardstick that identifies what is "normal". When illegal behaviour becomes normal it may not be regarded as deviant any longer; this is the lesson of history. The present historical situation indicates that we have reason to expect something similar to happen in relation to the sharing of files, as well as in other aspects of information technology law.

The Challenge of Transition

Copyright regulation is based on ownership and the reproduction of copies. Globally, it is solidly anchored at national, intergovernmental as well as supranational (EU) levels, has been broadened in terms of scope and criminalisation (INFOSOC), and has been strengthened in terms of enforcement (IPRED)—all as a legal response to regulations that did not function in the online milieu. It appears as though the legal construct that was developed in an analogue context is unable to incorporate or align with how the Internet is structured and with the conceptions possible in a digitised society. There is something completely different about ownership in the digital domain that does not work well in an environment where the phenomenon of making digital copies is not nearly as significant as making analogue copies.

Sweden has had the policies of an IT-savvy nation for almost two decades. The aim of these policies has been to develop infrastructure, and that information technology is good for "regional balance", for companies and job opportunities, for industry, and for the education system. Thus, on the one hand, there is and has been a strong political will to develop Sweden as a widely-connected IT nation, where the Internet is present in every home, and where everyone should be able to take part in, create and contribute to the web. On the other hand, traditional regulations and protectionist thinking is working to limit such behaviour and use of Internet. The development thinking of an industrial society has led to an infrastructure that has supported the development of norms that challenge the logic of industrial society, particularly bound to analogue reproduction and distribution of media content. This is the paradox that is being played out right now. The phenomenon is typical for a society in transition and well known from similar historical examples of societies changing from one organisational logic to another.

Some mental references belong to the passing paradigm, as do power and what can be called stakeholder interest, even within the sciences. This creates a time lag or mental delay before the new principles and norms become accepted and "mainstream". In the meantime, we have to live with contradictions, increasing social tension and economic inadequacies. Some people, being forerunners and norm-breakers, might suffer and be punished for their beliefs.

Since law is a reflection of society, legal science must understand how society changes, lest there be too strong a risk that the legal order becomes an institution that uses its powers to support the parties that act and are coming from the traditional order in society. Law will then play the role of an institution that distorts societal development to fit some interests before others, merely based on its paradigmatic kinship. The risk lies in the fact that the legal order can become a tool of power in a struggle between the paradigms, supporting one mentality before the other, and based only on which was introduced earliest. It is a task for the social sciences to question a given societal order and the "truths" that it rests upon. Law is never value neutral, something we as scientists has to reveal. The conceptions embedded in law may hinder a more fruitful transition to the new means of distribution and production, and distort the more genuine ways of stimulating creativity and cultural development, making the owners of media content that was created (or at least protected) in the latter half of the twentieth century the biggest beneficiaries. It is also social scientists' duty to help construct adequate concepts, labels, metaphors and tools for creating the new society. It is our task to look beyond the partly informational and partly deceptive metaphors of this construct. Digitisation challenges some of the preferred conceptions of our time. It is a time of conflict, and its outcome will shape creativity, and life, to come.

AUTHORS NOTES

The chapter touches on four central themes: Societal change and how to look at it; (copyright) laws' place in the transition; thought structures and their representations in law and debate; and Sweden as a case for the above. Many of the references in the chapter provide excellent further reading in relation to these central themes. In addition to those, here follow a few suggestions on additional reading.

For a grand take on societal change connected to information technology the modern classic trilogy of Manuel Castells' is worth looking into. *The Information Age: Economy, society and culture* is a grand synthesis managing to incorporate and tie together the reconstructions of identity, social movements, globalization, the decline of the national states, and the global criminal economy, with the transformation of work and employment the information technology revolution. See for instance the term *resistance identity*, as an identity produced by those actors who are in a position/condition of being excluded by the logic of domination, and *project identity* as a term for proactive movements that aim at transforming society as a whole, rather than merely establishing the conditions for their own survival in opposition to the dominant actors. (1997, pp. 10-12).

When it comes to a critical assessment of copyright in days of digitisation, the American law professor Lawrence Lessig is one of the leading authors. Lessig is known through a number of books on the nature of the Internet, its condition-changing force for the legal regulations, which quite naturally also regards copyright. He has extensively studied the interplay between legal regulations and Internet and its code, for instance in *Code and other laws of cyberspace* (1999), which he updated in *Code version 2.0*, (2006). Lessig has had a strong focus on culture and creativity, and what legal foundation that best would serves its preservation in a digitised world, and he drew attention to the potential harms of overregulation in *Free culture: how big media uses technology and the law to lock down culture and control creativity* (2004). He developed this critique to also include suggesting the possibilities of a hybrid economy, in *Remix: making art and commerce thrive in the hybrid economy* (2008). On the background of copyright, as it has developed in the twentieth century, see Said Vaidhyanathans *Copyrights and copywrongs: The rise of intellectual property and how it threatens creativity* (2001). Vaidhyanathan paints a bleak picture of the future and contemporary imbalance on how copyright functions as a regulative force in relation to creativity.

In a Swedish perspective, Johan Söderberg's *Allt mitt är ditt. Fildelning, upphovsrätt och försörjning* can be mentioned ("All mine is yours. File-sharing, copyright and making a living", author's translation), dealing with the historical and philosophical contexts of copyright in relation to the contemporary debate on file sharing and culture in Sweden. A case study that could be mentioned regards the recent economic development of the Swedish music industry, and was made at the Royal Institute of Technology in Sweden during 2009 by Johansson and Larsson (2009). Castells, M. (1996/2000). *The Information Age: Economy, society and culture, vol. 1: The rise of the network society*. 2nd edition, Blackwell Publishing. Castells, M. (1997/2004). *The Information Age: Economy, society and culture, vol. 2: The power of identity*. 2nd edition, Blackwell Publishing Castells, M. (1998/2000). *The Information Age: Economy, society and culture, vol. 3* (1998): *End of the millennium*. 2nd edition, Blackwell Publishing.

REFERENCES

Allwood, J. (1986). Mentalitet och språk – några reflektioner. In Sällström, P., (ed.) (1986) *Mentaliteter". Funderingar kring begreppet mentalitet och möjligheten av en forskning om mentalitet*, Åbo Akademi: Publications of the Research Institute of the Åbo Akademi Foundation.

Beck, U. (1995). Att uppfinna det politiska. Bidrag till en teori om reflexive modernisering, Uddevalla: Daidalos.

Brante, T. (1980). *Vetenskapens struktur och förändring*. Lund: Doxa.

Bryant, C. G. A. (1975). Kuhn, Paradigms and Sociology. *The British Journal of Sociology*, *26*(3), 354–359. doi:10.2307/589851

Burke, P. (1986). Strengths and weaknesses of the history of mentalities, in *History of European ideas*.

Burke, P. (1992/1990). *Annales-skolan. En introduktion*, Göteborg Bokförlaget Daidalos AB.

Cameron, L., & Low, G. (1999). *Researching and applying metaphor*. Cambridge University Press.

Christensen, A. (1994). *Hemrätt i hyreshuset. En rättsvetenskaplig studie av bostadshyresgästens besittningsskydd*. Stockholm: Juristförlaget.

Directive 2001/29/EC of the European Parliament and of the Council of 22 May 2001 *on the harmonization of certain aspects of copyright and related rights in the information society*, (Infosoc).

Ds 1999:36 The Swedish Penal Code.

Durkheim, E. (1933). *Emile Durkheim on the division of labour in society: Being a translation of his De la division du travail social*, with an estimate of his work by George Simpson. New York.

Eberhard, H. (2008). On the distinction between the concept of God and conceptions of God. *International Journal for Philosophy of Religion*, *64*, 63–73. doi:10.1007/s11153-008-9163-z

Eckberg, D. L., & Hill, L. Jr. (1979). The Paradigm Concept and Sociology: A critical review. *American Sociological Review*, *44*(6), 925–937. doi:10.2307/2094717

Ewerman, A., & Hydén, H. (1997). *IT och social förändring*. Byggforskningsrådet.

Fine, B. (2002). Economic Imperialism and the New Development Economics as Kuhnian Paradigm Shift? *World Development*, *30*(12), 2057–2070. doi:10.1016/S0305-750X(02)00122-5

Gerth, H. H., & Wright Mills, H. (1991). *(translated and edited) From Max Weber: Essays in Sociology*. London: Routledge.

Goodenough, O. R., & Decker, G. (2006), Why do People Steal Intellectual Property, in *The Gruter Institute Working Papers on Law, Economics and Evolutionary Biology,* Vol 4, Issue 1, Berkeley Electronic Press.

Horten, M. (2008). *The 'Telecoms Package' and the copyright amendments – a European legal framework to stop downloading, and monitor the Internet*. University of Westminster. http://www.laquadrature.net/wiki/images/0/0e/Telecom_package.paper.monica.horten.28.june.2008.v5.pdf (last visited 1 March 2010).

Horton, R. (1982). Tradition and modernity revisited, in Hollis and Lukes (eds.). *Rationality and relativism*, Mit Press.

Hugenholtz, B. (2000). Why the Copyright Directive is Unimportant, and Possibly Invalid, *European Intellectual Property Review, at 501–502.*

Hydén, H. (2002). *Normvetenskap*, Lund studies in Sociology of Law.

Kravets, D. (2008). Pirate Bay Says It Can't Be Sunk, Servers Scattered Worldwide, *Wired Blog Network*, 1 February 2008, available at http://blog.wired.com/27bstroke6/2008/02/the-pirate-bay.html.

Kuhn, T. S. (1970). *The Structure of Scientific Revolutions*. Chicago: University of Chicago Press.

Kullenberg, C. (2009). The Social Impact of IT: Surveillance and Resistance in Present-Day Conflicts How can activists and engineers work together? *FIfF-Kommunikation.* 37–40.

Lakoff, G., & Johnson, M. (1980, 2003). *Metaphors we live by*, Chicago and London: The University of Chicago Press.

Larsson, S. (2005). *Musikupphovsrätten i ett samhälle under förändring – Internet och fildelning ur ett rättssociologiskt perspektiv (Intellectual Property Rights in a Transitional Society – Internet and File Sharing from a Sociology of Law Perspective),* thesis for Master of Laws, University of Lund.

Larsson, S. (2008). *Between Daring and Deliberating, 3G as a sustainability issue in Swedish spatial planning*, Blekinge Institute of Technology, Licentiate Dissertation Series No. 2008:02, School of Technoculture, Humanities and Planning.

Larsson, S. (2009). The darling conceptions of your time, or: why Galileo Galilei sings so sadly in the chorus. In Eide, S. R. (ed.) *FREE BEER 1.0*, an anthology from the presenters at Free Society Conference 2008.

Larsson, S. (In press). The Path Dependence of European Copyright, *SCRIPT:ed.*

Larsson, S., & Hydén, H. (2008, July). *Different darling conceptions tied to different societal systems. The case of illegal file sharing in a brave new world.* Paper presented at Research Committee Sociology of Law annual meeting, Law and Justice in the Risk Society, Milan and Como, Italy.

Larsson, S. & Svensson, M. (forthcoming). Compliance or Obscurity? Online anonymity as a consequence of fighting illegal file sharing.

Lessig, L. (2008). *Remix: making art and commerce thrive in the hybrid economy*. New York: Penguin Press.

Litman, J. (2006). *Digital Copyright: Protecting Intellectual Property on the Internet, the Digital Millennium Copyright ACT, Copyright Lobbyists Conquer the Book Description, Prometheus books. Maine, H. (1861, reprint 1959). Ancient Law*. London: The World Classics.

Maroney, T. (2006). A Proposed Taxonomy of an Emerging Field. In *Law and Human Behavior* (*Vol. 30*). Law and Emotion.

Martins, H. (1972). The Kuhnian "Revolution" and its Implications for Sociology. In Nossiter, (Eds.), *Imagination and Precision in the Social Sciences. London: Faber & Faber, Morgan, G. (1999). Organisationsmetaforer*. Lund: Studentlitteratur. [*Images of Organization*, 1997, Sage Publications Inc.]

Patrick, A. O., & McBride, S. (2008). Showdown Looms over Pirated-Media Directory. Swedish prosecutors target organizers of pirate bay, a huge file-sharing guide, *The Wall Street Journal,* 11 January 2008.

Prop. 1999/2000:86 *Ett informationssamhälle för alla.*

Prop. 2004/05:110 Upphovsrätten i informationssamhället – genomförande av direktiv 2001/29/EG, m.m.

Renner, K. (1949). *The Institutions of Private Law and their Social Functions*. London: Routledge.

Santos, B. de Sousa (1995). *Toward a New Common Sense: Law, Science and Politics in the Paradigmatic Transition*. New York: Routledge.

SOU 1956:25 *Upphovsmannarätt till litterära och konstnärliga verk.* Lagförslag av Auktorrättskommittén.

SOU2003:35 Upphovsrätten i informationssamhället - genomförande av direktiv 2001/29/EG, m.m.

Sundqvist, G. (2001). *Bredbandspolitik – En tekniksociologisk analys av kommunala bredband.* STS Research reports no 2, Avdelningen för teknik- och vetenskapsstudier, Göteborgs universitet.

Svensson, M., & Larsson, S. (2009). *Social Norms and Intellectual Property. Online norms and the European legal development.* Research Report in Sociology of Law. Lund University.

The Swedish Copyright Act, Lag (1960:729) om upphovsrätt till litterära och konstnärliga verk.

Wallerstein, I. (1974). *The modern world-system.* New York: Academic P.

World Internet Institute report (2008). *Sverige i Världen 2008* [Sweden in the World 2008].

ADDITIONAL READING

Herrmann, E. (2008). On the distinction between the concept of God and conceptions of God. *International Journal for Philosophy of Religion, 64*, 63–73. doi:10.1007/s11153-008-9163-z

Johansson, D., & Larsson, M. (2009). *The Swedish music industry in graphs. Economic development report 2000-2008.* Stockholm: Royal Institute of Technology, and TrendMaze.

Lessig, L. (1999). *Code and other laws of cyberspace.* New York: Basic books.

Lessig, L. (2004). *Free culture: how big media uses technology and the law to lock down culture and control creativity.* New York: Penguin books.

Lessig, L. (2006). *Code version 2.0*, New York: Basic books cop.

Lessig, L. (2008). *Remix: making art and commerce thrive in the hybrid economy.* New York: Penguin Press.

Söderberg, J. (2008). *Allt mitt är ditt. Fildelning, upphovsrätt och försörjning.* Stockholm: Bokförlaget Atlas.

Vaidhyanathan, S. (2001). *Copyrights and copywrongs: The rise of intellectual property and how it threatens creativity.* New York, London: New York University Press.

KEY TERMS AND DEFINITIONS

Paradigm Shift: A label used to describe scientific progress in terms of "revolutions", developed by T S Kuhn. Here expanded to describe also societal transition in connection to "mentalities".

Mentality: Set of unspoken or unconscious assumptions, a structure of beliefs consistent within a culture or civilisation over a longer period of time.

Metaphor: A figure of speech in which a word or phrase that ordinarily designates one thing is used to designate another, and a fundamental part of the human conceptual system of thought and communication. Here used, together with "conception", as parts of the building blocks or mental grids that construct mentalities.

Conception: A thought structure, understanding, perception or logic.

ENDNOTES

1 The mentality concept fills an important function in terms of attention to the role that historic structures play in contemporary conflicts.

2 Among works exemplifying the *long durée,* Fernand Braudel remarked on Alphonse Dupront's study (Dupront, *Le Mythe de Croisade: essai de sociologie religieuse,*

1959, reprinted without the subtitle in 1997) of the long-standing idea in Western Europe of a crusade, which extended across diverse European societies far beyond the last days of the actual crusades, and among spheres of thought with a long life.

3 For a discussion, see Christopher G.A. Bryant (1975), and for an overview see Douglas Lee Eckberg and Lester Hill, Jr. (1979). For readers in Swedish, Thomas Brante (1980) provides an extensive elaboration of the issue.

4 The Portuguese sociologist of law, Boaventura de Sousa Santos, has used the metaphor of an overloaded camel being burdened by the load of laws (Santos 1995).

5 The Wall Street Journal published an article about Pirate Bay at the same time (Wall Street Journal, 11 January 2008).

6 Lag (2008:717) om signalspaning i försvarsunderrättelseverksamhet. Proposition 2006/07:63 En anpassad försvarsunderrättelseverksamhet. Ds 2005:30 En anpassad försvarsunderrättelseverksamhet. Betänkande 2009/10:FÖU3 Signalspaning.

7 Proposal for a Directive of the European Parliament and of the Council amending Directives 2002/21/EC on a common regulatory framework for electronic communications networks and services; 2002/19/EC on access to, and interconnection of, electronic communications networks and services; and 2002/20/EC on the authorisation of electronic communications networks and services.

8 HADOPI is the abbreviation for a French law officially titled *Loi favorisant la diffusion et la protection de la création sur Internet* or "law favouring the diffusion and protection of creation on the Internet", regulating and controlling the usage of the Internet in order to enforce the compliance to the copyright law. The abbreviation is taken from the acronym for the government agency created by the law.

9 Directive 2004/48/EC of the European Parliament and of the Council of 29 April 2004 on the Enforcement of Intellectual Property.

10 S M Kirkegaard, "Taking a sledgehammer to crack the nut: The EU Enforcement Directive" (2005) *Computer Law & Security Report*, Vol 21, Issue 6, at page 489.

11 What in Sweden is called Prop. 2008/09:67 Civilrättsliga sanktioner på immaterialrättens område - genomförande av direktiv 2004/48/EG.

12 Directive 2000/31/EC of the European Parliament and of the Council of 8 June 2000 on certain legal aspects of information society services, in particular electronic commerce, in the Internal Market ("Directive on Electronic Commerce")

13 For a discussion on "net neutrality", see C.T. Marsden, "Net Neutrality and Consumer Access to Content" (2007) SCRIPTed, Vol. 4, Issue 4, 407–435.

Section 5
Perspectives on Combating Cybercrime

Chapter 12
The Case of "Illicit Appropriation" in the Use of Technology

Eduardo H. Calvillo Gámez
Polytechnic University of San Luis Potosí, Mexico

Rodrigo Nieto-Gómez
Naval Postgraduate School, USA

ABSTRACT

In this chapter, the authors play the devil's advocate to those who favor strict government supervision over technology itself. The authors' argument is that technology is a "neutral" mean to an end, and that the use of technology to detract social deviations is dependent on public policy and social behavior. To elaborate their argument they propose the concept of "illicit appropriation", based on the Human Computer Interaction concept of appropriation. The authors argue that sometimes appropriation can be geared towards activities that can be considered as illicit, and in some cases criminal. They illustrate the use of illicit appropriation through a series of case studies of current events, in which they show that either a state or the individual can rely on illicit appropriation. The authors' final conclusion is that the use of technology to combat social deviations is not a technological problem, but a public policy issue, where a delicate balance has to be found between the enforcement of the law by technological means (approved by legislation), the user experience, the civil liberties of the individual and the checks and balances to the power of the state. This chapter is written from the expertise of the authors on Human Computer Interaction and Security Studies.

INTRODUCTION

Ramón de Campoamor y Campoosorio wrote the poem[1]:

En este mundo traidor
Nada es real ni es mentira
Todo es según el color del cristal
Con que se mira.

(In this treacherous world,
nothing is real or a lie;
everything is according to the color of the
glass you are looking through)

DOI: 10.4018/978-1-60960-094-5.ch012

In the everyday of the Mexican folklore, the introductory line of this poem has been forgotten, leaving simply the phrase: "Nothing is real or a lie, everything is according to the color of the glass you are looking through". Sometimes, the glass is being used to look into technology. Technology can be used to commit unspeakable acts, e.g. war, as well as to enhance human life, e.g. The Manhattan Project ended World War 2 and provided technological advances that we still use (Kelly, 2007). But technology is neither bad nor good as its value depends on the action that the technology is mediating. The role of technology to perform actions that can be deemed as good or evil depends on the eye of the beholder, as it was the case of the Manhattan Project, which helped humanity to end a war, but it also provoked the death of millions of civilians.

Arguably, the value of every human conduct is not left to interpretation. There are certain conducts that are deemed criminal (deviant acts) by the legislation of most democratic states such as pedophilia, human trafficking, kidnapping (or any other type of extortion), homicide, terrorism or copyright infringement. These deviant acts have benefited from the use of "new technologies" up to a point that they are considered as central for the commission of the illegal conduct. This prominent role of the technology has provoked a response from public actors that seems oriented to "punish" technology itself and limit its functionalities, and not only to prosecute and punish the crime.

However, it is when non-democratic States act in the same way to prosecute conducts that in their context are also considered deviations (e.g. China vs cyberporn or Iran vs Twitter), that the conflict between the neutrality of technology and the Nanny State becomes clearer. A biased criterion seems to dictate that it is valid for democratic states to restrict technology *in lieu* of the prosecution of criminal acts, but it is not valid when, outside the democratic world, those restrictions limit human rights or user experience.

In this chapter, we play the devil's advocate to those who favor strict government supervision over technology itself or that claim that the way to stop deviant acts is purely technological. We do not argue that some actions should be accepted and some others not. Instead, our argument is that technology is a "neutral" mean to an end, and that the use of technology to detract social deviations is dependent on public policy and social behavior. We also argue that technology designers should be aware that any new feature included to stop deviant acts might be used to commit those very same acts. We concur with Koops et al. that the goal of a technology-neutral regulation should be that "the effects of the ICT [Information & Communication Technologies] should be regulated, but not technology itself" (Koops, Lips, Prins, & Schellekens, 2006).

To elaborate our argument we propose the concept of "illicit appropriation", based on the Human Computer Interaction (HCI) concept of appropriation (e.g. Dix, 2007). Dix argues that the user of a technology appropriates its usage in a way not foreseen by the designer of the technology. We argue that sometimes that appropriation can be geared towards activities that can be considered as illicit, and in some cases criminal. We illustrate the use of illicit appropriation through a series of case studies of current events, in which we show that either a state or the individual can rely on illicit appropriation. In those examples, we argue that the use of technology to deter illicit behaviors is linked to civil liberties. One of our conclusions is that technology rarely creates new crimes, but it does ease the process of committing them. As one of the old mantras of information technology says: computers only make processes more efficient, in this case the process of committing a crime. Our final conclusion is that the use of technology to combat social deviations is not a technological problem, but a public policy issue, where a delicate balance has to be found between the enforcement of the law by technological means (approved by legislation), the user experience, the civil liberties of the individual and the checks and balances to the power of the state.

The chapter is organized as follows; firstly we review the concepts of appropriation and user experience as well as the concepts of democratic and non democratic states. Secondly, we introduce the concept of illicit appropriation. Thirdly, we present three case studies of illicit appropriation in democratic, non democratic states and, in order to avoid an excessively state-centric vision of the process of illicit appropriation, in the last case study we analyze it in the context of a criminal networked organization and their appropriation of everyday technologies for the commission of deviant acts. We close the chapter with conclusions.

TECHNOLOGY, USER EXPERIENCE AND APPROPRIATION

The role of technology is to improve human life. Humans have tried to improve their surroundings and daily interactions, both with the environment and fellow humans, through the invention of the written language, seed domestication, motors and computers just to mention few examples. In order to understand and design technology to have a deeper impact on daily life, a series of laws, guidelines and frameworks are usually devised to transfer technology from theory to praxis. These theories can go from the intrinsic design of the technology itself, e.g. would the metal melt if it is used to manufacture such engine? To the actual use of the technology by its intended audience, e.g. how long would it require to educate the user to learn how to drive? When the objectives are only to create a technology that solely deals with the core scientific aspects, disregarding the human user, it may not actually improve, but actually diminished, human life. For example, medical devices that produce harm because of a poor design (Thimbleby, 2008). Among the different aspects of technology, in this chapter we introduce illicit appropriation primarily using the front-end of Information Technologies (IT), but later on we extend the discussion to technology, including critical infrastructure. The focus of our discussion is on the interaction between user and technology and the utilization of technology in ways not defined by the designer.

Originally, the objective of any IT application was to be *useable* by its intended audience; that is, it could be used efficiently and effectively to perform a task and it would provide satisfaction to the user (Bevan, 1995). In other words, the application had to be able to let the user perform a task fast, correctly while being pleasant to use. New research in HCI suggests that the focus of the application should be to provide the user with a positive experience (Hassenzahl & Tractinsky, 2006). The role of satisfaction takes precedence over efficiency and effectiveness as it can be safely assumed that an application that does not let the user perform a task under these two conditions would provide a poor experience.

The concept of *User Experience* goes beyond the interface of the application; it is about the interaction of the user with the environment (Calvillo Gámez & Cairns, 2008). Designing to improve the user experience is grounded on the idea of understanding the users, the task and the context and making sure the application responds to all the needs of the user (Buxton, 2007). The designer needs to understand both the process and outcome of the interaction (Calvillo Gámez, Cairns, & Cox, 2010). The process is formed by those elements that make the interaction common among individuals. The user makes sense of the experience by anticipating, connecting, recounting, personalizing[2], interpreting and reflecting on the different elements that form the experience (McCarthy & Wright, 2004).

In making sense of the experience, the user reflects on the application used to perform the task. This changes the role of the application from a tool ready-at-hand to present-at-hand (Winograd & Flores, 1986; Heidegger, 1927). In other words, the application is no longer a medium to an end, but an end in itself. Present-at-hand is when the individual reflects on the tool, rather than on the task; it is usually associated to when the tool does not perform as expected, but it also gives the user the opportunity to find new uses for the tool not foreseen by its designer. That is, the user improves the experience by finding new uses to the application.

Using the application for something different than intended by its designer is known as *Appropriation* (e.g. Dourish, 2003; Dix, 2007). The various definitions are centered on the idea that the user appropriates the application by using it in another way as intended by its designer. It is important to distinguish between appropriation and personalization. Personalization is using the application for its intended use while adding non functional features, such as changing the picture of the desktop, colors, fonts, orders, and so on. Appropriation is not using the application for its original and intended use, such as using e-mails as a to-do list or creating a Grease Monkey script for Firefox to change its functionality. A simple and parochial example would be to use a shoe to drive a nail into the wall. It is important to notice that the appropriation of the application does not preclude the user from personalizing or using the application as expected. Furthermore, the user is likely to enhance the use of the application by adding the appropriated features (AF) and then sharing them via worth of mouth or blogs. For example, the GNU project (GNU Foundation) is based on appropriation and AF being incorporated into the new releases of the technology, evolving from AF to standard features. The designers release a basic GNU application, then the users adapt and change the application to fit their needs, and if enough users approve the introduced AF, then they are included in the new official release. Of course, to develop AF for any GNU application requires specialized skills. Firefox, on the other hand, provides support to add-on applications that although widely popular, are still not included in the standard version. In this case, the users can develop their own AF, or go to the add-on section of Firefox and add it. Firefox, just as GNU in general, are applications designed for appropriation; in particular appropriation by users with the technical ability to change the application. However, this is not a requirement for appropriation. A person using the inbox of the email as a to-do list, instead of its intended use of just receiving emails, does not require different skills as the rest of the users.

To summarize, appropriation is when the user interacts with the application in ways not defined by its designer and as an effort of the individual to improve the user experience. The AF proposed by the user are then shared and adopted by a wider population of users. The application, and the technology in general, are then used with the AF "on top" of the standard features. Before introducing what it means to illicitly appropriate an application, we first introduce the definition of what constitutes an illicit activity, and the caveat of what is illicit or not depending on the type of state on which this happens.

DEFINITION OF ILLICIT CONDUCTS IN DEMOCRATIC AND NON-DEMOCRATIC STATES

It is impossible to obtain a universal definition of what constitutes an illicit conduct, for the simple reason that there are as many definitions as there are sovereign nations in the world. It is a geographically dependent concept because it is linked to a jurisdiction, and every nation-state will define it differently depending on their legal tradition and local values. In that sense, an illicit conduct is any conduct deemed as such by the body that has the authority to create the law in that territory. We consider it to be a "positive law" definition: part of a law created by a proper authority, which defines a particular conduct as forbidden and punishable within the boundaries of that jurisdiction.

An exception to this criterion exists in the case of human rights, considered to be "inalienable rights of all members of the human family" (United Nations, 1948), as they derivate from natural law theory (Habermas, 1999). This means that no government has the right to curtail them, and because of that, most of the illicit conducts committed by state authorities do not have to be recognized in any law when they go against the fundamental rights of every citizen.

Or at least, that is the theory. In reality, as it is often the case with regard of International Law, "might makes right", and only the peer pressure of other nation-states or their projection of national power will allow some kind of punishment against the members of any government who commit one of the series of illicit conducts considered to be "crimes against humanity", e.g. genocide, enslavement, the deportation of populations, torture, enforced disappearance, among others (United Nations, 1998).

Crimes are the clearest example of an illicit conduct. While they are not the only ones, as many other conducts are considered to be against the law but not criminal (exceeding the driving speed limit or jaywalking are also illicit conducts, but in most countries they are not considered to be a crime). Criminal law tends to concentrate on the worst cases of illicit conducts, the ones that receive the harshest of punishments available in that jurisdiction. By worst it is usually meant those deviant conducts that are considered to hurt society as a whole, and not only the individual (that is why in criminal law cases, one of the parties is always "the state" or "the public" versus the presumed criminal).

Despite the fact that most of the English speaking countries derivate their laws from the common-law British tradition, a majority of them have today some kind of codification that contains a description of the conducts considered to be a crime, e.g. (Robinson & Dubber, 1999). In the rest of the world, a codified penal or criminal law is the norm. For those countries, the criminal code is written in the form of a catalog that contains a description of all the conducts deemed a crime (*actus reus*) and thus, they are in fact catalogs of illicit or deviant conducts.

The main difference between democratic and non-democratic states will then be, not in the way they structure their definitions of illicit conducts, but in their origin. In the former, a democratically elected parliament will be the source of all codification, while for the later, the origin of the definitions of the illicit conducts can be one person (a dictator) or a group of unelected individuals (a military junta, a "People's assembly" or the clergy of a religion), with little or no input from the majority of the population.

INTRODUCING ILLICIT APPROPRIATION

As discussed previously, appropriation is when the user interacts with the technology in ways not defined by its designer. Dix describes the association as the user "not playing by the rules" (Dix, 2007). From the point of view of HCI, designing for appropriation has to be encouraged as it provides a better user experience. It is for this reason that programs such as Firefox allow scripts or add-ons, or programs released under the GNU licensing scheme can be edited and configured to satisfy the needs of the user.

Dix continues explaining that there is another kind of appropriation --"subverted appropriation"--, where the user "uses something in a way it was not intended, not just because of something the designer didn't think about, but in order to thwart its intentions". For example, a user that installs a special kind of script to his browser, in order to avoid the advertisement content of a website (ad-blocker). In this case, this particular user is clearly subverting the intent of the web designer, who expects to make his profit from the ads revenue, but it is a valid appropriation for the designer of Firefox as it was created to use this type of script.

However, it is not always desirable that the user appropriates a given application; for example, running a Windows applications, such as the video game *Spores* (Wright, 2008), on a Linux distribution like Ubuntu. The web page Linux Outlaws (Spore on Linux Howto, 2009) provides a guideline into how to achieve this. *Spores* was designed to run only on Windows, but via introducing the AF suggested in the web site, it is possible to run it on Ubuntu. In this example, the instructions provided by the developer of the AF specifically tell the user to break the digital rights management (DRM) of the game. An early adopter of the technology may stumble upon the original source, while a late adopter may find the answer in a different website (e.g. Spores on WINE) where this warning is not issued. This may indicate that the original developer of the AF is doing illicitly, but that the late adopter may have not known this. However, both conducts would result in committing an illicit conduct. From the point of view of public policy, this is a problem.

This is what we define as *illicit appropriation*; it goes beyond subverted appropriation. It does not only break the designer's intentions, but also the law. On a scale, it is the "worst" kind of appropriation, as it is the only one that is punishable by law.

The case of the DVD Region Coding (RC) system highlights this battle between appropriation and public policy. DVDs incorporate a type of DRM that is supposed to prevent watching a DVD outside of its intended regional zone. This was a system intended to give the film industry the capacity to set different release dates in different regions of the world (DVD copy control Association, n.d.).

For example, to watch a DVD region 1, the user is supposed to be based in a region 1 country; thus, a DVD bought in Canada or the United States cannot be watched in Mexico. This is enforced by selling DVD players of only the approved region.

For the particular case of Mexico, this proved to be a bad geopolitical decision regarding the fight against copyright infringement. As the map in Figure 1 shows, this country was placed inside the region 4 that includes all Latin America, the Caribbean nations, Australia and New Zealand. This policy neglected the fact that Mexico's strongest international relations (human and commercial) are with the United States and take place inside the NAFTA zone (Domínguez & Fernández De Castro, 2009). This meant that in a very short period of time, Mexico had a vibrant "zone 1" DVD market, and a public eager to find ways to appropriate the DVD technology in order to circumvent these DRM issues.

This restriction can be circumvented by moving the DVD player outside of its original region; so buying a DVD player in the United States and moving it to Mexico would allow users in Mexico to watch a region 1 DVD. Because this is not illegal, it is a clear case of "subverted appropriation".

Other solutions are to use the computer to watch the DVD, and via software, modify the region of the DVD player software, or physically modify a DVD player to completely bypass the RC System. Both of these conducts would be considered to be an illicit conduct by the film industry.

All these limitations have a negative influence on the user experience. For this reason, few experienced users decided to appropriate the use and interaction of computer's DVD players. The DeCSS (DeCSS Central) library was introduced so that users could watch a DVD regardless of the RC. As a result, the Motion Picture Association of America presented a lawsuit against the developers of the library (DVD CCA Lawsuit(s)). The lawsuit was based on the idea that the developers were acting illicitly. Only one of the developers was on trail and he was found not-guilty (BBC Technology News, 2003). However, DeCSS did not become a standard feature of any computer, even though it was being used extensively all over the world. This was because it is believed

Figure 1. Map of the different RC currently in use in the Americas

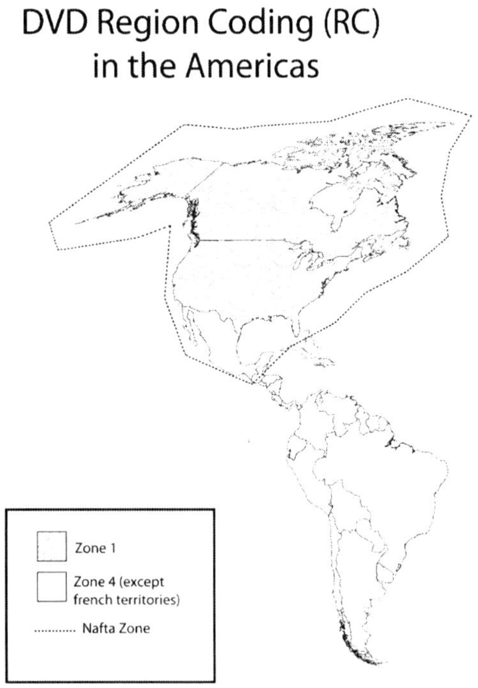

that DeCSS allows an illicit behavior of the users depending on the country on which they are based; although copyright infringement is not always equated with a criminal offense, it can be. The users can still download and use the application, but it is up to their legal framework if they are acting illegally or not.

Installing and running AF such as DeCSS can be seen as an effort of the user to thwart the intentions of the designer. One can assume that the designer of DVD players for computers are law abiding citizens that don't want to risk a potential lawsuit from media organizations; so they decide not to install the DeCSS library. The user that downloads and install the library is doing so to avoid something the designer, willingly, did not provide. Of course that in some countries DeCSS may actually break the law, where in other ones is just the posterior use of DeCSS that could result in breaking the law; for example, using DeCSS not only to watch a DVD, but to use it to copy a DVD without the intended locks.

We summarize then the process of appropriation in three different categories. Simple appropriation is desired from the point of view of user experience. The designer should allow the user to appropriate a given application in order to improve the experience of the user, as it was the case of the GNU applications or Firefox. Subverted appropriation is when the user thwarts the designer's intentions, as it was the case of the user of webpages who block commercial pop-ups, making it harder for the designer of the page to generate revenue. The third type of appropriation is when the user not only thwarts the designer's intentions, but also breaks the law. As it was discussed on the example of the DVD RC system, the conduct produced by the illicit appropriation can result in a criminal activity that is forbidden by law. See Figure 2 for a representation of the different types of appropriation.

This means that the concept is dependent on the existence of a previous legal declaration that a certain appropriation conduct is part of an illegal conduct. Even if the appropriation conduct subverts the intentions of the designer, then as long as it has not been declared illicit in a public law, it would only be considered to be a case of subverted appropriation, but not of illicit appropriation.

The core element of the process of appropriation is that the user appropriates the technology doing things in a way not foreseen, simple appropriation, or explicitly unwanted, subverted appropriation, by the designer, adapting the technology to his reality, outside of the designer's table. While the first kind of appropriation is seen as a positive phenomenon in design studies, the second kind, subverted appropriation, establishes a confrontational relation between the designer and the user, albeit still a legal one. Illicit appropriation differs from the other two in that, because of its

Figure 2. The three different types of appropriation. Simple appropriation is a goal of the designer so that the user can have a better experience while interacting with the application. Subverted appropriation is the user breaking the intentions of the designer. Illicit appropriation is when the user commits illicit or criminal activities with the application.

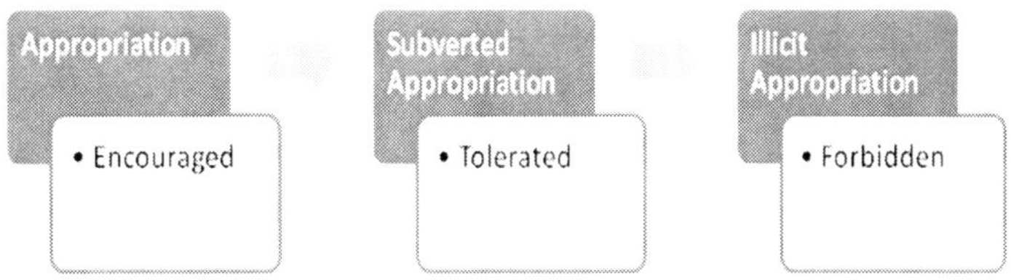

supposed deviant nature, it introduces a third actor, in addition to the designer and the user: the state.

Because some illicit appropriation processes have come to be considered as an intrinsic part of the commission of some crimes (cyber crimes being the clearest example) governments all over the world have decided to curtail or limit, not only the criminal conduct, but also the technology that facilitates the commission of the crime via the process of illicit appropriation. This limitation is done through legislation. The role of the state in protecting its citizens varies from country to country. In some countries, the state plays an active role enforcing the law thus providing regulations that limit the type of appropriation that a user can achieve with certain technology. In other cases, the role of the state is to deal with the consequences of the acts produced by illicit appropriation. And, in some cases even the state can be the perpetrator of illicit appropriation.

We will now study some of those cases were official actors in democratic and nondemocratic states alike, promulgate laws to fight some kind of illicit appropriation of technology, enforcing limits to the technology itself. Because the policymaking process in democratic and nondemocratic states differs considerably, we have separated the study of technology enforcement in these two groups.

We use the definition of democratic states provided by the Democracy Index (Economist Intelligence Unit, 2008), which is based on five indicators: electoral process and pluralism, functioning of government, political participation, political culture and civil liberties. For the Democracy Index of 2008, there are 30 full democracies in the world, 50 flawed democracies, 36 hybrid regimes, and 51 and authoritarian regimes.

Finally, in order to avoid an excessively state-centric vision of the process of illicit appropriation and law enforcement of technology, in the last section we analyze these subjects in the context of criminal networked organizations and their appropriation of everyday technologies for the commission of deviant acts.

CASE 1. TECH ENFORCEMENT IN NON DEMOCRATIC STATES

In this section we present two different examples of illicit appropriation in Non Democratic States. First we present the efforts of the Chinese government to filter pornography on the Internet, and then the role of Twitter during the 2009 election protests in Iran. It is important to remark that the classification of China and Iran as non democratic states is based on the Democracy Index of 2008 where China ranks 136 of 167 analyzed countries with a score of 3.04 points, and Iran ranks 145 with 2.83 points, both classified as authoritarian regimes.

The Green Dam of the People's Republic of China

The government of China has an official censoring policy for the Internet, contained in the Computer Information Network and Internet Security, Protection and Management Regulations approved by the State Council in December 1997 and promulgated by the Ministry of public security that same month (Ministry of Public Security of the State Council of the Chinese Government, 1997). The article 5 of the quoted regulations gives a list of the kind of information that cannot be transmitted using the Internet. Among them:

- inciting to overthrow the government or the socialist system,
- inciting division of the country (e.g. the independence of the Tibet province),
- making falsehoods or distorting the truth, spreading rumors,
- destroying the order of society,
- promoting feudal superstitions, sexually suggestive material, gambling, violence, murder,
- Terrorism or inciting others to criminal activity;
- openly insulting other people or distorting the truth to slander people;
- Injuring the reputation of state organs.

Most prominent among the censored type of information was that related to democracy and political parties, but not so strict on pornography (Zittrain & Edelman, 2003). The filtering of information is done on all the servers placed in the country through a series of firewalls. In other words, the filtering is done at the entrance of the Country.

On 2009, the Chinese government announced that it would require all personal computers sold in the country to include the "Green Dam" software (OpenNet Initiative, 2009) that would prevent all pornographic content from being displayed to the users. Green Dam checks a list of blocked websites, if the user tries to visit any of these websites, then the information is filtered (Wolchok, Yao, & Halderman, 2009). There are many complaints regarding the use of this software (Branigan, 2009), some people argue that it would allow the Chinese government to censor more information directly at the footstep of the users; others claim that the software is so poorly designed that it would put the computers with the software at risk (Wolchok, Yao, & Halderman, 2009). Some late reports claim that Green Dam will not be enforced (Taylor, 2009). However, even if installed or not, it is important to highlight the danger of illicit appropriation that can arise from Green Dam.

The original objective of Green Dam is to censor pornographic content. However, as the list that contains the blocked websites depends on a central database, it can be updated to include websites that contain different type of information. The person, or organization, in charge of updating the list could appropriate its use to censor information about democracy among other topics. Furthermore, the Green Dam would allow malicious users to use it as a gateway to gain access to the computer and its information. In both cases users would appropriate, subvertly, Green Dam to be used in a completely different way than intended; in the former case by blocking different types of websites than originally defined, in the later by using Green Dam to gain access to the computer. If said users are governmental organization, then this type of appropriation may not be deemed illicit depending on the legal framework of China. As we will see in the third case study, when similar activities are done by democratic governments, then they are not considered "as illicit".

Iran vs. Twitter during the 2009 Election Protests

In June 2009 Twitter, the micro-blogging service, delayed an upgrade to their system that would have taken down their servers for a whole day.

This was in the midst of the Iranian protests after their presidential elections. Before this could happen, the Department of State of the government of the United States asked the management of that company to keep the service running. Twitter had become the main communication system of the protesters, used not only to organize their rallies, but also to communicate to the rest of the world what was happening inside Iran's borders, closed to "traditional" means of international observation (Grossman, 2009). Although the Department of State confirmed that version, Twitter denied that they responded to the pressure of any government.

Keeping aside if the importance of the so-called "twitter revolution" was as significant as most of the media claims, it was clear that the US government was interested in keeping Twitter running; as it was seeing as neutral technology, in this case that means without the intervention of the Iranian government. The use of Twitter had become a matter of national interest for the US, beyond the financial strategy that this start-up company might have for its Iran operation.

Twitter was never supposed to be an instrument of civil disobedience and political participation. But as it is often the case with communication technology, the micro-blogging system was not supposed to be anything but a platform that would empower the user to use it in any way, shape or form that he or she decides. The open design of Twitter, which allow users to interact with it via SMS, instant messaging, Web and a search engine, as well as its broadcasting capabilities, allowed the Iranian protesters to appropriate the service. With it, they were able to organize protests, communicate events that were being censored by the state-controlled media broadcasting them to the outside world in a relatively anonymous and non-filtered way. Of course, the neutral and non-filtering of the information depends on the type of users with access to Twitter.

Within the context of nondemocratic states, the process of illicit appropriation is frequently portrayed in the media as a struggle for freedom against an oppressive regime. As it was the case in Iran.

CASE 2. TECH ENFORCEMENT IN DEMOCRATIC STATES: DIGITAL COPYRIGHT INFRINGEMENT AS AN ILLICIT APPROPRIATION PROCESS

It would be a mistake to consider that only non-democratic states have designed and enacted policies against the phenomenon of illicit appropriation. In fact, because of the nature of the political process in democracies, where the state is constrained by the law to do only what the law allows it, democracies have been a prolific source of legislation against illicit appropriation. In some cases, the state has gone as far as fighting the technology used during the appropriation process, and not only the conduct.

Piracy, as the illicit conduct of copyright infringement is commonly known, is the act of exploiting protected materials, against the wishes of the owner of the exclusive rights of that intellectual property, without compensation.

The beginning of the effort to protect intangible assets for their commercial exploitation can be found in medieval Europe. While the British crown had been issuing "letters patent" protecting several industrial designs and inventions since 1331, "some consider the first British patent [*and thus the first "modern" patent in the world*] to be the letter of protection granted to John of Shiedame in 1440" (Foster & Shook, 1993).

Regarding protection for literary works, the first copyright legislation in the world was passed also in the United Kingdom in April 1710. The statute of Anne, as it was known, "provided a legal protection of 14 years for works published after the commencement of the act" (Deazley, 2006). Its objective was "to encourage learned men to compose and write useful books" (The Statute of Anne, 1710).

Since then, intellectual and industrial protection have been the cornerstone of a system that tries to promote creativity by rewarding authors and designers with the exclusive right to exploit commercially their work, for a specific period of time.

Until the second half of the 20th century, all of these works were distributed in an analogous format (e.g. information in books, plans and designs in paper, music in LP's and then tape recordings, etc.) with little or no convergence at all among them. Because of that, all efforts to curtail copyright infringement were oriented to fight the distribution of tangible things, and to dismantle the industrial infrastructure needed to reproduce the protected material. While copyright infringement was a problem since the middle ages, it was one that was relatively contained.

But all that would change with the advent of the digital era. The history of digital technology and the compact disc might well be the most exciting case of simple and illicit appropriation. In 1977, Sony "showed a 30 cm digital record with a playing time of 150 minutes, and a few years later Philips demonstrated the 11.5 cm prototype of the Compact Disc to the international audio press". By 1982, the first commercial product was available in Japan and, one year later, in Europe (The history of the CD - The beginning, 2005).

The designers of the compact disc had in mind only the distribution of music in a digital format, in order to obtain a better sound quality. Instead, the public appropriated the technology in a process that was shaped by the realization that once the information –any information- was digitalized; it became "liberated" from the analogue physical support once contained it. It was turned into intangible digital media.

When that happened, a whole set of analogous and seemingly unrelated devices (LP's players, VCR's or even books) converged into one sole media, with a cost of reproduction close to zero, and independent of the constraints of physical reality. The advent of the World Wide Web some years later would mean that not only the cost of reproducing content became close to zero, but also that it was possible to distribute that content for "free".

Digitalization empowered the user to appropriate technology in a way that was impossible before, without the burden of manipulating the physical world to alter the substance of the media, as it was previously with an analog object. It also allowed the convergence of technology that is the norm today.

As Anderson points out, "the engineers brought us the technical infrastructure of the Internet and the Web -- TCP/IP -- and HTTP://-- but we were the ones who figured out what to do with it. Because the technology was free and open to all, we, the users, experimented with it and together we populated it with our content, our ideas, and ourselves. The technologists invented the pot, but we filled it" (Anderson, 2009). In other words, we appropriated digital technology beyond any designer's expectations.

As an unintended consequence, the system of copyright protection that was created in medieval times was put in jeopardy. The whole business model built around charging expensive prices for content has been confronted by the illicit appropriation process of piracy, which follows simple appropriation of digital technology. A user of digital music may produce a copy for her personal records, but that same copy could be distributed, or sold, without the consent of the copyright holder.

Since the 90s, public and private actors have employed different strategies to try to fight this new wave of digital copyright infringement, with very little success. Legislation has been one of the main tools to fight against this kind of illicit appropriation.

In 1998, the U.S. Congress passed the Digital millennium copyright act that criminalizes most conducts related to the illicit appropriation of digital technology. In 2001, the European Union passed the Copyright Directive aimed at the same objective. More recently, France passed in

September 2009 a second version of its "Hadopi" law (the first one was repealed by the constitutional tribunal), designed specifically to give the French state a graduated response capability to fight against illegal downloads.

But it's been the private sector, specifically the recording industry, who has been the most active actor in the fight against this kind of illicit appropriation. Using the legal framework of the Digital Millennium Copyright Act (and thus taking advantage of the actions of public actors), the Recording Industry Association of America (RIAA) sued the peer to peer service Napster in 1999, in what would become the most famous legal case regarding copyright infringement. Eventually, the RIAA won the legal battle against Napster, but lost the war against downloading as the main vehicle for delivering music. This law suit punished the technology, not the crime itself. Furthermore, the technology used in Napster is nowadays used in Skype; a type of simple and positive appropriation. Today, the illicit conduct of downloading copyrighted material is simpler than it was during the Napster years, and legal downloading is predicted to replace the compact disc as the main way of music distribution in 2010, with a completely new actor -- iTunes -- as the undisputed market leader (Whitney, 2009). iTunes used to distribute its files encoded with a type of lock that prevents the music to be copied in more devices, this has a direct impact on the user experience, as the user is unable to copy his music among his different computers or devices. In this case, the move of material from one device to another might be legal, but the solution to piracy may be an illicit appropriation of iTunes.

All these legislative and legal actions from the part of public and private actors have provoked a political reaction from the part of the -- until recently-- fringe sector of the electorate that wants to challenge the *status quo* regarding copyright regulation. From their point of view, open source efforts to create knowledge and innovation in an incremental way, cannot exist in the context of the traditional copyright paradigm, and have decided that the only way to oppose those policies is to become policymakers themselves. Founded in Sweden in 2006 with the aim of reforming copyright laws and fighting for privacy rights, the Pirate Party is now present in most European countries and has two seats in the European Parliament.

This means that as long as this political movement continues to be active, the classification of illegal downloading as a deviant conduct will be now permanently challenged in democracies with a "pirate" movement, proving once again the strong association between the fight against illicit appropriation and the policymaking process.

CASE 3. NON STATE ACTORS: ILLICIT APPROPRIATION DURING THE 9/11 ATTACKS

Terrorism can be the most hideous case of illicit appropriation. While there is no easy way to define terrorism, almost all authors agree that there are two key elements in any terrorist conduct: the use of targeted violence to provoke terror beyond the people that are directly targeted by such violence, and the intention to influence a political agenda.

The Report elaborated in March 2007 by the Independent Reviewer of Terrorism Legislation presented to the Parliament of the UK, concluded that, with some alteration needed, the definition in the Terrorism Act 2000 "is useful and broadly fit for purpose" (Independent Reviewer of Terrorism Legislation, 2007). In consequence, and given the fact that we are not interested in continuing here the debate on a universal definition of terrorism, we use the official definition of the Terrorism Act 2000:

"Terrorism Act 2000"

Section 1.

(1) In this Act "terrorism" means the use or threat of action where-

(a) the action falls within subsection (2),
(b) the use or threat is designed to influence the government [or an international governmental organisation] or to intimidate the public or a section of the public, and
(c) the use or threat is made for the purpose of advancing a political, religious [, racial] or ideological cause.

(2) Action falls within this subsection if it-
(a) involves serious violence against a person,
(b) involves serious damage to property,
(c) endangers a person's life, other than that of the person committing the action,
(d) creates a serious risk to the health or safety of the public or a section of the public, or
(e) is designed seriously to interfere with or seriously to disrupt an electronic system.

(3) The use or threat of action falling within subsection (2) which involves the use of firearms or explosives is terrorism whether or not subsection (1)(b) is satisfied."

The real target of the terrorist action is not those who suffer the violence, but the people who witness it. Terrorism is always in a direct need of an audience to scare.

Nevertheless, new technologies have permitted a kind of escalation in the objectives of terrorist organizations, that today seek not only to impress the public that witness their action, but also to inflict massive casualties and economical loses to their opponent (OECD, 2003). The Madrid bombings of 2004 (191 casualties) or the London bombings of 2005 (52 casualties) are both good examples of the new scope of terrorist attacks.

The most successful non state-sponsored terrorist attack in history is –by far-- the attack of 9/11 in US soil (2,293 casualties). Al-Qaeda in particular is known for designing its operations with the objective of maximizing "the casualties and costs that they produce" (Jackson, 2009).

Even if those acts accomplished the "traditional" objective of terrorizing a witnessing audience, the 9/11 attacks had a magnitude that made them much more that "just a terrorist attack". They were equivalent to the level of an act of war: The number of casualties was bigger than during the Perl Harbor attack, it took place at the heart of two of the most important cities in the United States and the attack was directed against the civilian population.

The attacks succeeded in putting a halt to all commercial air traffic in US airspace for days to come for the first time in US history, and made impossible to continue the operations of the US financial markets. They also destroyed the emergency management office of New York City, crippled the first response capabilities of the state of New York, partially shattered the national headquarters of the US Armed Forces and made necessary the evacuation of most government buildings, placing government continuity at risk (National Commission on Terrorist Attacks Upon the United States, 2004). The associated costs of the attacks have been calculated in literally hundreds of billions of dollars (Makinen, 2002).

One reason for the level of success that Al-Qaeda had during 9/11 is that these attacks took terrorism to the realm of illicit appropriation in an unprecedented way. They were not only able to sabotage a component of the Critical Infrastructure of the United States (the transportation sector), but they illicitly appropriated part of that infrastructure and turned it against the population it serves, transforming it from a tool of transport into a tool of terror.

The "Marsh report" of 1997 is considered to be the cornerstone of the field of Critical Infrastructure Protection in the United States. It defines Critical Infrastructure as "a network of independent, mostly privately owned, man-made systems that function collaboratively and synergistically to produce and distribute a continuous flow of essential goods and services" (Marsh, 1997). This infrastructure is vital to the well-being of any developed nation. The National Infrastructure Protection Plan of 2009 of the United States organizes this infrastructure around 18 interconnected sectors (National Infrastructure Protection Plan: Partnering to enhance protection and resiliency, 2009).

While "normal" terrorism traditionally tries "only" to sabotage a system to inflict casualties to their perceived enemy, destroying their infrastructure (e.g. the Pan Am 103 bombing of 88; 270 casualties) during 9/11, the four hijacked planes were not only sabotaged, but they were also illicitly appropriated, in order to transform them from a transportation tool into a weapon of mass destruction. The intentions of the aeronautic designers were completely subverted as they never intended for the planes to be used to demolish buildings and kill its occupants; breaking every law that governs and regulates civilian aviation.

The terrorists understood the rules that allow the planes to be used "as intended". They attended flying schools inside the United States where they learned the basic knowledge needed to fly the planes (they learned how to use the technology), and they also studied the limitations of the security system at the airports, inside the cabin of the planes, and in the way air traffic control took place in the United States prior to those attacks.

Only then, the Hamburg cell was ready to appropriate illicitly part of the air transportation sector of the United States turning this technology against its legitimate users.

A terrorist attack that appropriates part of a technology, will be always more dangerous than a plot that only seeks to destroy an infrastructure. The reason is simple: when the terrorist illicitly appropriates part of a critical infrastructure, he's not only in the position to inflict damage by destroying such infrastructure, but he is able to appropriate some key elements of that technology, to create a deadlier system. That is what happened in 9/11, where instead of "only" the 246 fatalities in the four planes (excluding the 19 hijackers), the terrorists were able to kill as many as 10 times more innocent victims, using everyday technology against the civilian population.

Theoretically, there are a lot of other scenarios where the illicit appropriation of a part of a critical infrastructure might result in high costs both in terms of money and more importantly, in human fatalities: e.g. Planes (or trains) can be used as bombs; water infrastructure, as well as the food supply chain, can be appropriated as a pathogen delivery system (Wein & Liu, 2005); the chemical industry or the hazmats storage facilities can be appropriated to be turned into a dirty bomb (Rosoff & von Winterfeldt, 2005), etc.

Cyber terrorism might present some of the worst case scenarios related to illicit appropriation of critical infrastructure systems by radical actors. The appropriation of the SCADA (*Supervisory Control And Data Acquisition*) systems that control most critical infrastructure would allow them to remotely turn the technology that any modern society needs to function, against that same society, e.g. opening the levees to flood a city or in the worst-case scenario of all, hacking the command-and-control center of a nuclear power plant (Greenberg, 2007).

It can be considered that the whole Homeland security policy in the US is a big and costly "illicit appropriation prevention" system of systems. It tries to force the user of those different systems to use them only in the way they were designed, with no appropriation whatsoever.

Consider the TSA (Transportation Security Administration), part of the Department of Homeland Security of the United States, which was created after 9/11 in order to centralize in one federal agency the aviation security systems at the airports, which prior to the attacks were managed by private companies (TSA is responsible also for the security in all other transportation systems, but most of its resources go to aviation security).

The objective of this relatively new agency is to create and manage a mixed system of people and technology that will prevent the illicit appropriation of any part of the transportation sector in the United States, and in any country from where planes fly to the US.

Therefore, the example of the DHS shows that there are valid cases when a ban on appropriation is justified for critical infrastructure protection concerns. Based on this, we can define the endeavors of the "red team", a team that tries to think like the adversary in order to defeat the defense systems of an infrastructure, that take place in any terrorism prevention program have the objective of creating a system that: a) cannot be sabotaged and b) cannot be appropriated.

But one should not forget that in all cases, the priority of the actions and the policy involved should be to avoid and prosecute an illicit conduct and not the illicit appropriation by itself. More than punishing the illicit use of a technology, the objective should be punishing homicide, terrorism, kidnapping, the destruction of private property, etc.

In consequence, the policy to prevent illicit appropriation in the context of Critical Infrastructure Protection should be oriented to the prevention of the appropriation process and not to the prosecution of deviant acts already committed. That should be the task of the justice system.

If someone uses a hammer to kill a person (illicit appropriation), society should punish the homicide… not the use of the hammer.

CONCLUSION

In this chapter we have presented the case of illicit appropriation. Our main argument is that, for good or evil, technology is a mean to an end; not an end on itself. Based on these arguments, we present the key conclusions of our discussion as follows:

- The use of technology is dependent on the individual's intentions.
- The individual can use the technology for something completely foreign to what the designer of the technology intended via appropriation.
- Appropriation is an effort of the individual to improve the user experience.
- What constitutes an illicit conduct is defined by the state, except when the state curtails human rights.
- Illicit conducts are combated by the state via laws, which are defined by policy makers.
- When using technology to combat an illegal conduct, designers have to work hand and hand with policy makers.
- Determent of Illicit conducts is not a technological issue; however, technology solutions can be used to combat them when they are the consequence of a solid public policy.
- Illicit appropriation should be prosecuted when:
    - a) the appropriation goes beyond the normal "use" of technology, and
    - b) puts in danger non users.
- Policy makers have to take into account the magnitude of the conduct. It is not the same illicit appropriation for the sake of using a hacked MP3 player, that illicitly appropriating part of the transportation system of a country to commit a terrorist attack. Thus, illicit appropriation should not be a crime, but the conduct that required such appropriation.
- Policy makers have to consider the use of technology to prevent illicit appropriation, but technology responses should not be considered the silver bullet that would replace sound public policy.

There is a trend in legislative bodies around the world to enact legislation against technology itself, when confronted by illicit conducts with a strong technological component. While in some cases it is necessary to regulate a new kind of technology to protect the life of other users (e.g. the transit regulations needed after the invention of the car or to maintain the integrity of security systems), most of the time these laws demonize technology, punishing the appropriation of technology by law abiding citizens and diminish the quality of the user experience, with very little effects against the illicit conduct itself.

In fact, because the illicit appropriation of technology often requires some technical skill, the person appropriating the technology will have more access to the tools and knowledge needed to circumvent the legal interdiction, while the honest user is left with a second class technology (e.g. A "pirate" who downloads an mp3 with no Digital Rights Management has none of the limitations that an honest user has when buying a song with DRM and trying to use it in his or her mp3 player).

It is our argument that, in general terms, it is better and more efficient for the state to fight the illicit conducts, instead of the technology used to commit them. Furthermore, because technology is a neutral mean to an end, legislation in democratic states should be "tech-friendly" in order to facilitate the fight of an illicit conduct. A "tech-friendly" strategy would give more technological tools to the state to fight illicit conducts, while maintaining the quality of the user experience.

A last example: In Mexico, Nazario Norberto, congressman of the Revolutionary Democratic Party (PRD) drafted a bill in Feb 2010 to limit and regulate the use of twitter in Mexico because, in his opinion, people are using this social tool to commit drug related crimes and avoid sobriety checkpoints (SIC) (Miller, 2010). His approach is the anti-technology approach. The legislator is trying to get a law approved that would limit the way twitter users interact with this social media. One could even ask if there is any illicit conduct being committed at all, by writing in twitter the emplacement of a sobriety checkpoint: Is it not protected speech? And what is the difference of that conduct and warning a friend by telephone or the DJ of a nightclub informing his clientele via the loudspeakers?

Instead, a tech-friendly policy would understand the technology and adapt it to the public interest. If the checkpoint placement strategy is redesigned, it could use the technology instead of confronting it: Checkpoints placed in high-speed roads and then broadcasted over twitter would make those roads safer, getting rid of the drunk drivers. This would also give the police department of the city the ability to canalize high risk traffic to the streets reported as "free of roadblocks", concentrating latter more of its agents there. Ultimately, the police department could access twitter and learn where drunk drivers are expecting a roadblock and move it to "surprise" them, and even broadcast using the same hashtag (keyword to find information in twitter) to reach the public in a new way, and compel them in real time to not drink and dive, proposing them alternatives like free or subsidized taxis.

A tech-friendly policy regarding this issue wouldn't try to cripple the neutral technology. Instead, it would adapt the operation of the authority (in this case, the police department) to take advantage of the altered reality.

Non democratic states have understood the power of policies that don't oppose technology faster than democratic states, illicitly appropriating social media technology: "both Twitter and Facebook give Iran's secret services superb platforms for gathering open source intelligence about the future revolutionaries, revealing how they are connected to each other. These details are now being shared voluntarily, without any external pressure. Once regimes used torture to get this kind of data; now it's freely available on Facebook" (Morozov, 2009).

This sole fact underscores the need for democratic states to maintain and adopt tech friendly policies designed to respect human rights, civil liberties and maintain the quality of the user experience, not only to provide an example and to keep the moral high ground, but also to keep technology neutral, a condition for the knowledge economy (Drucker, 1992) that gives democratic nations a powerful competitive advantage over non democratic ones.

Technology has to be seen by the legislative branch of government as a transversal factor in the functionality of any organization. A factor that helps the organization in processing and presenting the information more efficiently, but more importantly, a factor that is dependent on the process that the technology is trying to optimize. If the process of committing a crime or illegal conduct is properly regulated, then the technology used to pursue it would not need to be punished. At the same time, if the organization understands properly a process that is trying to enforce, then the "positive" use of the technology would be a consequence. It should not be the case, in either direction, to focus first on the technology and then on the process.

As we argued at the beginning of the chapter, we have played devil's advocate. This was done in order to raise awareness of the need to create interdisciplinary teams, and prepare professionals that understand both technology and policy, to combat deviant acts. Hence, this would provide an efficient use of technology to facilitate humanity.

REFERENCES

Anderson, C. (2009). *Free: The Future of a Radical Price*. New York: Hyperion.

BBC Technology News. (2003, January 7). Teenager wins DVD court battle. http://news.bbc.co.uk/1/hi/technology/2635293.stm. London, UK.

Bevan, N. (1995). Measuring usabilty as quality of use. *Software Quality Journal*, *4*(2), 115–130. doi:10.1007/BF00402715

Branigan, T. (2009, July 2). China's Green Dam internet fitering system will go ahead, official says. *Guardian*, pp. http://www.guardian.co.uk/world/2009/jul/02/china-green-dam.

Buxton, W. (2007). *Sketching user experience: Getting the design right and the right design*. Morgan Kaufmann.

Calvillo Gámez, E. H., & Cairns, P. (2008). Pulling the strings. A theory of Puppetry for the Gaming Experience. In S. Günzel, M. Liebe, & D. Mersch (Ed.), *Conference Proceedigns of the Philosophy of Computer Games* (pp. 308-323). Potsdam, Germany: Potsdam University Press.

Calvillo Gámez, E. H., Cairns, P., & Cox, A. L. (2010). Assessing the Core Elements of the Gaming Experience. In Bernhaupt, R. (Ed.), *Evaluating Experience in Games*. London, UK: Springer. doi:10.1007/978-1-84882-963-3_4

Deazley, R. (2006). *Rethinking Copyright: History, Theory, Language*. Cheltenham: Edward Elgar Publishing Limited.

DeCSS Central. (n.d.). *DeCSS Central*. Retrieved from http://www.lemuria.org/DeCSS/

DeCSS Central. (n.d.). *DVD CCA Lawsuit(s)*. Retrieved from http://www.lemuria.org/DeCSS/cca.html

Dix, A. (2007). Designing for Appropriation. *Proceedings of the BCS HCI 2007 Conference, People and Computers XXI. 2*. London, UK: BCS-eWik.

Domínguez, J. I., & Fernández De Castro, R. (2009). *United States and Mexico: Between Partnership and Conflict*. Abingdon, Oxfordshire: Routledge.

Dourish, P. (2003). The appropriation of interactive technologies: Some lessons from placeless documents. *Compuer Supported Cooperative Work (CSCW)*, 465-490.

Drucker, P. (1992). *The age of discontinuity*. New York: Harper & Row.

DVD copy control Association. (n.d.). *Frequently Asked Questions (FAQ)*. Retrieved september 3, 2009, from www.dvdcca.org: http://www.dvdcca.org/faq.html

Economist Intelligence Unit. (2008). *Index of Democracy 2008.* London: The Economist.

Foster, F. H., & Shook, R. L. (1993). *Patents, copyrights & trademarks*. New York: John Wiley & Sons.

Foundation, G. N. U. (n.d.). *GNU Operating System*. Retrieved from http://www.gnu.org

Greenberg, A. (2007, August 22). *America's Hackable Backbone.* Retrieved August 29, 2009, from Forbes: http://www.forbes.com/2007/08/22/scada-hackers-infrastructure-tech-security-cx_ag_0822hack.html

Grossman, L. (2009, June 17). Iran Protests: Twitter, the medium of the movement. Washington, DC: *Time Magazine*.

Habermas, J. (1999). *Between Facts and Norms: Contributions to a Discourse Theory of Law and Democracy*. Cambridge, MA: MIT Press.

Hassenzahl, M., & Tractinsky, N. (2006). User experience -- a research agenda. *Behaviour & Information Technology*, *25*(2), 91–97. doi:10.1080/01449290500330331

Heidegger, M. (1927). *Sein und Zeit*. Tübingen: Max Niemeyer.

Independent Reviewer of Terrorism Legislation. (2007). *The Definition of Terrorism*. London: The Home Department.

Jackson, B. A. (2009). Organizational Decision-making by Terrorist Groups. In P. K. Davis, & K. Cragin, *Social Science for Counterterrorism: Putting the Pieces Together.* Santa Monica: RAND Corporation.

Kelly, C. C. (2007). *The Manhattan Project: The Birth of the Atomic Bomb in the Words of Its Creators, Eyewitnesses and Historians*. New York: Black Dog and Leventhal.

Koops, B., Lips, M., Prins, C., & Schellekens, M. (2006). *Starting Points for ICT Regulation: Deconstructing Prevalent Policy One-Liners (Information Technology and Law)* (*Vol. 9*). Asser Press.

Linux Outlaws. (2009). *Spore on Linux Howto*. Retrieved from http://linuxoutlaws.com/spore

Makinen, G. (2002). *The Economic Effects of 9/11: A Retrospective Assessment*. Washington, DC: Congressional Research Service.

Marsh, R. T. (1997). *Critical Foundations: Protecting America's Infrastructures*. Washington, DC: The White House.

McCarthy, J., & Wright, P. (2004). *Technology as Experience*. Cambridge, USA: MIT Press.

Miller, M. E. (2010, 02 02). *"Twitteros" are Mexico's Latest Outlaws.* Retrieved 02 02, 2010, from CBS News: http://www.cbsnews.com/stories/2010/ 02/02/world/main6166446.shtml

Ministry of Public Security of the State Council of the Chinese Government. (1997). *Computer Information Network and Internet Security, Protection and Management (translation of the US Embassy.* Beijing.

Morozov, E. (2009). Iran: downside to the "twitter revolution". *Disent* (fall), 10-15.

National Commission on Terrorist Attacks Upon the United States. (2004). *The 9/11 Comission report*. Washington, DC: US Congress.

2009*National Infrastructure Protection Plan: Partnering to enhance protection and resiliency.* Washington, DC: Department of Homeland Security.

OECD. (2003). *Emerging Risks in the 21st century*. Paris: OECD.

OpenNet Initiative. (2009). *China's Green Dam: The Implications of Government Control Encroaching on the Home PC.* OpenNet Initiative Bulletin.

Robinson, P. H., & Dubber, D. M. (1999, March 12). *An Introduction to the model Penal Code.* Retrieved September 1, 2009, from PennLaw. University of Pennsylvania Law School: http://www.law.upenn.edu/fac/ phrobins/intromodpencode.pdf

Rosoff, H., & von Winterfeldt, D. (2005). *A Risk and Economic Analysis of Dirty Bomb Attacks on the Ports of Los Angeles and Long Beach.* Los Angeles: Center for Risk and Economic Analysis of Terrorism Events. University of Southern California.

Taylor, M. (2009, August 13). China drops Green Dam web filtering system. *Guardian*, pp. http://www.guardian.co.uk/technology/2009/ aug/13/china-drops-web-censorship.

The history of the CD - The beginning. (2005). Retrieved september 23, 2009, from Philips Research: http://www.research.philips.com/technologies/projects/cd/index.html

The Statute of Anne. (1710, April 10). Retrieved juin 7, 2009, from The Avalon Project. Yale Law School: http://avalon.law.yale.edu/1 8th_century/anne_1710.asp

Thimbleby, H. (2008). Feature: Ignorance of interaction programming is killing people. *interactions, 15* (5), 52-57.

United Nations. (1948, December 10). *The Universal Declaration of Human Rights.* Retrieved September 22, 2009, from Un.org: http://www.un.org/en/documents/udhr/

United Nations. (1998, July 17). *Rome Statute of the International Criminal Court.* Retrieved september 12, 2009, from UN.org: http://www.un.org/children/conflict/key documents/english/romestatuteofthe7.html

Wein, L. M., & Liu, Y. (2005). Analyzing a bioterror attack on the food supply: The case of botulinum toxin in milk. *Proceedings of the National Academy of Sciences of the United States of America, 12*(28).

Whitney, L. (2009, August 18). *iTunes reps 1 in every 4 songs sold in U.S.* Retrieved September 18, 2009, from Cnet News: http://news.cnet.com/8301-13579_3 -10311907-37.html

WINE HQ. (n.d.). *Spores on Wine.* Retrieved from http://appdb.winehq.org/objectManager.php?sClass=application&iId=8185

Winograd, T., & Flores, F. (1986). *Understanding computers and cognition: a new foundation for design.* Ablex Publishing Corporation.

Wolchok, S., Yao, R., & Halderman, J. (2009). *Analysis of the Green Dam Censorware System.* http://www.cse.umich.edu/~jhalderm/pub/gd/.

Wright, W. (2008). *Spores.* Maxis Electronic Arts.

Zittrain, J., & Edelman, B. (2003). *Empirical Analysis of Internet Filtering in China. Harvard Law School, The Berkman Center for Internet and Society.* Cambridge, USA: Harvard Law School.

ADDITIONAL READING

Calvillo Gámez, E. H. (2009). *On the Core Elements of the Experience of Playing Video Games: Studying the Gaming Experience.* Germany: LAP Lambert Academic Publishing.

De Sola Pool, E. (1983). *Technologies of Freedom.* Cambridge, USA: Belknap Press.

Lewis, T. G. (2006). *Critical Infrastructure Protection in Homeland Security.* New Jersey: Wiley. doi:10.1002/0471789542

Neuman, W. R., McKnight, L. W., & Solomon, R. J. (1999). *The Gordian Knot: Political Gridlock on the Information Highway*. Cambridge, MA: MIT Press.

Nieto-Gómez, R. (2009). *La Homeland Security des États-Unis et ses répercussions géopolitiques sur la construction de la "Sécurité du Territoire National" au Mexique (PhD dissertation).* Paris: Institut Francais de Geopolitique --Université Paris8--.

Orwell, G. (1949). *1984*. London: Secker and Warburg.

Vinge, V. (2006). *Rainbows End*. New York: Tor Books.

KEY TERMS AND DEFINITIONS

Appropriation: When the user interacts with the technology in ways not defined by its designer.

Subvert Appropriation: When the user thwarts the intention of the designer to use the technology in a way not defined by the designer.

Illicit Appropriation: When the user breaks the law by appropriating the technology.

User Experience: The process and outcome of an individual interacting with an application on a particular environment.

Illicit Conducts: A kind of behavior that is deemed illegal by the legislation of a specific country.

Homeland Security: A concerted national effort to prevent terrorist attacks within the United States, reduce America's vulnerability to terrorism, and minimize the damage and recover from attacks that do occur.

Deviant Acts: Actions that violate the norms of a region, including but not limited to formal legislation.

ENDNOTES

1. Poem and translation obtained from http://daxdavis.com/nothing-is-real-or-a-lie/
2. McCarthy & Wright use the word "appropriation", however, this word is used in this article with a different meaning as the one intended by them. To avoid confusions, appropriation is changed for personalization without changing the intended meaning of the sense making process.

Chapter 13
Designing Usable Security Feedback for Web-Filtering Systems

Ricardo Mendoza-González
Aguascalientes Institute of Technology, Mexico

Jaime Muñoz Arteaga
Autonomous University of Aguascalientes, Mexico

Francisco Álvarez Rodríguez
Autonomous University of Aguascalientes, Mexico

ABSTRACT

Currently, many powerful applications designed to combat social deviations are available, like the web-filtering systems, which measure the content of a website before submitting it to the user, notifying whether the content of the website presents (or not) information related to pornography, violence, racism, among others, and prompting the user to not accessing the site, or even blocking access to the website. Nevertheless, frequently the feedback of these systems is not well-designed, which may confuse users and lead to mistakes, disappointments, and misunderstandings. In order to reduce this concern, a method is provided to developers with guidance in designing usable security notifications to be incorporated in web-filtering systems. The method is structured through a library of user interface design patterns which integrates essential concepts of security and usability. The authors show the effectiveness of the patterns by using an illustrative example as a proof-of-concept together with a preliminary study.

INTRODUCTION

In this chapter we present a non-exhaustive collection of design patterns intended to facilitate the way some security aspects (notifications) are conveyed to the end user through the UI (User Interface). The design solutions offered by the patterns proposed allow achieving an appropriate security feedback, which could be used by most of the web-filtering systems available.

In the same way, web-filtering systems can be used by users with different experience; therefore the notifications showed by these systems must meet the basic requirements of usability and

DOI: 10.4018/978-1-60960-094-5.ch013

universal design. The security feedback designed with the proposed patterns, could make possible the correct interpretation about security feedback (related to the content of a particular website) showed by web-filtering systems, regardless of the users experience and backgrounds (experts, advanced, and beginners).

We believe that the implementation of the proposed patterns allows incorporating easily the offered solutions into the design process, generating positive results throughout the development cycle. See Figure 1.

Additionally, Figure 1 refers to a theoretical contribution which consists in incorporating the end user into the design/improvement process of the security feedback, by means of measuring methods for obtaining the user's perception about the usability of a given UI, cf. (Mendoza, 2009 b; 2009 c; Muñoz, 2008).

We also consider the combination of visual and auditive notifications to enhance the comprehension of the security feedback designed by means of the proposed patterns. For which we create a basic model to exemplify the presentation of information security feedback to the user when a non-appropriate content is detected in a web-site. Our model is divided into three basic stages (see Figure 2): First, an additional notification form is triggered to notify user about content filtering issues, we suggest sonification because audio may be more disruptive than an image, but visual notifications can be possibly augmented with any other kind of feedback. Then, the visual feedback is effectively designed based on the design patterns that are explicitly based on *HCISEC* criteria. Finally, the feedback with options and suggestions is constructed and showed.

The principal objectives of this chapter are presented following:

- To reduce the risk of accidental pornography views by children while they use internet (doing homework, sending e-mails, etc.).
- To provide a starting point for developers to create adequate security information feedback used by web-filtering systems.
- To reduce the propagation of websites with content related to social deviations, by means of an appropriate security feedback for web-filtering systems.
- To promote the inclusion of additional feedback to reinforce the visual notifications to improve users' perception to security alerts.

Figure 1. Implementation of the proposed patterns through the development life cycle

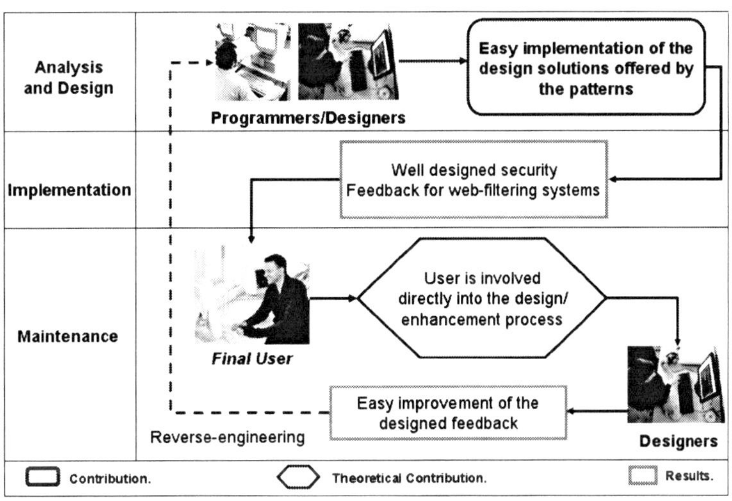

Figure 2. The suggested three steps for feedback showing

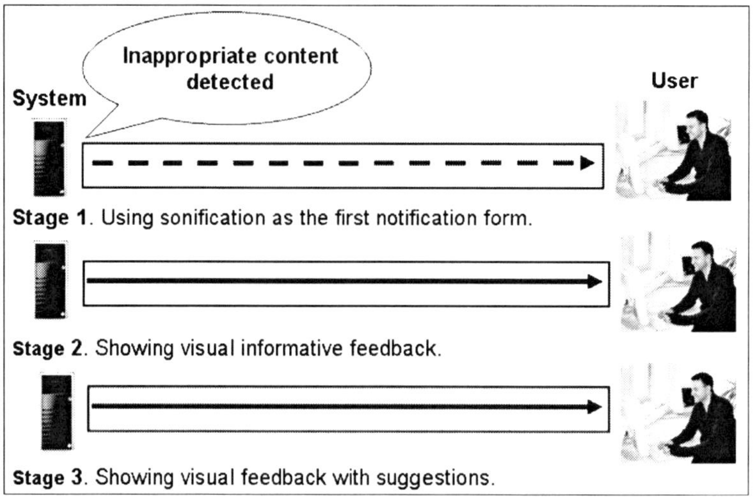

- To complement previous efforts to find equilibrium between usability and security for interactive web applications.

According to Rosemberg (2004), in a sociological order, many polls have shown that parents are deeply concerned about what their children may encounter on the Internet. Some of these concerns are directly related to the easy access by anybody to sexuality, bad values, and commercialism. The parents also fear the potential for invading a family's privacy and for adults taking advantage of children. Thus, if the Internet is unsafe for kids, as many parents clearly believe, and the growth of e-commerce requires making it a safe place, then filtering as part of an overall business strategy makes sense. Also filtering is used to avoid illegal political campaigning, illegal downloads, smear campaigns, anti-government propaganda, among others. Even Federal governments use web-filtering to combat child pornography.

However it is important to mention that some countries under authoritarian regimes use the web filtering technology to restrict the access to international information (e.g. *The Google News China Controversy* – China is practicing true censorship by excluding non-Government approved, non Popular Republic of China's government, news resources).

Putting aside the negative aspects, web-filtering is very important, and developers of these technologies must be particularly careful in designing their user interfaces as more and more users use them, being one of the main motivations for this research.

BACKGROUND

According to Guermazi (2007), web-filtering is used to prevent access to undesirable web pages. A number of technologies for web filtering have emerged recently to analyze the content of websites. The most of these technologies are complex systems based in techniques of Artificial Intelligence (e.g. Data-mining Algorithms, Neural Networks, among others), which are oriented to detect content related with human deviations like violence, hate propagation, drugs, pornography, among others in a particular websites, and to protect children and other unsuspecting users from exposure to such materials (Lee, 2003).

Many times these systems are not directly used by the users, but there is interaction between them by means of the security feedback showed through the UI of the websites. The term *User Feedback* is often referred to as any form of communication directed from a system to the user. Bearing in mind previous definition, it is possible to infer that the term *Information Security Feedback* is any information related to the system's security conveyed to the end user through the UI (Mendoza, 2009 b).

It is very important that this feedback is presented through a well-designed user interface. Recent works, cf. (Chiasson, 2007 a; 2007 b; 2006; Dhamija, 2007; Cranor, 2003) conclude that a well-designed user interface could contribute to an increase in the correct interpretation of security notifications, avoiding misunderstandings. In the same way, (Braz, 2007; Cranor & Garfinkel, 2005) establish that many times excessive cognitive burden during the use of an on-line system derives from a bad-designed user interface. In general terms, these and other problems are directly related with the user interface design, and appear when the usability aspects are ignored during the development process of a system. The standard ISO 9241-11 defines usability as an "Extent to which a product can be used by specified users to achieve specified goals with effectiveness, efficiency and satisfaction in a specified context of use" (ISO, 1998, pp. 1-2). Another important aspect related to usability, is undoubtedly the *Universal Design*, which also known as *Design for All*, focuses on UI design without barriers to some disabled users. Therefore it is necessary that the UI support the use of alternative interaction devices appropriate to the physical capabilities of each user, including a variety of feedback forms (Center for Universal Design, 2008).

Nevertheless, little is known about the best manner to integrate human sensorial channels to perceive some security alerts. The most popular feedback is the use of visual alerts. Other novel –yet relatively unexplored– techniques exist, such as olfactory, gustative, tactile, and auditive.

After visual, auditive feedback are probably the most advanced in terms of research work. In addition, the use of sonification has the following advantages (Garcia-Ruiz, 2007):

- Sonification, in theory, should permit to assign a specific sound to a specific attack.
- A particular sound may be identified in a set of auditive alarms.
- Sonification combined with visual notifications permits an efficient sensorial correlation.

Many researchers believe that by combining different forms of feedback it is possible to increase the likelihood of success of the security notifications. UI that combine two or more feedback forms are called *Multimodal Interfaces* and the Visual-Auditive multimodal form is currently one of the most effective and popular due to the good correlation between both (auditory and visual) human sensory channels.

As mentioned earlier, regardless the feedback form used by any system its UI needs to be well designed. Currently, many resources are available to facilitate the design of a UI, like the Security Human Computer Interaction *HCISEC* (Johnston, 2003), which is focused in the design of user interfaces for security applications. To achieve a successful application of the *HCISEC* it is necessary to consider the design criteria proposed by Johnston (2003). These criteria facilitate developing usable interfaces used in a security environment:

- *Visibility of system status*: The UI must inform the user about the internal state of the system (e.g., using messages to indicate that a security feature is active, etc.). The warning or error messages must be detailed, but specific, including a suggested corrective action for some security problem, and links to obtain additional information or external assistance.

- *Aesthetic and minimalist design*: Only relevant security information should be displayed. The user must not be saturated with information and options, and the UI must avoid the use of technical terms as much as possible. The security UI must be simple and easy to use, maintaining a minimalist design.
- *Satisfaction*: The security activities must be easy to realize and understand. Without the use of technical terms in the information showed to the user, in some cases, it is convenient to use humor situations or figures to present important security concepts to the user in an entertaining manner.
- *Convey features*: The UI needs to convey the available security features to the user clearly and appropriately; a good way to do it is by using figures or pictures.
- *Learnability*: The UI needs to be as non-threatening and easy to learn as possible; it may be accomplished using real-world metaphors, or pictures of keys and padlocks. The meaning of these metaphors may be incorporated to the security interface indicating users how to easily use the specific security features.
- *Trust*: It is essential for the user to trust the system. This is particularly important in a security environment. The successful application of the previous criteria should typically result in a trusted environment. The traditional concept of trust can be adapted for the *HCISEC* to "The belief, or willingness to believe, of a user in the security of a computer system" (Johnston, 2003, p. 680). The degree of trust that users have in a system will determine how they use it. For example, a user that does not trust a web site will not supply their credit card details.

The objective of *HCISEC* is to facilitate the design of a user interface. However most of the current guidelines are not specific enough, which limits its application in real scenarios. Related with this topic, Braz (2007) emphasizes the need of specific design guidelines for the development of usable secure systems. The guides must include techniques and strategies to facilitate the use and application of the knowledge during the development cycle.

An interesting and effective alternative to reduce the need previously mentioned corresponds to the application of design patterns. Several authors like (Buschmann, 1999; Kienzle & Elder, 2002; Sanz, 2005) share the idea that a pattern represents a proven solution for a recurrent problem within a certain environment. Within the scope of security's system design, the use of security patterns could be also considered. In a security context patterns contains experience like solutions to security problems. Similar to design patters, security patterns propose solutions that impacts in systems' implementation (Sanz, 2005). According to Kienzle and Elder (2002), some of the security patterns will take the form of design patterns, but not all security patterns are design patterns. "Because of the popularity of design patterns in the software engineering community, the natural inclination is to assume that anything going by the name security patterns should be described using a UML diagram and include sample source code.

MAIN FOCUS OF THE CHAPTER

Problems, Issues, and Controversies

We firmly believe that poorly designed feedback used by web-filtering systems may cause many setbacks to internet users, like the *involuntary* visits to websites whose content is related to human social deviations (hate propagation, bullying, flagellation, animal cruelty, among others), which may be linked to the increasing popularity and spread of such sites.

Another problem with an inadequately designed feedback in web-filtering systems is that it often denies the proper access to professional users to the content and information provided by websites related to: Medicine, forensics, armed forces, among others. Many times the web-filtering systems do not associate the websites' content with the users' professions. The system must provide users with options into an adequate feedback for deciding whether or not accesses to a particular website, but maintaining a low level of permissiveness to try to keep children protected from inappropriate content.

Bearing in mind the above, we emphasize the lack of a specification (under a uniform format) oriented to easily conceive usable feedback for web-filtering systems. We attack this challenge proposing a specification based on the design patterns paradigm, and related to design basics of usability and security.

Solutions and Recommendations

Basically, our proposal consist of a collection of design patterns divided into three essential levels, which are oriented to represent the basic aspects to handle user interfaces for web-filtering systems (See Figure 3):

- *Informative Feedback*: This level includes the interactive patterns useful to present information about: available security features, the correct way to use these features, detection of threats, and internal status of web-filtering systems. We incorporate auditive feedback in this level to complement visual notifications.
- *Interaction Feedback*: This level brings together the interaction forms useful to establish the feedback's navigation and operation.
- *Interactive Feedback*: This level includes the interactive patterns to specify the security feedback needed to convey information to the end user when the elements of the interface are handled by means of the mouse or the keyboard.

This distinction of levels contributes to facilitate an adequate design of basic, but very usable and comprehensible feedback. In the same way, the use of patterns facilitates the incorporation of design solutions into the development life cycle, cf. (Mendoza, 2009 c; Muñoz, 2009; Muñoz, 2008).

In order to present a general view of our classification, we define some of the interactive patterns presented in Figure 3. According to the challenge presented in the preceding section we consider several possible recurrent design prob-

Figure 3. Collection of patterns

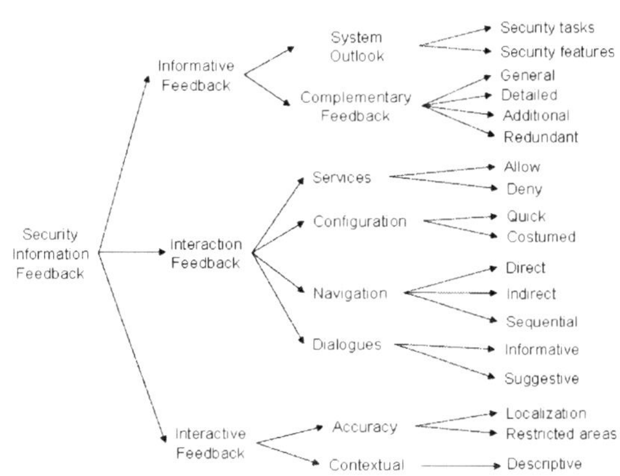

Figure 4. A fragment of the specification of design patterns

Informative Feedback

System Outlook

Pattern's Name: Security tasks

- **Problem**: How to inform users about the security tasks performed by the web-filtering system?
- **Context**: The interface must inform users about the security activities realized by the system. These notifications must be showed avoiding complicate sentences, and the use of technical terms.
- **Motivation**: Users' confidence can be increased by being properly notified about the security of the system.
- **Solution**: Users can be notified using simple text messages without technical terms. These messages should be reinforced with images, pictures or others visual notifications.

Complementary Feedback

Pattern's Name: Additional

- **Problem**: How to minimize the chances of users' misinterpretations about security notifications showed by the web-filtering?
- **Context**: The interface must notify users about the internal state of the system in a quick and efficient manner.
- **Motivation**: If users are properly notified in time on any security issue, the number of errors, confusions, misunderstandings, etc., could be reduced
- **Solution**. A good alternative to quickly alert users is to complement the visual notifications with some additional feedback form.

Note: This research promotes the use of sonification as a reinforcement of visual feedback.

Interaction Feedback

Services

Pattern's Name: Allow

- **Problem**: How to give users more control over the web-filtering?
- **Context**: Sometimes, professional users need access to websites whose content can be detected by the system as inappropriate.
- **Motivation**: If the web-filtering system provides options that allow users to decide whether or not accesses to a website, user could have a feeling of more control over the system.
- **Solution**: Including as part of the feedback dialogues with options to allow users to make the decision whether or not access to a website. It is necessary to strengthen these dialogues with specific questions to identify if the user is a minor, avoiding as much as possible to disclose inappropriate content to children.

Configuration

Pattern's Name: Costumed

- **Problem**: How to provide configuration options to professional users?
- **Context**: Sometimes the web-filtering system could be used by experienced and professional users which will need to modify specific security aspects.
- **Motivation**: If the web-filtering systems allow users to configure certain basic aspects in order to improve their interaction, then, the system acceptance by users could be increased.
- **Solution**: The web-filtering system must to show just the needed feedback to allow the configuration of basic security features, even for expert users. Critical security aspects must to be established only by system's managers.

Navigation

Pattern's Name: Sequential

- **Problem** How to facilitate the access to security features showed in previous screens?
- **Context**: The system must allow navigation between screens in an agile and easy manner, considering a logical and continuous sequence.
- **Motivation**: An easy and quick access to the all the elements and features of the system represents a very favorable experience during the user-system interaction.
- **Solution**: Including a simple navigator as part of the feedback showed through the UI.

Dialogues

Pattern's Name: Suggestive

- **Problem**: How to facilitate the interpretation of a security alert?
- **Context**: The web-filtering system must provide users with suggested actions to follow to mitigate mistakes, disappointments, and misunderstandings, when an inappropriate content is detected in some website.
- **Motivation**: A suggested action represents a contribution which will assist in decision making.
- **Solution**: By means of specific messages, avoiding the use of technical terms or irrelevant information, users could be notified about the internal state of the web-filtering system. The messages must include suggested actions to follow in case of eventuality.

Interaction Feedback

Accuracy

Pattern's Name: Restricted areas

- **Problem**: How to notify users about the restricted areas in the web-filtering system?
- **Context**: The system must inform users the limits established for specific elements in the UI. e.g. Disabled options, unavailable configuration options, among others. This notification must to be made in a very kind manner.
- **Motivation**: An adequate feedback assists in the introduction of restrictive messages.
- **Solution**: Presenting changes in the shape of the mouse's cursor is a good alternative to notify about the restricted elements of the interface. Additionally, messages with a touch of humor could be used (in moderation).

Contextual

Pattern's Name: Descripted

- **Problem**: How to provide basic information about specific feedback elements?
- **Context**: The web-filtering system must inform users about the functions of each elements of the feedback showed.
- **Motivation**: These notifications could reduce errors related with security, and contribute to establish a trusted environment.
- **Solution**: Using concise messages visible only when the user points the mouse over a specific feedback element.

lems, which are based on the *HCISEC* criteria. Then a suggested specific solution is offered by the patterns, complemented by the problem context and motivation for the application of each pattern. A fragment of our patterns' specification is presented in Figure 4.

In dealing with the problems and lacks presented in the preceding section, we believe that the application of our proposal during the creation of web-filtering systems could to provide several benefits to users and developers, such as those presented in Figure 5.

Figure 5. Some benefits obtained by applying the proposed patterns

Users	Developers
• The usable feedback designed through our patterns could to reduce the users' misunderstandings, disappointments, mistakes, among others setbacks, when some issue is detected by the web-filtering system.	• The use of the proposed patterns permits a better use of the knowledge between the members of the development team.
• The feedback created using the proposed patterns is oriented to properly notify users, reducing the possibility of *involuntary* access to websites with inappropriate content.	• The collection of patterns assists in the design of feedback oriented to promote a trustworthy environment between the system and user.
• The feedback designed through the patterns contributes to keeping children protected from inappropriate content showed in websites. By means of usable and adequate notifications, including authentication feedback.	• The design solutions offered by the patterns can be incorporated easily during the different stages of the development life cycle.
• The patterns suggest to incorporate authentication feedback (Questions related to the user's profession, age, educational level, etc.), which could to contribute in associating the content of websites with the profession of users.	• The collection of patterns proposed is very flexible, allowing adding or removing patterns, changing them, or even modifying the structure of the classification, considering the specific requirements of a particular web-filtering system.
• The patterns suggest the use of additional feedback forms to strengthen visual notifications, particularly with sonification. This combination of feedback allows a very good sensory correlation, and better understanding of security alerts, even for inexperienced users.	• Reduce the lack of useful specifications to design *usable-secure* applications (e.g. web-filtering systems).

CASE STUDY

We use a case study as a proof-of-concept, which is conducted according to the testing methodologies presented in (Chiasson, 2007 a; 2007 b; 2006). In this study we evaluate the usability of the security feedback showed by twelve web-filtering systems, and then we suggest a number of general design enhancements using the proposed patterns and presenting it through a prototype.

Apparatus (Web-Filtering Systems)

We select ten web-filtering systems from TopTenReviews (http://www.toptenreviews.com/), a well-known web systems ranking site. For our case study we took the top ten, listed under "2009 Internet Filter Software Review Product Comparisons." The remaining two are web-filtering tools (*K9 Web Protection* and *VIMA Web Filter*), which broadly speaking, are aimed at filter content related to pornography in websites, and are not listed by TopTenReviews.

We following briefly describe the selected web-filtering tools bearing in mind some technical aspects and reviews revealed by TopTenReviews site (http://www.toptenreviews.com/).

Note: We use the free trial versions of the following web-filtering tools to perform this study.

- *Net Nanny Parental Controls 6* (http://www.netnanny.com/): This web-filtering tool combines cutting-edge technology with non-technical settings and reports for informed, proactive parents who may or may not be computer or internet savvy. According to Internet-Filter-Review According to TopTenReviews/Internet-Filter-Review (http://internet-filter-review.toptenreviews.com/), this web filtering system stay at the forefront of internet filtering, offering some unique and compelling features not found in other programs.
- *Safe Eyes* (http://www.internetsafety.com): Safe Eyes is a very popular web-filtering tool because offering coverage on multiple computers. In addition, provides the flex-

ibility to treat different user groups differently in terms of internet privileges. This approach makes it much easier to manage filtering needs for multiple people in a family or business

- *CYBERsitter 10* (http://www.cybersitter.com/): CYBERsitter allows the customization of filtering with almost any program that connects to the internet, including instant messaging programs and email accounts. A particular feature of this tool is that it is one of the few available products to include a basic email–filtering option. Other programs don't offer nearly as many options for email, completely blocking email instead of filtering it.
- *CyberPatrol* (http://www.cyberpatrol.com/): This web-filtering system is relatively easy to use, quite customizable, and effective. CyberPatrol offers a host of good features and filtering settings, but lacks a few advanced features found in some of the higher-performing internet filter solutions. Additionally, this web-filtering tool offers a quality internet filtering program with some compelling features.
- *MaxProtect* (http://www.maxsoftware.com/): This tool is very comprehensive and useful. The system offers: Internet filtering, desktop filtering, and gives the administrator control over the applications and time spent on the computer. Nevertheless, this product is it only filters through keyword and URL blacklists, and on occasion fails during the overblocking (websites that should not have been blocked) and underblocking (websites that should have been blocked, but were not).
- *McAfee Parental Controls* (http://www.mcafeestore.com/): This tool it is very effective and reliable, nevertheless it is complex to use and interpret. In addition, its setup process is complex.
- *FilterPak* (http://www.familyisafety.com/): This tool allows blocking URL's, IP addresses, and keyword searches. The system allows override and reactivates the "parental control". This tool works with any ISP Including AOL or local providers, and it is compatible with high speed access and won't slow down the connection. It's customizable supporting multiple family profiles.
- *imView* (http://www.imview.com/): This web-filtering tool is a monitoring type program intended to be used by parents to protect their children from getting into trouble online. It can record most online activities such as websites visited, keywords searched, and it can record keystrokes. It can also block specific websites and it can be set to block websites that contain certain keywords in their URL. All reports, screenshots and activity can be accessed remotely from any computer and the program can send the parent/monitor keyword triggered email alerts.
- *Norton Parental Controls* (www.symantec.com/norton/internet-security): This product was created by the well known Symantec Company, nevertheless was not as easy to use as some of other products. Certain important functions are not where users would expect them to be. Another weakness of this tool is the underblocking (websites that should have been blocked, but were not), many websites that should have been blocked were allowed through. Most foreign objectionable websites were allowed right through.
- *Parental Controls* (http://www.webroot.com/): This web-filtering system was designed to ensure the safety of the children as they surf the internet. Parental Controls allows parents to monitor which sites are accessible to their children, and follow up later on to see which sites their children

visited. Users (Parents) can choose to receive this information in a daily, weekly or monthly report; or they can go at their own convenience. Parents also have the ability to set up profiles for each of the computers users to enact predefined levels of internet restriction.

- *K9 Web Protection* (http://www1.k9web-protection.com): The web filtering technology used by this tool divides Internet content into 60 distinct categories. These categories are stored in a master database called "Blue Coat", through which webmasters maintain and update more than 15 million ratings of websites and domains. A Web site will fall into one or more of these categories, based on the content of the site. This system allows the software configuration to block or allow specific categories.
- *VIMA Web Filter* (http://www.vimatech.com/): This tool analyze images on more than 150 parameters, the VIMA WebFilter objectionable image and website filtering software can carry out effective real-time and/or off-line filtering of pornographic content and websites. Because of the extent of the analysis it performs, VIMA WebFilter is able to differentiate between a nude picture on a pornographic site and a nude picture in a work of art, among other things. This particular feature is one that is inadequate or completely lacking in other filters. This web-filtering system also minimizes the number of objectionable images that get through to the user, while also maximizing the amount of acceptable material that does get through. In addition, VIMA WebFilter offers tremendous flexibility and can be customized and personalized to meet different filtering criteria.

Usability Evaluation

We use a fragment of the measurement model presented in (Mendoza, 2009 a), which consist of a collection of non-exhaustive metrics assigning a question-metric to each of the *HCISEC* criteria (requirements) outlined in section Background. The set of metrics is presented in Figure 6. We use the same metric "Positive response= 10. Negative response= 0" for this Question-Requirement set. The numeric value assigned to the responses is just a reference point. A particular *importance* level must be assigned by developers or designers, depending on the objectives of the web-filtering system.

Then, we apply the simple formula shown in Figure 7 to represent the satisfiability of requirements suggested by this method.

Figure 6. Metrics for usability evaluation

Metrics
Requirement 1: Visibility of system status **Question**: UI includes elements to inform user about the internal state of the system?
Requirement 2: Aesthetic and minimalist design **Question**: The notifications are presented to the user avoiding irrelevant and complex information?
Requirement 3: Satisfaction **Question**: Security and information options are presented in a user friendly manner avoiding the use of technical terms?
Requirement 4: Convey features **Question**: UI includes graphic elements to complement the text messages, in order to clarify the information in a better way?
Requirement 5: Learnability **Question**: UI includes elements that make the website easier to use?
Requirement 6: Trust **Question**: UI includes elements that help increase the trust of the user in the system?

Figure 7. (1) β_i denote the value of the ith requirement (see Figure 6) for usability. The importance value for the ith usability requirement is denoted by φ_i (the developers must provide the importance value).

$$y = \sum_{\forall i,\, i \in Metrics} \beta_i \phi_i$$

According to the measurement model, we considered the following values for the metrics (see Figure 4): Positive response=10. Negative response=0. The level of *importance* could vary between 1 and 10, however, for our purposes we gave an *importance* value of 10 to all requirements (e.g. φ_i = 10, for all i). As per the requirement score, for the considered web-filtering systems we gave the following values (see Figure 8).

Data Analysis

Figure 6 depicts the scores given by the metrics using Equation (1). The web-filtering systems depicted in Figure 8 are ordered as follows: Web-filtering systems 1 to 10 are the top web content filter tools from TopTenReviews, the next two correspond to additional web-filtering systems which are not listed in the TopTen ranking.

Observe that the metrics' feedback provide consistent scores which are comparatively in line with the TopTenReviews rankings. We do our analysis under the assumption that technical review from TopTenReviews, and the popularity of the web-filtering systems could reflect accurately the levels of usability.

In addition, the scores given by the metrics allow us to observe that most of the evaluated tools have a relatively well-designed feedback, but present certain shortcomings in important areas

Figure 8. Metrics' feedback given by Equation (1)

Metrics / **Web-Filtering Tools**	Visibility of system status	Aesthetic design	Satisfaction	Convey features	Learnability	Trust
1. *Net Nanny 6*	10	10	0	10	10	10
2. *Safe Eyes*	10	0	10	10	10	10
3. *CYBERsitter 10*	10	10	10	0	10	0
4. *CyberPatrol*	10	0	10	10	10	0
5. *MaxProtect*	0	10	0	0	10	0
6. *McAfee Parental Controls*	10	0	0	0	10	0
7. *FilterPak*	0	10	0	0	10	0
8. *imView*	0	0	0	0	10	10
9. *Norton Parental Controls*	10	0	0	10	0	0
10. *Parental Controls*	0	0	0	10	10	0
11. *K9 Web Protection*	10	0	0	0	10	10
12. *VIMA Web Filter*	0	0	0	0	10	0

to be covered according to the design criteria of HCISEC.

We particularly found that the most of the analyzed web-filtering tools do not includes logotypes or other UI elements that continuously promote or reinforce a trustworthy environment between the user and the application.

On the other hand we also note that some web-filtering systems have excellent designed interfaces and feedback in their configuration and control center. Nevertheless the security feedback is inexistent during the internet-surfing, and the tools merely deny the access to a site identified as a threat without any explanation to users.

Bearing in mind the above, it is possible to suggest a number of design enhancements to the security feedback used by the web-filtering systems analyzed. We took the security feedback of the web-filtering tool with the lowest score (*VIMA Web Filter*) as a basis for implementing the solutions offered by the patterns and thus provide a general example of design suggestions. These suggestions could improve even the web-filtering systems with higher scores.

Suggested Design Improvements

As mentioned above, the improvement suggestions for the security feedback presented by *VIMA Web Filter* were incorporated into a prototype entirely generated from the solutions and specification offered by the proposed design patterns (see Figure 4).

We use the Java Studio Creator 2 Update 1 Free version (http://developers.sun.com/jscreator/downloads/) to generate the prototype used for this study.

To provide a general view of the suggested enhancements to the feedback of *VIMA Web Filter*, we present the following graphical examples (see Figures. 4 to 6) and the corresponding explanation.

- *Pattern's Name*: Suggestive, Additional, Security Tasks, Sequential, and Allow.
- *Problem*: How to facilitate the interpretation of a security alert and provide more control over the system?
- *Solution*: By means of a traffic lights, a sound alarm, and a specific message (without technical terms or irrelevant information) the users will be notified about the content of a particular website. The messages include a suggested action to reduce mistakes or misunderstood, as well as a link to obtain additional information.

This (Figure 9) shows the appearance of the security feedback when inappropriate content is detected by we-filtering system, in this case, we incorporate Sonification to encourage the visual feedback. A "Beagle-barking" sound is generated (We choose this specific sound because represents the logo of *VIMA Web-Filter*), red colour is used in the traffic lights (Application of the design patterns *Additional*, and *Security Tasks*). Immediately, the interface presents a dialogue-message with options "Continue", "No Access", and "More Info" (Application of the design patterns *Suggestive*). At the same time, the dialogue shows the Beagle-Image logo at the top right corner of the page.

- *Pattern's Name*: Informative, Suggestive, Sequential, and Security Tasks.
- *Problem*: How to appropriately inform users about a denial of access to a particular web-site?
- *Solution*: By means of a no-complex concise message (avoiding technical terms or irrelevant information) the users will be notified about the denial of access to a particular website.

Figure 10 shows the suggested appearance of a notification for denial of access to a restricted website. In this case, we incorporate a traffic lights again, the logotypes of *VIMA Web-Filter* and *Image Beagle* to reinforce the trust in the web-filtering system. Then red colour is used in

Figure 9. Appearance of the suggested security feedback. All trademarks and logotypes referenced in this figure are property of their respective owners.

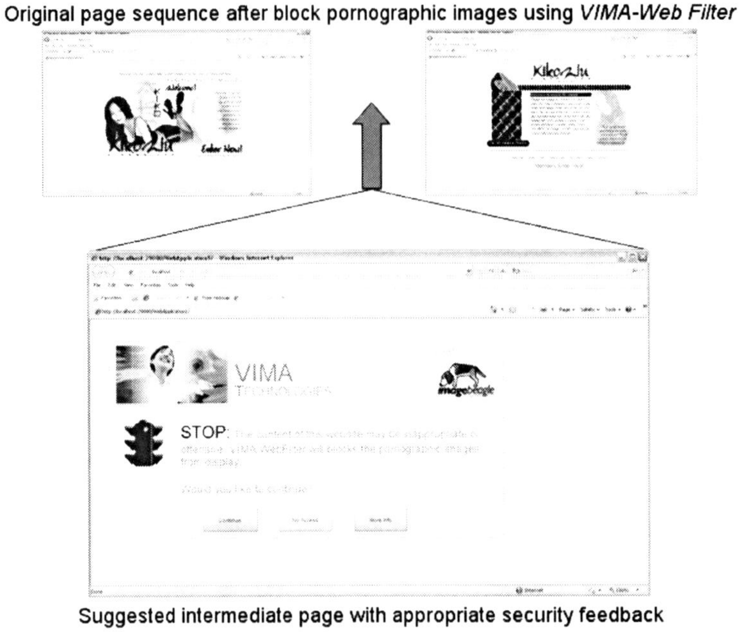

Figure 10. Suggested feedback to inform users about the denial of access. All trademarks and logotypes referenced in this figure are property of their respective owners.

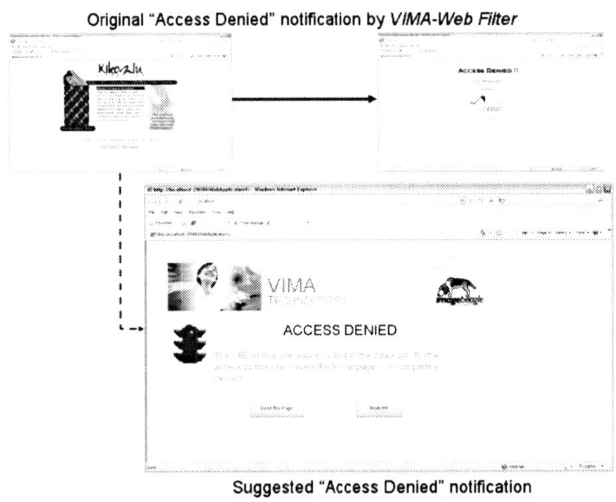

the traffic lights (Application of the design patterns *Additional*, and *Security Tasks*). UI displays a message informing the cause of the denial of access. In addition, this message is complemented by a link to obtain additional information. We kept the same minimalist design for all the security feedback trying to promote a trustworthy environment.

- *Pattern's Name*: Redundant, Security Features, and Security Tasks.

Figure 11. Continuous notifications about the system's operation and internal state. All trademarks and logotypes referenced in this figure are property of their respective owners.

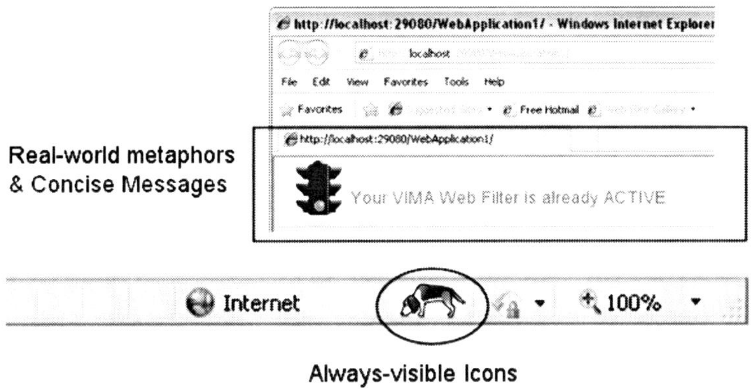

- *Problem*: How to appropriately maintain informed users about the web-filtering operation?
- *Solution*: Using always-visible notifications (in restraint) like logos, images, short messages, colours, real-world metaphors, among others, the users will be correctly informed about the system tasks and internal state (Enable or Disable).

We show in Figure 11 two examples of feedback to maintain notified users about the internal state of the system. The first example shows the use of a simple text message complemented by traffic lights (real-world metaphor) (Application of the design patterns *Redundant*, and *Security Features*). In the second example we suggest the incorporation of the company icon-logo at the bottom of the browser, this type notifications is widely used by web applications. These simple but important aspects could contribute to increase the user trust in the web-filtering tool.

FUTURE RESEARCH DIRECTIONS

The technologies for facilitating humanity represent a very wide field of research, which is of utmost importance to explore. We believe that is necessary that more Researchers, laboratories, Universities, and even Governments engage in this challenge. There are plenty of issues to address as future work related with the book's theme, but we emphasize those particularly related to the subject of this chapter:

- To generate most complete specifications to design suitable and usable feedback exclusively used by web-filtering systems.
- To present the specify patterns in a formal fashion, e.g. using languages like: Pattern Language Markup Language (http://www.hcipatterns.org/PLML1), or Balanced Pattern Specification Language (BPSL) presented in (Taibi, 2006).
- To create most sophisticated prototypes using the patterns specifications.
- To perform usability studies using standard questionnaires to obtain appropriate user feedback about the prototypes' usability.
- Continuously improving the specifications and the prototypes using the user's feedback.
- To incorporate in the prototypes additional feedback forms to strengthen visual notifications (e.g. haptic feedback, olfactory feedback, tactil feedback, among others).

- To promote this topic in conferences and journals of Human-Computer Interaction and Usable Security.

CONCLUSION

In this chapter we present an alternative to design basic but very usable security feedback particularly used by web-filtering systems. This proposal consists of a non-exhaustive collection of patterns based on the *HCISEC* design criteria (Johnston, 2003), which are oriented to facilitate the incorporation basic usability concepts during the user interface design process.

According to the objectives of this research we conclude that a well designed feedback may contribute to reduce the risk of accidental pornography views and thus also reduce the propagation of websites with content related to social deviations.

In particular, the feedback designed with our patterns includes auditive notifications that could quickly notify users, even faster than images, about the content of a particular website. We believe that sonification may allow Parents to have more control over the children's internet surfing, e.g. even if the parents are in an adjoining room to the children's computer-room; they could hear the audio alert generated by the web-filtering system.

Similarly, the solutions offered by the patterns promote a well designed security feedback for web-filtering systems, which represents a further effort to bridge the gap between usability and security.

In general terms, it is important to mention that the study performed reflect a good incorporation of essential usability and security aspects to achieve a trustworthy environment through the conveyed information by web-filtering systems.

Furthermore, the patterns facilitate the design of security feedback easy to understand and interpret by users with different levels of experience and backgrounds (experts, advanced, and beginners).

We believe that design patterns represent a very good alternative to convey theoretical usability and ergonomic concepts to physic secure applications. In addition, this process allows not only impact final user with design results, but also user integration in the design process by means of UI elements to collect users' feedback about usability and effectiveness of security notifications. Conversion of this data to UI enhancements becomes easier by design patterns consideration.

REFERENCES

Basili, V. R., Caldiera, G., & Dieter Rombach, H. (2004).The goal question metrics approach. Retrieved April 15, 2007, from http://wwwagse.informatik.uni-kl.de/pubs/ repository/basili94b/encyclo.gqm.pdf

Bellovin, S. M. (2006). On the brittleness of software and the infeasibility of security metrics. *IEEE Security & Privacy,* 1540-7993, p. 96.

Braz, C., Seffah, A., & M'Raihi, D. (2007). Designing a trade-off between usability and security: A metrics based-Model. In Baranauskas, M., Palanque, P., Abascal, J., Junqueira Barbosa, S.(Eds.) *11th IFIP TC 13 Conference on Human-Computer Interaction, INTERACT 2007* (pp. 114–126). Rio de Janeiro. Lecture Notes in Computer Science, Springer.

(1996). InBuschmann, F., Maunier, R., Rohnert, H., Sommerlad, P., & Stal, M. (Eds.). Pattern-oriented software architecture: *Vol. 1. A system of patterns*. Hoboken, NJ: Wiley.

Chiasson, S., Biddle, R., & van Oorschot, P. C. (2007). A second look at the usability of click-based graphical passwords. *In Symposium on Usable Privacy and Security* (1-12). Pittsburgh, Pennsylvania, ACM Press.

Chiasson, S., van Oorschot, P. C., & Biddle, R. (2006). A usability study and critique of two password managers. In *15th USENIX Security Symposium* (pp. 1-16), Vancouver, B.C., Canada, IEEE, Press.

Chiasson, S., van Oorschot, P. C., & Biddle, R. (2007). Graphical password authentication using cued click points. In Biskup, J., & López, J. (Eds.). *12th European Symposium On Research In Computer Security* (pp. 359-374). Dresden, Germany. Lecture Notes in Computer Science. New York: Springer.

Cranor, L. F. (2003). Designing a privacy preference specification interface: A case study. In *Workshop on Human-Computer Interaction and Security Systems, ACM CHI 2003*. Fort Lauderdale, ACM Press.

Cranor, L. F., & Garfinkel, S. (Eds.). (2005). *Security and usability: Designing secure systems that people can use*. Sebastopol, CA: O'Reilly Media.

Dhamija, R. (2007). Security usability studies: Risk, roles and ethics. In *Workshop on Security User Studies, ACM CHI 2007*. San Jose, CA: ACM Press.

García-Ruiz, M., Edwards, A., Aquino-Santos, R., Vargas Martin, M., & Mendoza-González, R. (2007). Using sonification to teach network intrusion detection: A preliminary usability study. In Montgomerie, C. & Seale, J. (Eds.), *World Conference on Educational Multimedia, Hypermedia and Telecommunications, EDMEDIA*'07. (pp. 849-857). Chesapeake, VA: AACE.

Guermazi, R., Hammami, M., & Ben Hamadou, A. (2007). Combining classifiers for web violent content detection and filtering. In Shi, Y., Albada, G.D. van, Dongarra, J. & Sloot, P.M.A. (Eds.), *7th International Conference on Computational Science, ICCS 2007* (pp. 773-780). Beijing, China. Lecture Notes in computer Science.New York: Springer.

ISO International Organization for Standardization. (1998). *International ISO standard* 9241-11, Ergonomic requirements for office work with visual display terminals (VDTs) - Part 11: Guidance on usability. 1st edition, Switzerland.

Johnston, J., Eloff, J., & Labuschagne, L. (2003). Security and human computer interfaces. *Computers & Security*, *22*(8), 675–684. doi:10.1016/S0167-4048(03)00006-3

Kienzle, D. M., & Elder, M. E. (2002). Security patterns for web application development. *Final Technical Report*. Retrieved March 10, 2009, from http://www.scrypt.net/~celer/ securitypatterns/final%20report.pdf

Lee, P. Y., Hui, S. C., & Fong, A. C. M. (2003). A structural and content-based analysis for web filtering. *Internet Research: Electronic Networking Applications and Policy*, *13*(1), 27–37. doi:10.1108/10662240310458350

Mendoza González, R., Muñoz Arteaga, J., Vargas Martin, M., Álvarez Rodríguez, F. J., & González Calleros, J. (2009). a pattern methodology to specify usable security in websites. In *3rd International Workshop on Secure Systems Methodologies using Patterns, SPattern'09*. Linz, Austria. IEEE Press.

Mendoza González, R., Vargas Martin, M., Muñoz Arteaga, J., Álvarez Rodríguez, F. J., & Garcia-Ruiz, M. A. (2009). A measurement model for secure and usable e-commerce websites. In *22nd IEEE Canadian Conference on Electrical and Computer Engineering, CCECE/CCGEI 2009* (pp. 77-82). St. John's, Newfoundland and Labrador, Canada. IEEE Press.

Mendoza González, R., Vargas Martin, M., Muñoz Arteaga, J., Álvarez Rodríguez, F. J., & Ochoa Ortíz Zezzatti, C. A. (2009). Web service-security specification based on usability criteria and pattern approach. *Journal of Computers*, *4*(8), 705–712.

Muñoz, J., Mendoza, R., Vargas Martin, M., Vanderdonckt, J., & Álvarez, F. (in press). a methodology for designing information security feedback based on user interface patterns. *Advances in Engineering Software*.

Muñoz, J., Mendoza, R., Vargas Martin, M., Vanderdonckt, J., Álvarez, F., & González Calleros, J. (2008). A method to design information security feedback Using patterns and HCI-security criteria. In Lopez Jaquero, V., Montero Simarro, F., Molina Masso, J.P. & Vanderdonckt, J. (Eds.), *7th International Conference on Computer-Aided Design of User Interfaces, CADUI 2008* (pp. 283-294). Albacete, Spain. Lecture Notes in Computer Science, Springer.

Nielsen, J. (2001). Usability metrics. Retrieved September 2, 2007, from http://www.useit.com/alertbox/20010121.html

Sanz, D., Díaz, P., & Aedo, I. (2005). Security modeling in information web systems. In Díaz, P., Montero, S., & Aedo, I. (Eds.), *Web Engineering and Design Patterns*. Pearson Education. (In Spanish)

Taibi, T. (2006). Formalising design patterns composition. *IEE Proceedings. Software*, *153*(3). doi:10.1049/ip-sen:20050072

UD Center for Universal Design. (2008). The Universal Design Principles. *North Carolina State University*. Retrieved May 15, 2009 from http://www.design.ncsu.edu/cud/ about_ud/udprin ciples.htm

ADDITIONAL READING

Atoyan, H., Duquet, J., & Robert, J. (2006). Trust in new decision aid systems. In *18th International Conference of the Association Francophone d'Interaction Homme-Machine* IHM'06, Montreal, (pp. 115–122). New York: ACM Press.

Baumann, K., & Thomas, B. (2001). *User interface design of electronic appliances*. New York: CRC Press.

Beirekdar, A., Vanderdonckt, J., & Noirhomme-Fraiture, M. (2002). A framework and a language for usability automatic evaluation of web sites by static analysis of HTML source code. In Kolski, C. & Vanderdonckt, J. (Eds.), *4th International Conference on Computer-Aided Design of User Interfaces*, CADUI'2002, (pp. 337-348). Dordrecht: Kluwer Academics Publisher.

Berry, B., Hobby, L. D., McCrickard, S., North, C., & Pérez-Quiñones, M. A. (2006). Making a Case for HCI: Exploring Benefits of Visualization for Case Studies. In *World Conference on Educational Multimedia, Hypermedia & Telecommunications*, EDMEDIA'06. Virginia: AACE.

Bevan, N. (2005). Cost benefits framework and case studies. In Bias, R. G., & Mayhew, D. J. (Eds.), *Cost-Justifying Usability: An Update for the Internet Age*. San Francisco: Morgan Kauffman. doi:10.1016/B978-012095811-5/50020-1

Bias, R. G., & Mayhew, D. J. (1994). *Cost-justifying usability*. San Francisco: Morgan Kauffman.

Bouillon, L., Limbourg, Q., Vanderdonckt, J., & Michotte, B. (2005). Reverse engineering of web pages based on derivations and transformations. In *3rd Latin American Web Congress LA-Web'05*, (pp. 3- 13). Buenos Aires, Argentina. IEEE Computer Society Press.

Braz, C., & Jean-Marc, R. (2006). Security and usability: The case of the user authentication methods. In *18th International Conference of the Association Francophone d'Interaction Homme-Machine IHM'06*, (pp. 199-203) Montreal. New York: ACM Press.

Brown, A., Vargas Martin, M., Kapralos, B., Garcia-Ruiz, M., & Green, M. (2009). Towards music assisted intrusion detection, (Poster), *IEEE Symposium on Security and Privacy*, Oakland, CA.

Calero, C., Moraha, M. A., & Piattini, M. (2008). *Handbook of research on web information systems quality*. Hershey, PA: IGI Global.

Chemseddine, M., & Noirhomme-Fraiture, M. (2008). Complex and dynamic data representation by sonification. In Forbrig, P., Paternò, F., Pejtersen, A.M. (eds.). *IFIP International Federation for Information Processing*, Volume 272; *Human-Computer Interaction Symposium*. (pp. 195–200), Boston, Mass. Berlin, Germany: Springer.

Chong, Lee J. & McCrickard, S. (2007). Towards extreme(ly) usable software: Exploring tensions between usability and agile software development. In *Agile 2007 Conference*, (pp. 59 – 71). Washington D.C.: IEEE Press

Cranor, L. F. (2003). Designing a privacy preference specification interface: A case study. In *Workshop on Human-Computer Interaction and Security Systems*, Fort Lauderdale. New York: ACM Press.

Dix, A., Finlay, J., Abowd, G., & Beale, R. (1998). *Human-computer interaction*. London, UK: Prentice Hall.

Folmer, E., & Bosch, J. (2004). Architecting for usability; a survey. *Journal of Systems and Software*, *70*(1), 61–78. doi:10.1016/S0164-1212(02)00159-0

Gamma, E., Helm, R., Johnson, R., & Vlissides, J. M. (1994). *Design patterns: Elements of reusable object-oriented software*. New York: Addison-Wesley.

Jacko, J. A., & Sears, A. (2003). *The human computer interaction handbook: Fundamentals, evolving technologies, and emerging applications*. New Jersey: Lawrence Erlbaum.

Lund, A. M. (1997). Another approach to justifying the cost of usability. *Interactions (New York, N.Y.)*, *4*(3), 48–56. doi:10.1145/255392.255403

Martínez-Ruiz, F., Muñoz Arteaga, J., Vanderdonckt, J., González-Calleros, J., & Mendoza, R. (2006). A first draft of a model-driven method for designing graphical user interfaces of rich internet applications. In *4th Latin American Web Congress*, LA-Web'06, Universidad de las Américas UDLA, Cholula, Mexico.

Martínez-Ruiz, F., Muñoz Arteaga, J., Vanderdonckt, J., González-Calleros, J., & Mendoza, R. (2006). In *the IASTED International Conference Communication, Network, and Information Security*, MIT, Cambridge, Massachusetts.

Muñoz, J., Mendoza, R., Álvarez, F., & Vargas Martin, M. (2006). A classification of security patterns for the transactions between a requester, an intermediary, and a web-service.

Muñoz, J., Mendoza, R., Álvarez, F., Vargas Martin, M., & Ochoa, A. (2007). Integration of auditive and visual feedback in the design of interfaces for security applications. In *Workshop on Perspectives, Challenges and Opportunities for Human-Computer Interaction in Latin America*, CLIHC'07, Rio de Janeiro, Brazil.

Muñoz-Arteaga, J., Mendoza-Gonzalez, R., & Vanderdonckt, J. (2008). A classification of security feedback design patterns for interactive web services. In Heikkinen, S., Jorstad, I., & Tapus N. (Eds.) *3rd International Conference on Internet Monitoring and Protection* ICIMP'08, (pp. 166-171). Bucharest,CA: IEEE Computer Society Press.

Nielsen, J. (1997). *Measuring the usability of reading on the web*. Retrieved January 10, 2009 from www.useit.com/ alertbox/readingmetrics.html

Nielsen, J., & Landauer, T. K. (1993). A mathematical model of the finding of usability problems. In Arnold, B., van der Veer, G., & White, T. (Eds.) *INTERACT '93 and CHI '93 conference on Human factors in computing systems*, (pp. 206-213). Amsterdam, The Netherlands: ACM Press.

Preece, J., Rogers, Y., Sharp, H., Benyon, D., Holland, S., & Carey, T. (1994). *Human computer interaction*. Boston: Addison-Wesley.

Rosemberg, R. S. (2004). *The social impact of computers*. London: Elsevier Academic Press.

Sommerville, I. (2004). *Software engineering* (7th ed.). Boston: Pearson/Addison-Wesley.

Stephanidis, C. (2005). *Universal access in health telematics: A design code of practice, Thematic Network (Working Group) "Information Society for All"*. Berlin, Germany: Springer.

KEY TERMS AND DEFINITIONS

Feedback: This term is often referred to as to any form of communication directed from a system to the user.

Design Patterns: Broadly speaking design patterns represents a proven solution for a recurrent problem within a certain environment.

Usability: The standard ISO 9241-11 defines usability like "Extent to which a product can be used by specified users to achieve specified goals with effectiveness, efficiency and satisfaction in a specified context of use" (ISO, 1998, pp. 1-2).

HCISEC: According to (Jonhston, 2003) HCISEC is the part of a user interface which is responsible for establishing the common ground between a user and the security features of a system. HCISEC is human computer interaction applied in the area of computer security.

Web-Filtering: Technologies oriented to analyze the content of websites to prevent access to undesirable web pages.

Universal Design: Focuses on UI design without barriers to some disabled users.

Multimodal Interfaces: UI that combine two or more feedback forms, e.g. A user interfaces that shows visual notifications combined with auditive notifications (Sonification).

Sonification: Several authors define Sonification (a.k.a. Auralization) as the use of non-speech audio to convey information or perceptualize data. Brad Gover from Waterloo University (http://audiolab.uwaterloo.ca/~bradg/auralization.html) defines auralization as the process of rendering audible, by physical or mathematical modeling, the sound field of a source in a space, in such a way as to simulate the binaural listening experience at a given position in the modeled space.

Chapter 14
Human Factors Required for Building NLP Systems

Daout Daoud
Princess Sumaya University for Technology, Jordan

Samir Abou El-Seoud
Princess Sumaya University for Technology, Jordan

ABSTRACT

The chapter illustrates the importance of human factors required for building natural language processing (NLP) systems. The authors will examine different NL interface style and processing and correlate them with human factors such as: domain, interface, text style and medium of communication. They verified our assumption by presenting a NLP system which was built as a proof-of- concept. However, because of semantics and the very nature of language, the authors discuss our concern with possible abuse by unscrupulous persons who would attempt to exploit NLP systems for reasons other than legitimate information exchange.

INTRODUCTION

Techniques of automatic natural language processing (NLP) have been under development since the earliest computing machines, and in recent years these techniques have proven to be robust, reliable and efficient enough to lead to commercial products in many areas. The applications include machine translation, natural language interfaces and the stylistic analysis of texts but NLP techniques have also been applied to other computing tasks besides these.

DOI: 10.4018/978-1-60960-094-5.ch014

A natural language (NL) interface accepts user inputs in natural language allowing interaction with some system, typically a retrieval system, which then results in sufficient responses to input NL text or query statements. However, these systems can also be used subsequently to facilitate communication with the outside world. Hence, a natural language interface should be able to translate unrestrained natural language statements into appropriate actions for the system, and there should be safeguards should the system be used to communicate freely with others.

This type of unrestricted NL interface is an attractive choice because, if it could be built, it would offer many advantages. Firstly, it does not require any learning and training, because its structure and vocabulary are already recognizable by the user. Secondly, natural language allows users to encode composite meanings. Thirdly, this type of interface is text-based, making it suitable for all types of devices and medium. In contrast, form-based or graphical user interfaces need more sophisticated and specific resources.

Incorporating a NL interface requires translating ambiguous user inputs into clear intermediate representations. Two main problems are associated with building such systems: the first one is handling linguistic knowledge and the second one is handling domain knowledge.

A study of the current workplace shows that the deployed NL interface systems are rare and most of them are only prototypes. This problem is not related to the openness or restrictiveness of the domain. Although most task oriented activities are domain-specific, we do not yet find any restricted NL interface-based operational systems especially for poorly informatized languages such as Arabic. Not all languages have received equal investment in linguistic resources and tool development (Riloff, Schafer et al. 2002). As an example, most of the research published on IE discusses problems related to English, which is a resource-rich language. While some of the existing English-based systems performance is comparable to that of human experts, by comparison, Natural Language Processing (NLP) in the Arabic language is still in its initial stage (Hammo et al. 2002).

NL-based systems have the reputation of high development cost and low quality. Our goal in this chapter is to show that the most important issue in building NL based systems is the incorporation of human factors in the development, regardless of the target language in terms of resources richness, or type or complexity of the domain, or even cleanliness of the input text. If this approach is combined with treating a NLP project as an engineering problem, and not only as a traditional linguistic problem, it is almost guaranteed to produce a system with industrial quality and high extensibility, with the minimum resources possible.

The key success factors in building NL systems are understanding how people encode their thoughts, and finding the right representation to model the concerned domain knowledge. Furthermore, to help explore Socially Aware Language Understanding, it is important to identify a set of tasks and task performance goals, target groups of language users, and a set of situations, and then evaluate the effectiveness of meeting increasingly challenging goals under different approaches and styles of computationally mediated language usage.

In this chapter, we will first give background information about natural language interface styles and processing. Then, we present the human factors that involved in building a NLP system. Methods used in incorporating human factors are also discussed. Finally, we will discuss, as a case study, a system that can handle spontaneous Arabic SMS text and show human factors were key success elements.

NLP APPLIED TO SOCIAL AND PSYCHOLOGICAL ISSUES

Disorders such as Cerebral Palsy and Lou Gehrig's disease produce severe physical disabilities that make normal communication impossible. Therefore, developing syntactic prediction systems should contribute to the development of intervention devices that preserves the user's ability to use language freely while maintaining speed and ease. One can combine natural language processing (NLP) and popular syntax theories to devise such prediction system to model the syntax rules that specify how words can be combined. This allows the syntactic predictor to make rule-based, linguistic determinations about what words can follow those already processed. It can be used with flexible abbreviation systems to eliminate possible expansions for personalized abbreviations. The syntactic predictor could also be used with other devices to reduce the effort required of the user by predicting what word forms he or she is likely to type next. In modeling linguistic knowledge, this system provides a more natural solution to the communication problem than many systems currently in use (Van Dyke, 1991)

As computing capacity grows and resources become better defined, a major problem NLP faces is how to filter unwanted messages either by peers or by persons with normal communicative abilities. This is a problem because in some societies, persons with different abilities are routinely discriminated and face physical and verbal abuse. For this reason, NLP becomes even more important in that it needs to also protect users from cyberbullying, teasing, soliciting sex or other inappropriate acts.

Rehabilitation engineering endeavors to integrate technology into vocational, educational, and independent living settings in order to increase the independence of persons with physical or sensory disabilities. There are numerous sub-areas in this field: robotics research for artificial limbs, production of sensory aids like hearing aids, developing physical therapy techniques, and devising marketing strategies to promote this technology. Augmentative and Alternative Communication (AAC) derives its motivation from the sub-field of Rehabilitation Engineering. Word prediction for disabled users is an example of how natural language processing (NLP) techniques and modern syntax theories can be used to improve the communication devices currently available to disabled users.

Communication devices are technological interventions between the disabled user and the world he or she hopes to communicate in. Unfortunately, the world consists of many people who wish to take advantage of physically or cognitively impaired persons. That is why NLP is a sword that can cut both ways. The history of their development includes intervention strategies that were not able to take the needs and desires of the user into consideration because of technological limitations. Often they placed a cognitive or physical burden on the user because they were not intuitive or were complicated to operate. The primary consideration has changed from finding any way possible for these people to communicate, to finding the communication aid that is best suited to each individual, while still providing safeguards against abuse. The emphasis is now on developing intervention devices which feature easy training, flexible use, and allow a reasonable communication rate, although future work must deal with filtering undesirable persons who wish to contact communicatively impaired persons.

The typical user for the system developed by (Van Dyke, 1991) is cognitively intact and therefore has the mental capability and desire to use language the same way a non-disabled individual would. The user's disability affects his or her motor capability and muscular control in a way that produces limited dexterity. These users are typically non-speaking, and have difficulty typing, writing, or even controlling a joy-stick to select letters. In the worst case the user is limited to using a single-switch interface which makes communication very slow.

Two types of disorders that typically produce this condition are developmental, such as Cerebral Palsy, or degenerative, like Lou Gehrig's disease. Cerebral Palsy (CP) is diagnosed at infancy in children whose normal muscular control is deficient (Griffith, 1985). The child usually exhibits unusual body postures, purposeless body movements, and poor coordination and balance. Although some children with CP may suffer from mental retardations, many have a high intelligence despite their muscular disabilities. However, it is important to mention that although many of these children have average or even high intellectual capacity, emotionally speaking, communicating with persons who wish to harm them emotionally can actually result in more harm than good.

The clinical name for Lou Gehrig's disease is Amyotrophic Lateral Sclerosis (ALS). This disorder afflicts adults late in their lives, meaning that there is no previous language impairment or cognitive disability (Griffith, 1985) to hinder communication. Patients suffer from muscle twitching and weakness, beginning in the hands and spreading to the arms and legs, or from the stiffening of muscle groups usually in the extremities. They will often lose control of muscles that perform swallowing and communicative functions.

Some stroke victims could also benefit from this technology; however, often they will have more severe linguistic impairments which make using this system inappropriate. Whatever the ailment, if the user can be characterized as linguistically, or cognitively, intact but with deficient motor skills, the device I have developed has the potential to facilitate their communication.

By modeling syntactic knowledge in the computer, one can produce a system that can improve existing communication devices (Van Dyke, 1991). The improvement provided is a more natural one for the user because it comes from the information humans use anyway when they communicate. It is not an ad hoc solution to the communication problems these people face, it is a solution motivated by the nature of the problem: an inability to use language in a "natural" unconstrained way. If we can make the machine use language the way a human does, then rather than being hindered by the technology the user's disabilities force him to use, both machine and human can cooperate to enhance the disabled person's communication.

We could conclude that word prediction for disabled represents a successful application of natural language processing (NLP) to the problem of augmentative communication. A syntactic predictor can be used to speed communication rate because the system draws on the same rules for creating a sentence that the disabled user exercises as he or she forms sentences. Because of this, the computer is able to intelligently anticipate the form of the word the user will type next. The difficulty of applying this technology, however, is how much language is inferred or has a secondary meaning that is not desirable. This creates an additional challenge for NLP.

The work described by (Van Dyke, 1991) aimed at making augmentative communication devices more efficient and "usable" for the disabled user. The work has focused on how syntax can be used to eliminate the possible expansions of a creative abbreviation entered at run-time. Using a parallel parsing strategy, Van Dyke, has found it possible to reduce the effort required of the user because he or she is offered only the grammatically appropriate words as abbreviation expansions. A result of this is that the user is less likely to be confused by the words the computer offers as choices since they are always syntactically relevant to the situation. Other ways that the list of possibilities can be reduced and relevancy be maintained could come from applying other kinds of linguistic knowledge of the sort humans use to understand language. For example, discourse tracking is a kind of pragmatics that could be used to give the system knowledge like "since we have been talking about eating breakfast, it is probably the case that "tbl" stands for "table" and not "tablet". However, such abbreviations can come to mean still other things that are inconvenient for users. Semantics could also be used to reflect the fact that if the user has used the verb "drink" then we expect the following NP to be some inanimate, consumable object rather than a person's name or things like "table".

This sort of information would add to the power that syntactic prediction gives the system and eventually the user will have an extremely small and precise set of words to choose from.

More work could also be done at the syntax level, in the form of adding to the kinds of structures the grammar is able to handle. For example, currently the grammar cannot handle coordination, ellipsis, or topicalization; all of which are reasonably common in spoken language. The appendix at the end of this work includes a test suite that demonstrates the coverage of the grammar as it stands at this writing. From this it is easy to see where additions to the grammar could be made.

The implementation described by Van Dyke could also be improved through a more critical analysis of the Degree phrase, especially regarding the lexicalizations of the degree head and the relationship of the head to the elements it subcategorizes for. In particular, it would be useful to reanalyze the status of the quantifier phrase and constituents that can occur in specified position.

In (Ernst 1991), a positional interpretation of these items is given that may allow a more exact specification of word order in the pre-head noun positions of DP's. The task will be to find an explanation for these kinds of phrases that does not sacrifice capturing the generalities between them.

In (Demasco, McCoy et al. 1991), a technique that applies Natural Language Processing (NLP) to Augmentative Communication called compansion has been developed to increase the communication rate of physically disabled individuals. Compansion takes as input a compressed message (i.e., uninflected content words) from the disabled individual, and generates a syntactically and semantically well-formed sentence. The developed system has a vocabulary of over 1000 words. It handles most tenses, produces a variety of sentence constructions, and has the capability to infer the verb or subject in some situations. The structure of the semantic parser has been radically overhauled. It relies on a much more flexible knowledge structure than the template matching that it previously employed. In addition, a new version of the generator is being employed which provides a greater number of possible sentence constructions. Areas for further improvement include developing a syntactic pre-processor that will determine modification bindings as well as determining the scope of conjunctions by examining the order of the input. Also, it is intended to allow for more complex sentence constructions, make use of discourse information, and to expand the coverage of the system to include metaphorical expressions.

The work by (Demasco, McCoy et al. 1989) discusses the use of natural language processing (NLP) techniques in the design of language interfaces for augmentative communication systems. It presents a new technique called sentence companion that utilizes both natural language understanding techniques and generation techniques to allow a user to express full grammatical sentences with compressed content word sequences. Companion is a term derived from the two words compression and expansion. In general, the technique allows the user to input compressed sequences of linguistic units. The system will dynamically expand these units into a more "appropriate" representation. Sentence compansion works at the sentence level. The user inputs a sequence of content words, and the system produces a grammatically correct sentence that expresses the user's meaning. It currently assumes that the user's vocabulary set consist solely of words. The major goals of augmentative communication research are to improve the interface between the user and the system. This interface can be viewed as consisting of two components. The physical interface transfers user motor acts into meaningful system input. The language interface provides the user with a set of linguistic units to select from via the physical interface. Improvements to this component of the interface can be obtained from a wide variety of techniques.

At the simplest level, the language interface can be viewed as a structured collection of linguistic units (also called vocabulary sets). Early devices implemented vocabulary sets as simple two dimensional matrices that corresponded to a physical keyboard or to a visual keyboard (e.g., scanning). An extension to this structure that is commonly used today is the incorporation levels. A system with levels incorporates mechanisms to map different matrices to the same physical interface, thus allowing much larger vocabulary sets.

A great deal of effort has been devoted towards further improvement of the language interface. One of the earliest techniques used was optimization of vocabulary sets. Optimization is based on the nonuniform access time to specific units in the vocabulary set. For example, in a simple row column scanning array, the access time to an item is proportional to its distance from the upper left hand corner of the display. (Foulds, Baletsa et al. 1975) utilized this characteristic to design an optimized alphabetic vocabulary set.

Optimization is possible because the human language is structured and rule governed. The keyboard optimization discussed above relied on the fact that in general, letters occur with varying frequencies. Because 'e' occurs more frequently than 'z', it should be more accessible to the user.

Letter sequence frequency has also been exploited in other ways such as in linguistic prediction. Predictive systems rely not only on general information about language, but also on information obtained from the user. A predictive system will dynamically alter the structure of the language interface by offering the user next probable selections based on his/her previous selections (Baletsa, Foulds et al. 1976). Prediction is distinguished from optimization by two important characteristics:

- the system is dynamic (and in theory more responsive to the user's needs),
- the system relies on more linguistic information.

To continue making progress in the design of better language interfaces, it is necessary to utilize techniques via NLP that take greater advantage of the structure of human language and derive more information from the device user. This greater knowledge of human language at a more embedded level is important, particularly to ensure that these children do not have to face any discrimination, bullying or unwanted contact.

Natural language processing (NLP) techniques can be distinguished from the current methods used in AAC by two major factors. First, most techniques currently used (e.g., prediction) rely heavily on language statistics. While statistics are extremely useful, NLP's emphasis on language rules provide the potential to create more flexible rule based systems. This flexibility, however, can lead to abuse on a syntactic level as the rule based and semantically based communication can be at odds with one another. Second, because language statistics are primarily limited to letter constructs (e.g., bigram, trigram, etc.), many of the currently available techniques are limited to word production. Again, persons just slightly misspelling words can send obscene messages, bully, or tease these children with special needs. Current NLP research is active at all levels of linguistic complexity. Phonology and morphology address how words are formed. Syntax and semantics address how words are put together into meaningful sentences. Finally, pragmatics deals with conversational information that transcends the simple meaning of constituent sentences.

The need for systems with natural language interfaces has become increasingly essential as more and more people access information through their web browsers and mobile phones. This increase in the use of the web to access information on research, commerce, business, finance, sports, health, etc., requires employing rapid, reliable and accurate means of interacting with those systems.

Most of these systems are web-based, despite the fact that the mobile medium is becoming more popular among people, specifically in the developing countries.

Some other applications of Natural Language Processing (NLP) to the Field of Augmentative and Alternative Communication (AAC) is the field of study concerned with providing devices or techniques to augment the communicative ability of a person whose disability makes it difficult to speak in an understandable fashion. A variety of AAC devices and techniques exist today. Some are non-electronic word boards containing words and phrases in standard orthography and/or iconic representations. For example, the work done by (Demasco and McCoy 1992) targets a population of AAC users who do not have language impairments. One can imagine that such a user would like their device to output well-formed sentences (in fact, naturally "think" in terms of such sentences). However, because of the nature of their physical impairment and the time that it takes to compose well-formed sentences, users of current devices must often settle for output that is not as desirable. In this work they have token advantage of the regularities in natural language production to allow the system to act in an intelligent fashion. This intelligent fashion makes it easier for the user to cause the device to output well-formed text, but saves the user time and/or effort in entering that text.

The fields of Natural Language Processing (NLP) and Computational Linguistics attempt to capture regularities in natural (i.e., human) languages in an effort to enable a machine to communicate effectively with a human conversational partner (Grishman 1986; Allen 1987; Gazdar and Mellish 1989; Allen 1995). Three major areas of research in these fields (*syntax*, *semantics*, and *pragmatics*) deal with regularities of language at different levels. Various techniques have been developed within each which will be useful for application to various AAC technologies.

Syntax: The syntax of a language captures how the words can be put together in order to form sentences that "look correct in the language" (Allen 1987). Syntax is intended to capture tructural constraints imposed by language which are independent of meaning. Syntactic knowledge is being successfully applied in a number of AAC projects. For example, several word prediction systems use syntactic information to limit the words predicted to those which could follow the words given so far in a syntactically valid sentence (Swiffin, Arnott et al. 1987; Van Dyke 1991; Van Dyke, McCoy et al. 1992). To some extent, many grammar checkers available today and systems aimed toward language tutoring (e.g. Newell, Arnott et al. 1992; Wright, Beattie et al. 1992; Suri and McCoy 1993; Suri and McCoy 1993)) also use syntactic information, though there is still great room for improvement.

Semantics: The area of semantics deals with the regularity of language which comes from the meanings of individual words and how the individual words in a sentence form a meaningful whole. A problem in semantics is the fact that many words in English have several meanings (e.g., "bank" may refer to the edge of a river or to a financial institution). In similar fashion, the word "pussy" can refer to a cat or female genetalia. For this reason selectional restrictions are useful but cannot in itself solve the problems of abuse as slight purposeful misspellings can provide a different semantic outcome. In Computational Linguistics the use of selectional restrictions (Katz and Fodor 1963), case frames (Fillmore 1968; Fillmore 1977), and preference semantics (Wilks 1975) is based on the idea that the meanings of the words in a sentence are mutually constraining and predictive (Small and Rieger 1982). When the words of a sentence are taken as a whole, the meanings of the individual words can become clear. However, when the words can be taken out of context, they can have a completely different meaning. That is why much more work has to be done in the area of protecting users.of NLP.

Pragmatics: Pragmatic information refers to the broad context in which language and communication takes place (Joshi, Webber et al. 1981; Levinson 1983; Allen 1987). Situational context and previous exchanges produce conversational expectations about what is to come next. Natural language processing has concerned itself with developing computational mechanisms for capturing these same expectations in a computer.

The goal of the work "Natural Language Processing Principles for Improving Deaf Writing" by (McCoy and Suri 1991) is to develop a computer tool to correct the written English of deaf writers. The envisioned program will accept a written document, analyze the document looking for errors, and provide corrective advice when an error is found. Essentially, the system will act as an English tutor that will help deaf individuals improve their English writing skills. The design of this program is based on the belief that English should be viewed as a second language for many deaf people and that errors in the written English of deaf writers arise due to language transfer (LT) coupled with the individual's relative lack of exposure to English. The envisioned program has been designed so that it will take advantage of the writer's knowledge of American Sign Language (ASL) in identifying errors and suggesting corrections.

Linguists have found evidence of language transfer between spoken languages. Based on these findings, it is reasonable to expect language transfer from ASL (a visual-gestural language) to English (a spoken language). The work by (McCoy and Suri 1991) is driven by analysis of writing samples based on the research in language transfer. This analysis leads to taxonomy of errors.

The system itself consists of two phases. In the first phase, the system identifies errors. To do this, it relies on a grammar of English which has been augmented with a set of error rules which capture the errors in our taxonomy. The second phase of processing generates a correction.

The majority of the work to date has been devoted to analyzing writing samples from deaf individuals with an ASL background in order to develop the error taxonomy. The work by (McCoy and Suri 1991) have implemented a prototype system that contains an extensive grammar of English which has been augmented with a number of error rules (both syntactic and semantic). The system can use these rules in order to identify a subset of the sentence level errors identified by the sample analysis.

The major focus by (McCoy and Suri 1991)is the correction of errors that result from language transfer at the discourse level. It has been documented that much of the language instruction of the deaf has concentrated on the sentence level and thus deaf students may reach the point where their writing lacks discourse cohesion even though the individual sentences are grammatically correct. The correction of discourse level errors is particularly useful for many deaf writers. The work by (McCoy and Suri 1991)investigates models which can identify such discourse errors. Work has begun on the correction phase of processing which must be tailored to the individual user.

The work "Progress in the Application of Natural Language Processing (NLP) to Information Retrieval Tasks" by (Smeaton 1992) examined and reviewed some progress in using the lexical, syntactic, semantic and discourse levels of the language analysis for tasks like automatic and semi-automatic indexing of text, text retrieval, text abstracting and summarization, thesaurus generation from text corpus and conceptual information retrieval. The author also included his own work on the application of syntactic analysis to the matching and ranking of phrases using structured representations of texts. The paper also discussed the prospects for gains in terms of overall retrieval effectiveness or quality.

The visually handicapped have consistently progressed over the last decades in their efforts towards inclusion in the mainstream. Integration in education and professional life in particular was possible due to the deployment of computers in every day life, without which it would not have been possible, or at least not to the same degree. In the wake of this integration, more and more blind students are attempting post-secondary studies. This situation has created many new challenges and new needs specifically related to the in depth study of documents in a reasonable time so as to produce assignment submissions and research papers.

Over the last decades, more and more visually handicapped students have attempted post-secondary studies. This situation has created many new challenges. One of them is the need to study text and electronic documents in depth and in a reasonable time. Blind students cannot flip through the pages of a book, skim through the text or use a highlighter.

On a regular basis, post-secondary students, especially those in Liberal Arts, have to do research on specific topics. Their task consists in consulting a wide variety of books, documents and Web sites and producing an essay, anything from a few pages to a full-fledged thesis. To be able to better understand the difficulties that visually handicapped students face when trying to write a paper, we need to give a quick description of the tools they use to read and write.

The visual handicap can be divided into two broad categories, the partially sighted and the totally blind. For the totally blind, one way to make up for the absence of sight is the use of speech. The way computer technology is used in this case is to provide spoken output as screen readers that use an internal speech synthesizer. Words appearing on the screen are read aloud. The major problem with using speech as a medium is that it provides a very small working window due to the constraints of short term memory and linearity of speech. The work in (Miller 1956) provided evidence that people can remember about 7 chunks (terms) in short-term memory. The work in (Cowan 2001) goes even further and suggested as little as 3 to 5 terms.

The other way to compensate for the loss of visual ability is through the sense of touch. For two centuries now, the blind have been able to read by moving their fingers across raised dots on thick paper. The appearance of computer based 'Braille displays' in which each dot can be raised or lowered through electromechanical devices has caused a huge leap in the accessibility to electronic documents. The major constraint here is that the user can only 'see' a 40-character window.

To write a paper, a student normally scans a substantial amount of documents quickly and easily using fast reading techniques; develops an outline while reading one or more of these documents more systematically, highlighting and taking notes; reviews and rearranges his notes and uses them to create a draft of his document; refines the document iteratively until it is complete; at all times, refers back to previous readings and versions. At first sight, all these steps may seem easy, even mundane. However, for the visually handicapped, several problems exist. To name only a few:

- The amount of material to be read is, by itself, often a challenge – even for a sighted student.
- Reading difficulties go from very mild to very severe for handicapped students. This can be due to many factors such as a lack of reading and summarization skills, visual handicaps, dyslexia,...
- Note taking is an art in itself. The classic index card method facilitates sorting of notes but is tedious and can hardly be used by students with visual handicaps.
- Contrary to sighted users who can see a full screen of information, the blind have no overview of documents. Everything is always seen through a 40 character window for Braille users, or a window of less than 8 spoken words for speech users.

What is more, the above problems are encountered while performing all major steps in writing compositions: Research, Analysis, Outlining and Composition.

As witnessed by the CSUN series of conferences (e.g. [CSUN's 19th and 20th Annual International Conferences "Technology and Persons with Disabilities", 2004 and 2005 resp., LA, USA]), the topic of hardware and software applications to help the visually disabled has received a lot of attention. However, the resulting software applications are either geared at reading or writing, but not both. In addition, most of the features provided are text annotation tools and very little NLP techniques are used. As many of these systems are commercially available, no paper describing their inner working seems to be available.

In fact, NLP techniques might be used to improve the reading and writing tasks of the visually handicapped include only word prediction [http://freedomscientific.com/LSG/products/wynn.asp] and [www.kurzweiledu.com] and homonym checking [www.texthelp.com]. Homonym support provides auditory and visual reinforcement of commonly confused like-sounding words. To avoid the confusion between homophones, the program color codes confusable words and lists possible alternatives with audible definitions and sample sentences. Word prediction allows the application to predict the most likely word to be typed given the previous context. The user types a letter and the program offers a list of the most likely words beginning with that letter. If the required word is on the list, it can be quickly selected. If the word is not on the list, typing the next letter will bring up a different choice and so on. Again, as these systems are commercial, it is not clear if a language model is used, or if a simple dictionary look-up is performed.

The work by (Chelin, Kosseim et al. 2006) propose a solution in the form of an experimental prototype and show how natural language processing (NLP) techniques can profitably assist blind students in meeting their academic objectives. The techniques used include the automatic creation of indices, passage retrieval and the use of WordNet for query rewriting. The paper also presents a technology application of a practically usable software. The system was evaluated quantitatively and qualitatively. The evaluation was very encouraging and supports further investigation. It has also been discussed how Natural Language Processing (NLP) techniques can assist blind students in meeting their academic objectives. The paper presents a prototype system built called ESCAS as an information probing and gathering environment or more practically as a text editor with features to assist in reading and writing documents. Its goal is to reduce the time it takes a student to do research on a specific topic and ultimately produce a paper in a time frame close to their sighted friends. The purpose here is to investigate where and how far NLP techniques can facilitate the access of students to information pertinent to their research. These techniques will convey additional information about the content of a document and faster access to relevant positions within the document. Text annotation tools, such as the facility to create and annotate outlines, bookmarks, notes... are the typical tools offered to help the visually handicapped. These tools facilitate the access and the composition of documents, but cannot help in manipulating their content. Various NLP techniques have been used to zoom in on material within an input document that is particularly relevant to the subject being developed by the user. In the case of sighted users, this can be achieved by skimming or fast reading through the material. The intent was not only to replace skimming with what NLP can offer but also go beyond skimming and implement such functionality that could be of help even to those who have no handicaps.

The developed system ESCAS can be seen as an information probing and gathering environment that offers features based on NLP techniques. The purpose here is to investigate where and how far existing NLP techniques can facilitate the access of students to information pertinent to their research. These techniques convey additional information about the content of a document and faster access to relevant positions within the document. The evaluation of the system was more qualitative, than formal. Future work should include a formal evaluation of the system with several students, and compares their use of the system.

Since the advent of highly active antiretroviral therapy (HAART), HIV/AIDS has transitioned to a chronic illness for those who have access to treatment. The medication list may be captured in structured format, but many of the important nuances of the medication regimen, the history of past regimens, patient reactions to the regimen, and adherence behavior are typically captured in narrative. Moreover, given the demographics of persons with HIV/AIDS in the US, HIV/AIDS is frequently accompanied by complex psychological and social problems that also require complex management in conjunction with managing the HAART regimen. Consequently, many data crucial to providing optimal HIV/AIDS treatment are generally captured in narrative formats that are not suitable for further computer processing. Natural Language Processing (NLP) is an applicable tool for encoding clinical information from free text format in clinical settings. MedLEE (Medical Language Extraction and Encoding) (Friedman 2000) is an a comprehensive extraction, structuring, and encoding NLP system in use at New York Presbyterian Hospital that has been extensively evaluated.

Electronic health record (EHR) systems approach the problem of data capture from clinicians in one or both of two general ways. The first consists of using domain specific user interfaces for capturing data in a structured, and possibly coded format. The second consists of collecting and storing clinical information in computer systems in text format. "Plain text" capture is easier and cheaper to implement, but in order to "unlock" the information for uses such as computer-based decision support or outcome measurement, it must be structured and coded after the fact (Barrows, M. et al. 2000). Using structured, coded approaches alone may result in loss of significant clinical information typically contained in clinical narratives. Natural language processing (NLP) offers a strategy for integrating approaches to create structured reports from narratives for further computer processing.

In recent years, significant progress has been achieved toward increased structured data entry using standardized health care terminologies. Concurrently, the value of narrative as the clinician's rich description of the encounter and source of vital information has been reaffirmed. Natural language processing (NLP) offers a strategy for integrating these approaches to provide structured reports for further computer processing.

As part of a larger project aimed at using narrative data to enrich the online medical record as well as using NLP to improve the quality of information in the EHR, (Hyuna, Bakkena et al. 2004) analyzed a small sample of documents in a corpus of progress notes to identify potential challenges associated with using NLP for HIV/AIDS clinic notes and to characterized the specific sublanguage in the domain of HIV/AIDS care. They provided illustrative examples of five types of challenges.

(Hyuna, Bakkena et al. 2004) randomly selected clinic notes of 80 patients from a corpus of de-identified notes from 707 HIV/AIDS patients written by physicians and nurse practitioners during the provision of HIV/AIDS care in a West coast academic health sciences center. A small sample was then processed to identify potential issues and solutions (e.g., preprocessing to replace abbreviations vs. additions to lexicon) prior to analyzing the corpus using NLP. The authors manipulated the data using PERL in order to pre- process for MedLEE parsing. They manually analyzed the parsed output and compared it to the original notes and then classified four types of ***generic issues*** and ***HIV/AIDS domain specific issues*** if discrepancies were found, e.g. loss of meaning or misinterpretation of meaning, in order to identify the best way to deal with those issues. Here below are their results:

1. **Generic Issues**
    - Preprocessing issues
        - Usual abbreviations, but currently are not in the MedLEE lexicon, e.g. *abd, cor, psych, Rx, T, DOE*

        - Unusual abbreviations, but common in clinic notes, e.g. *cont (meaning, 'continue'), gen (generally/generalized)*
        - Abbreviations with more than one meaning, e.g. TM (Temporomandibular; Tympanic Membrane)
    - Lexical issues
        - General words, e.g. *health, consult, encourage, family.*
        - HIV/AIDS specific words, e.g. *viral load, antivirals.*
    - Grammar issues
        - Distribution of modifiers – modifier not distributed across a conjunction, e.g. *Now amylase or lipase normal.*
        - Sentential connectives – component words recognized, but not joined together by connectives, e.g. *Had transfusion as scheduled last week due to anemia.*
        - Zeroing – a phrase contained in a previous statement is dropped, so MedLEE could not parse appropriately, e.g. *CD4 = 162. 11% decrease.*
    - Ambiguous records
        - *PAP & pelvic done today* ('Pelvic' is ambiguous since it can be a body site or an acronym of specific procedure on the pelvic region).
        - *2. full set of labs today* ('full set of labs' is not clear).
2. **HIV/AIDS Specific Issues**: A large portion of psychological and health management sentences were not parsed appropriately, e.g., *Does not have a plan to commit suicide but feels very hopeless, Review viral load and discuss therapy.*

The AIDS ambulatory clinic domain has a unique sublanguage, which contains terms that are not used in other clinical environments, e.g., 'viral load.' The authors' approach for improving MedLEE processing intended to markup documents during pre-processing provided the correct interpretations of phrases that are unique to the sublanguage. This work will be challenging, but the identification of specific patterns may help to automate the analysis process, by using these patterns in rule-based or machine learning methods. The authors envision the use of this NLP system to improve the quality of clinical documentation and ultimately, the quality of HIV/AIDS care.

For the operation of mobile robots, the design of the man-machine interface has a major impact on the acceptance by the user. This is especially valid when dealing with users of minor or no training, which can be found in almost any application area, ranging from transport or construction tasks to assistive technology for elderly and disabled. The work by [Drews, P. and Fromm, P., 1997] introduces a natural language processing (NLP) software which transforms a natural language sentence into a series of commands for a mobile service robot. Both the language processor and the command set of the controller may be modified by the user to adapt it to different application areas. The language processor is based on a modified finite state transducer network. The grammar/word pattern extracted by the network is processed by a transformation module which converts the contained information into the controller command. A context memory and a question generator allow the processing of incomplete instructions and command modifications. A prototype application is presented which allows severely disabled persons bound to a wheelchair to control the chair and their home environment by natural language. The work is specifically directed at the needs of people who have lost the use of speech through motor impairment such as ALS (Lou Gehrigs disease).

The paper by (Copestake and Flickinger 1998) discuss how research on language as a form of joint action yields insights into the types of natural language processing (NLP) techniques that can be used in AAC based on text-to-speech devices. Most work on NLP techniques for such systems has concentrated on word and phrase prediction or on facilitating retrieval of complete preconstructed messages. A prototype system has been developed, which incorporates prediction and message retrieval, and has been in daily use by a person with ALS for several years. The system runs on a standard laptop PC. There are considerable advantages for the user in this approach, as opposed to dedicated AAC hardware, since it allows the use of the same computer for email, Web access and so on. Investigation of data logged from this system, and from audio-taping a few hours of conversations involving the user, has suggested that a wider range of NLP techniques might be utilized. Communication using an AAC device has much in common with written media. Often an AAC user is able to use written forms of communication such as electronic mail without appreciable handicap. The fundamental deficiency in AAC for people whose disability is physical rather than linguistic or cognitive is not that communication as such is impaired but that problems arise because the expectations of face-to-face conversation are violated, in particular with respect to timing. In fact, (electronic) written media have some advantages over spoken media which are potentially shared by AAC: specifically the possibility of recording and indexing by content and the ability to reuse material, both from previous communications and from external sources. Of course, face-to-face AAC does have more immediacy than normal written communication - the AAC user can use non-verbal signals (although the ability to do this may be greatly reduced in ALS) and the other participants in the conversation will be speaking normally. Many AAC systems allow for retrieval of fixed text, such as "Please could you" to preface a request. The paper by (Copestake and Flickinger 1998) improved on this by developing a series of flexible `templates' for use in an AAC system. The role of NLP in this process is to predict the closed class words and otherwise ensure that the output is grammatical, as well as doing the contextual prediction. This is related to work on companion (Demasco and McCoy 1992), but the work by (Copestake and Flickinger 1998) aimed to accomplish this for an arbitrarily large vocabulary. Parameters such as urgency also require variation in speech intonation and volume. All this will take considerable research in NLP. This approach has been described in (Copestake 1997).

To conclude, because of the inherent differences in media, it is unrealistic to expect that conversation with an AAC user can be the same as face-to-face conversation between non-AAC-users, however successful work on incorporating NLP techniques eventually is. However, communication using an AAC device can be made more effective through speeding up user input by taking advantage of conventions, and also by reusing text. Furthermore, AAC devices can benefit from the additional functionality of PCs, particularly in serving as a memory aid, and in allowing the retrieval of text from a range of sources.

The goal of the work done by (Copestake and Flickinger 1998) was to maximize the immediacy of conversations using an AAC device while still allowing the AAC-user self-determination. The paper briefly discusses three complementary aspects of using NLP in AAC devices: speeding up AAC input by making use of language conventions, maximizing utility of stored material and managing conversations to reduce the effect of delays.

As we mentioned before, some people have disabilities which make it difficult for them to speak in an understandable fashion. The field of Augmentative and Alternative Communication (AAC) is concerned with developing methods to augment the communicative ability of such people. In addition to problems that make speaking difficult AAC users often have difficulties in coordinating extremities (so typing on a standard keyboard may be impossible and access to large keys is often very slow. Cognitive difficulties may also be Present. The field of AAC is concerned with developing methods that provide access to communicative material under reasonable time and cognitive constraints.

Over the past couple of years, the Applied Science and Engineering Laboratories (ASEL) at the University of Delaware and the duPont Hospital for Children has been involved with applying natural language processing (NLP) technologies to the field of (AAC). One of the major projects at ASEL (The COMPANSION project) has been concerned with the application of primarily lexical semantics and sentence generation technology to expand telegraphic input into full sentences (Demasco, McCoy et al. 1989; Demasco, McCoy et al. 1991; Demasco and McCoy 1992). While the Compansion prototype is vieed as a promising and successful application of NLP to ACC, it raises some questions when viewed as a practical ACC system. While this project has shown some very promising results its direct application to a communication device is somewhat questionable (primarily because of the computational power necessary to make the technique fast). The paper by (McCoy 1997) described some of the problems with bringing Compansion to a standard communication device and introduced some work being done in conjunction with the Prentke Romich Company (PRC) (a well known communication device manufacturer) on developing a pared-down version of Compansion for people with cognitive impairments. Characteristics of the language used by a particular population being studied has permitted to apply some simple NLP techniques which are proving to be sufficiently robust for this task.

(LeBlanc, Geiger et al. 2006)show that Natural Language Paradigm (NLP) has proven effective in increasing spontaneous verbalizations for children with autism. This study investigated the use of NLP with older adults with cognitive impairments served at a leisure-based adult day program for seniors. Three individuals with limited spontaneous use of functional language participated in a multiple baseline design across participants. Data were collected on appropriate and inappropriate vocalizations with appropriate vocalizations coded as prompted or unprompted during baseline and treatment sessions. All participants experienced increases in appropriate speech during NLP with variable response patterns. Additionally, the two participants with substantial inappropriate vocalization showed decreases in inappropriate speech. Implications for intervention in day programs are discussed.

The Natural Language Paradigm (NLP) is a leisure-based behavioral intervention developed by (Koegel, O'Dell et al. 1987)to increase the language of non-verbal children with autism. The procedure incorporates many strategies for promoting generalization described by (Stokes and Baer 1977) including teaching in natural contexts, incorporating naturally occ responsiveness to social and environmental stimuli, positive affect) (Koegel, Koegel et al. 1999). Motivation is enhanced by using a loose shaping criterion to increase the probability of reinforcement during early language attempts and by incorporating choice and learner-initiated activities (Koegel, Koegel et al. 1992; Charlop-Christy, LeBlanc et al. 1999; Koegel, Koegel et al. 1999).

Several studies have found NLP effective for teaching language to children with autism (Koegel, O'Dell et al. 1987; Laski, Charlop et al. 1988; Camarata 1996; Koegel, Camarata et al. 1998)even with children for whom traditional vocal imitation training proved ineffective. (Laski, Charlop et al. 1988) trained parents to implement NLP with their children with autism with resulting increases in generalized speech. Parents found the intervention enjoyable and thought their children enjoyed it as well. Importantly, parents did not need to implement the procedures perfectly for them to be effective.

NLP may prove appealing for other populations with language impairments because of the use of preferred stimuli delivered contingent upon vocalizations and the ease of implementation in natural contexts (Charlop-Christy, LeBlanc et al. 1999). To date, no researcher has investigated the use of NLP with older adults with cognitive impairments, though cognitive decline is often accompanied by decreases in language (Dijkstra, Bourgeois et al. 2004). Though few studies in behavioral gerontology have directly targeted conversational skills, an exception is (Bourgeois 1993) who found that photographic memory aid books resulted in increased on-topic statements for individuals with dementia. Additional interventions in this area are needed. NLP may prove particularly appropriate for leisure-based day programs for older adults because the intervention is designed to incorporate preferred leisure activities. This study evaluated the effects of NLP on the speech of three adults with cognitive impairments and limited functional language.

An explicit goal of NLP is enhancing learner motivation (i.e., willingness to participate in instruction, responsiveness to social and environmental stimuli, positive affect) (Koegel, Koegel et al. 1999). Motivation is enhanced by using a loose shaping criterion to increase the probability of reinforcement during early language attempts and by incorporating choice and learner-initiated activities (Koegel, Koegel et al. 1992; Charlop-Christy, LeBlanc et al. 1999; Koegel, Koegel et al. 1999).

NATURAL LANGUAGE INTERFACE STYLES

Free Natural Language Interface

Evidently, natural language is considered the simplest technique of human-machine interaction. It allows a wide range of expressions by helping the user to specify as many parameters as needed in a single request. The defining characteristics of a natural language interface are that users need not explicitly learn the lexicon and syntax of the system, so that they are able to express what they want in the language they are used to (Long 1994).

Natural Language Interfaces were first used as a means for querying databases "Natural Language Database Interfaces (NLDBIS) " (Androutsopoulos, Ritchie et al. 1995). The main idea was that, for a user with no deep computer science knowledge, it is easier to query the database in natural language instead of using SQL expressions. Moreover, natural language expressions are often shorter than SQL ones, and there are cases when it is difficult to formalize expressions like "some", "a few", "often" etc. Most natural language interfaces to database systems have been prototypes, built by the research community.

The first NLDBIS was the LUNAR system built by Woods in 1973 for NASA. It concerned chemical analyses of moon rocks. Two databases were implemented, including the chemical analyses and the literature references. It managed to handle 78% of requests without error, a figure that rose to 90% when dictionary errors were corrected. However, ungrammatical sentences were not handled well and the NLP component was not very flexible.

SHRDLU was an early natural language understanding computer system, developed by Terry Winograd at MIT for his Ph.D. Thesis 1968-1970. SHRDLU allowed user interaction using English terms. The user instructed SHRDLU to move various objects around in a small "blocks world" containing various basic objects: blocks, cones, balls, etc.

By the late 1970's, LIFER/LADDER was one of the first good database Natural Language Processing (NLP) systems. It was designed as a natural language interface to a database of information about US Navy Ships. This system used a semantic grammar to parse questions and query a distributed database.

Natural Language interfaces to databases were commercially available in the late 1970s, but largely died out by the 1990s: porting to new databases and especially to new domains requires very specialized skills and is essentially too expensive (automatic porting was attempted but never successfully developed) (Copestake, 2003).

Those early prototype systems showed that the analysis and understanding of natural language input is a highly complex process, requiring linguistic knowledge (morphology, syntax, semantics, and pragmatics) as well as a well elaborated knowledge-base and a very complex dataflow control between the components (Vertan, 2004).

Until now, such systems require careful use, because they strongly depend on the correctness of their input, which, however, often contains typing errors, grammatical re-formulations, or slightly ungrammatical everyday idioms.

Traditional techniques for handling these difficulties involved spelling correctors and very large grammars and lexicons, but these techniques turned out to be unsuccessful.

The second reason why natural language interfaces are not in common use today is the large amount of time it has traditionally taken to construct a natural language interface. Current technology is such that each interface must be constructed on a case by case basis for each application. Thus, only applications that can justify such large expenditure of manpower are candidates for possible applications. However, given the quality of the resulting system, the effort has not proven to be cost effective.

Controlled/Restricted Natural Language Interface

A controlled natural language has a limited subset of the vocabulary and syntax of a full natural language. This allows ambiguity to be reduced and processing time to be kept within reasonable bounds (Long, 1994). To retain the properties of ease-of-use and ease of remembering, the limitations of the system must somehow be conveyed to the users without requiring them to learn the rules explicitly.

However, some of these systems impose limitations on the vocabulary and syntax to the extent that it is no longer a controlled natural language, but rather a command language.

In this technique, the system imposes on users the way they should encode their thoughts and utterances. It does not allow them to express their requests naturally, so that they have to remember the allowed lexicon and syntax of the controlled language. As an example, Google SMS[1] imposes a controlled language on its offered services. For example, to get movie showtimes theater listings and movie details, users should follow the following language model (see, for example, Figure 1):

- If you're looking for showtimes of a movie that's currently playing, enter the movie's title followed by your location (a zip code or city and state).

Figure 1. A sample of a controlled NL interface

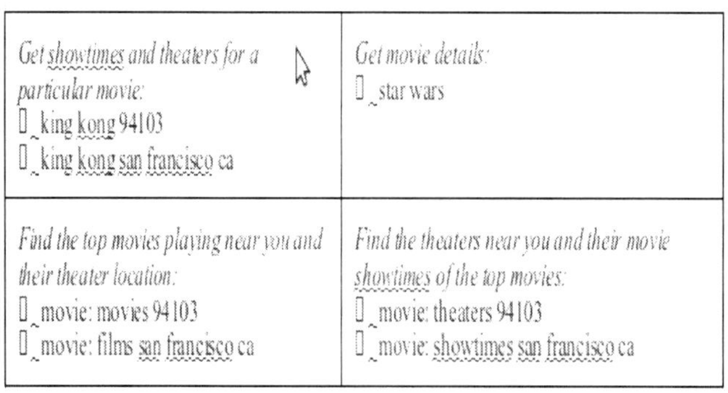

- For a listing of theaters near you and the showtimes of their top movies, enter 'movie: theaters' or 'movie: showtimes' followed by your location.
- For a listing of movies playing near you and their theater location, enter 'movie: movies' or 'movie: films' followed by your location.
- For movie details such as running time, genre, MPA rating, and critics' rating, simply enter the movie title.
- Get local listings in addition to your movie showtimes when you type 'theaters' or 'films' followed by your location.

Google SMS provides similar instructions for each offered service and users have to memorize them in order to use them correctly.

These systems might work well for simple tasks or under controlled environments where users know the task well and can be trained on the controlled language interface.

On the other hand, when the task becomes more complicated with many parameters to express, it is difficult to have a working system based on a controlled language without learning what structures are acceptable, which is impractical for public services.

For example, PENG is a machine-oriented controlled natural language that has been designed for non-specialists to write precise specification texts in a seemingly informal notation. To guarantee the efficient usage of this controlled natural language, a text editor with an intelligent feedback mechanism has been developed that guides the writing process and guarantees well-formed linguistic structures that can be translated unambiguously into first-order logic (Schwitter 2004).

Another controlled language system is the KANT System (Knowledge-based, Accurate Natural-language Translation), which has been primarily targeted towards the translation of technical documents of heavy equipments (Mitamura, 1999).

(Pool 2005) evaluated 32 controlled natural languages which are currently available and came to the following conclusions:

In many cases, controlled natural languages have not been designed for Web-scale use. Most often, they have been designed within commercial organizations for the improvement of the processing of the language used in specialized functions performed in those same organizations. Utterances in that context are usually constrainable grammatically and lexically at relatively low cost, for three reasons. First, the domain is limited (e.g., heavy machinery maintenance). Second, the functions performed by utterances within the domain are limited (e.g., instructions). Third, a small and stable set of authors do the documentation, so that any investment they make in learning to comply with a controlled language is amortized over a large body of production. These advantages may account for the relative popularity of controlled natural languages as proprietary industrial tools.

Menu-Based Natural Language Interface (MBNLI)

To avoid the problem of training users which is associated with controlled natural language interfaces, menu-based natural language interfaces (MBNLI) have been introduced. They help users understand the capabilities of the system, thus avoiding any user negative or false expectation (Thompson, Ross et al., 1983).

Menu-based natural language interfaces propose a broad variety of natural expressions: the user has a choice among a large number of alternatives at any place in the input, but the input language is still controlled in the sense that any choice has a well-defined (syntactic) structure with an associated precise semantic interpretation. Additionally, choosing from a menu eliminates typos. As the quality of a natural language processing system depends strongly on the formal quality of the input, these kinds of systems have been regarded as the ideal cost/effect compromise for human-computer interaction (Vertan and Hahn 2003).

The first fully implemented interface, NL-Menu, was developed in the eighties. It was a menu-based natural language understanding system. Rather than requiring the user to type his input into the system, input to NLMenu is made by selecting items from a set of dynamically changing menus. Active menus and items are determined by a predictive left-corner parser that accesses a semantic grammar and lexicon (Tennant, Ross et al. 1983).

ROSY (RObust SYntactic analysis, 1987) was developed at the University of Saarbrücken and was meant as an extension of NLMenu, taking into account the particularities of the German language. With the addition of a morphological processor, the natural impression of the generated sentences was strongly improved. Several other interfaces based on the same principles were developed afterwards, mainly for database query tasks (Vertan and Hahn, 2003).

One of the main disadvantages of these systems is the use of an extensively controlled language. Additionally, it is not flexible and does not allow spontaneous input with obvious limitations on the menu design and updating of the dictionary. Finally, this approach is not adequate for all types of devices and communication modes and overwhelms users and frustrates them with lengthy interactions.

Natural Language Dialogue Interfaces (NLDI)

A "Dialogue Interface" conducts a chat-like conversation with users, to extract exactly what they are looking for or trying to accomplish.

Natural language dialog has been used in many areas, such as for call-center/routing applications, e-mail routing information retrieval and database access, and for telephony banking (Chai, Horvath et al., 2002).

A good example of NLDI is the HappyAssistant dialogue system described by (Chai, Horvath et al. 2002). Another example of NLDI systems is the Subjex.com search engine, which establishes two-way conversations with the users. Here's a sample dialogue:

User: WinZip

Subjex> *Would you like me to search for WinZip?*
User: Yes
Subjex> *Here are some results. Should I look for* 2000 *or* WinZip computing *or something else?*
User: Something else
Subjex> *What specific information about WinZip do you require?*
User: Download sites

The approach of NLDI goes far beyond the keywords approach. It actually asks questions back to the user. This simple differentiation has a profound impact on the resulting user experience. A two-way conversation leads to fewer and highly relevant (targeted) search results; without forcing the user to wade through potentially millions of results or necessitate the learning of advanced search "Boolean" techniques using keywords. Keyword search give users only one opportunity to input what they are searching for before beginning another search. User queries are generally very short which naturally leads to a large number of documents being returned (Kruschwitz and Al-Bakour 2005). Hence, adopting this interaction style is efficient in avoiding irrelevant information and in providing more precise results to users. Furthermore, this short and uncomplicated users' input will cause a more accurate processing of input text.

Specifically, in e-commerce environments, NLDI has also been proven to be more efficient in finding information than the keywords approach or MBNLI (Chai, Horvath et al., 2002).

In a similar manner with other NL interactions, the quality and complexity of such systems depends on the nature of the natural sublanguage used in the dialogues. Allowing unrestricted natural language dialogue in unrestricted domains would appear to require full human conversational competence, which does not seem feasible in the foreseeable future.

Additionally, this approach might work well with certain medium of communications such as Internet or PSTN, but, it is inappropriate for paid SMS-based services in which the users are charged for each message.

NLP APPROACHES CURRENTLY USED

Different approaches are used to process natural language text. Strategies and techniques relevant for our goal of building NL-based interfaces for e-commerce include word-based, syntax-based and semantic-based processing.

Information Retrieval (IR) Techniques (Bag of Words)

Information retrieval involves returning a set of documents in response to a user query. The user specifies his information needs by providing a set of keywords and the information system retrieves the documents which best approximate the user query. Hence, the goal of this approach is to find relevant documents in a large collection, in response to a user's query expressed as a sequence of words (Guarino, Masolo et al., 1999).

Search engines are a form of IR and one of the most essential tools on the Internet: they help find Web sites relating to a particular subject or topic using the bag of words approach.. Search engines are basically huge databases containing millions of records that include the URL of a particular Web page along with information relating to the content of the Web. Search engines crawl the Web and log the words from the web pages they find in their databases. The most popular search engines include: Google, AltaVista, Excite, Hotbot, Infoseek, Lycos, Webcrawler, Yahoo, and so on.

Information retrieval systems are usually evaluated based on two metrics – precision and recall. Precision refers to the ratio of relevant (or correct) documents returned to the total number of documents returned. Recall refers to the number of relevant documents returned out of the total number of relevant documents available in the document collection being searched.

Retrieval based on keywords is often of fairly low quality. There are two possible reasons. First, the user query may be composed of too few terms which usually implies that the query context is poorly characterized. This is frequently the case, for instance, in the Web. This problem is dealt with through transformations of the query such as query expansion and user relevance feedback. Second, the set of keywords generated for a given document may fail to summarize its semantic content properly. This problem is dealt with through transformations in the text such as identification of noun groups to be used as keywords, stemming,

Figure 2. Keywords-based matching in IR-based e-commerce systems

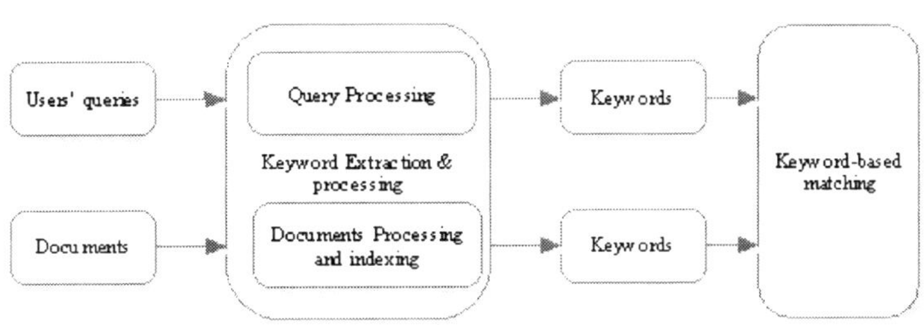

and the use of a thesaurus. Additionally, for reasons of efficiency, text compression can be employed.

In the case of e-commerce applications this approach is unlikely to allow precise matching, because keywords based processing lacks the understanding of semantic meaning of the search words. (Chai, Horvath et al. 2002).

Additionally, bag of words processing is not adequate for handling numeric values such as price which are essential in e-commerce applications. Furthermore, given the dynamic nature of e-commerce contents, a more frequent crawling and indexing is required, as compared to traditional content.

Despite a large body of research on keyword matching, the effectiveness of retrieval systems is still rather low. Although, it might work for search engines and large documents retrievals, it does not suit e-commerce systems. The types of dissimilarities between queries and documents cannot be solved using this approach.

Syntactic-Based Processing

In this type of systems, more processing is performed based on the syntactic structure of a sentence. The output of this processing is expressed usually by a parse tree. This approach is used mainly in traditional machine translation systems. However, for NL-based system that accepts users' queries and generate responses, different complexity factors (Appelt and Israel, 1999) must be considered and studied before selecting the best level of syntactic analysis.

Different levels of text processing are used, for example, in the MUC events competing groups with a wide range of theoretical orientations converging on a common approach: domain-dependent templates for representing the critical patterns of concepts and a limited amount of syntactic processing to find appropriate phrases that fill slots in the templates (Sowa, 1999). This was illustrated by abandoning TACITUS which was a system that spent most of its time on syntactic analysis that were irrelevant to the ultimate goal. SRI replaced it with FASTUS, a system that is triggered by keywords, finds phrase patterns without attempting to link them into a formal parse tree, and matches the phrases to the slots in the templates (Sowa, 1999).

Partial syntactic analysis of a sentence is called shallow parsing as an alternative to full sentence parsing. While earlier work in this direction concentrated on manual construction of rules, most of the recent work has been motivated by the observation that shallow syntactic information can be extracted using local information — by examining the pattern itself, its nearby context and the local part-of-speech information (Li and Roth, 2001). Learning and statistical methods has been used recently to recognize shallow parsing patterns such as syntactic phrase or words that participate in syntactic relationships.

On the other hand, to perform syntactic-based matching, both the words and the syntactic rela-

Figure 3. A typical syntactic-based information system

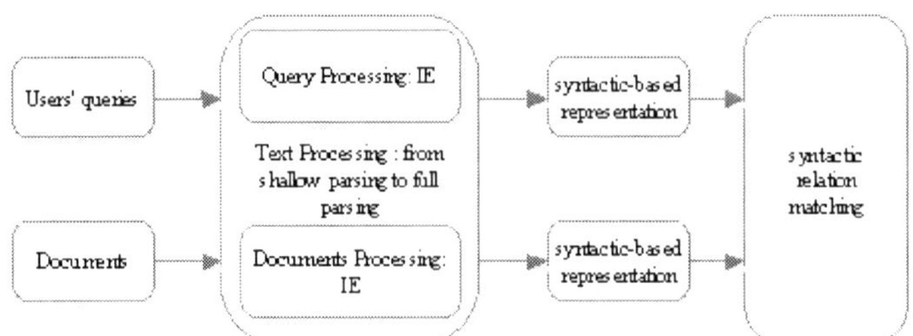

tions between them in the query are matched with words and relations in the document. Determining the syntactic relations requires full parsing of the sentence to identify the syntactic relations. However, reports show that using syntactic relation matching produces no improvement, over using just keywords. Indeed even more (Smeaton, O'Donnell et al., 1995) obtained worse results from relation matching using a tree-matching procedure than from keyword matching. One possible reason why syntactic relation matching has not yielded better results is the difficulty of identifying syntactic relations accurately (Khoo, 1997).

Additionally, it is possible to use syntactic processing (shallow or full) to identify keywords and fill predefined templates. However, further processing and matching could be based on keywords only.

Semantic-Based Processing

Semantic-based systems process text with the objective of extracting content or meaning. These types of systems rely on semantic grammars, in which the sentence is not described by syntactic relations, but by using the semantic classes that categorize occurring words (Kittredge, 1982).

Semantic processing requires these topics to be addressed: knowledge representation, words meaning and the extension of knowledge representation with synonyms and related words (Engels and Bernt Bremdal, 2000).

The LIFER/LADDER (1977) (Templeton and Burger 1983) system was the first to use a semantic grammar (that is, it used labels such as "SHIP" and "ATTRIBUTE" rather than syntactic labels such as noun and verb). This means that it was closely tied to the domain for which it was engineered.

The Conceptual Dependency Theory of Schank (Schank, 1975) was the first attempt to apply knowledge representation to language processing (Engels and Bernt Bremdal, 2000). He wanted a representation that was unambiguous and unique. His aim was to express the meaning of any sentence in any language. The representations were intended to be language-independent. For any two sentences that are identical in meaning, regardless of language, there should be only one representation.

The Conceptual Dependency formalism was used to build the first commercially deployed IE (Information Extraction) system ATRANS (Lytinen and Gershman, 1986), which handled money transfer telexes. This system used semantics and did not require complete syntactic analysis of text and exploited the fact that the content of money transfer telexes is highly predictable. The text is processed by identifying actors (originating customer, originating bank, receiving bank, etc.) in order to fill in a template that was used, after human verification, to initiate automatic money transfers. It took something like thirty man-years to make it work (Schank, 1991).

Figure 4. a typical semantic-based systems

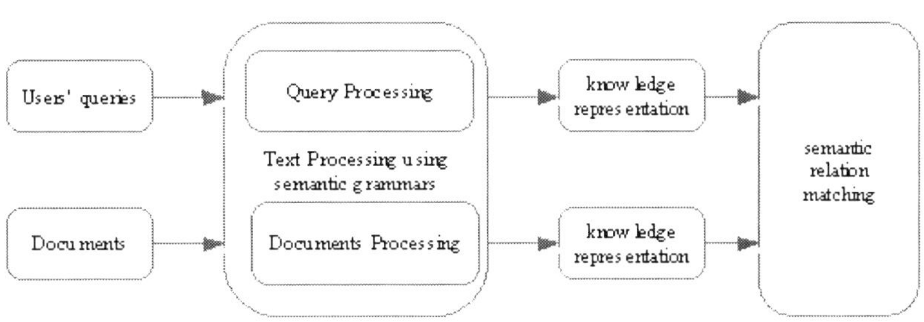

The main problem with semantic-based systems is that they are domain-specific. Hence, porting to other domains is often not straightforward. The development of such systems requires extensive domain knowledge. On the other hand, they have higher recall and precision than syntactic-based systems.

HUMAN FACTORS BASED DEVELOPMENT OF NLP SYSTEMS

For Natural Language Processing to be able to contribute to the full range of dynamic situations in which language is used, it needs to recognize, interpret, and respond appropriately in all the 'contexts' in which language is encountered, not just formal, well-written genres. This requires additional levels of interpretation beyond standard semantics, and can be thought of as self-adapting personal language processing, which will incorporate all the sets of features which convey meaning based on linguistic and paralinguistic cues that humans use in their social-communicative interactions, whether in speech or in every day written communications (e.g., email, text messaging, or instant messaging).

NLP system, just like any other system, should be designed with the user in mind. The structure of such systems is influenced by different human factors:

The Domain

It is noticeable that in restricted domains of knowledge, among certain groups of people and in particular types of texts, people have their own way of encoding their thoughts. Such restrictions can be said to reduce the degree of lexical and syntactic variation in text (Lehrberger, 1982). These specific languages are called either sublanguages or restricted or specialized languages.

Examples of sublanguages are the languages of weather bulletins, aircraft maintenance manuals, scientific articles about pharmacology, hospital radiology reports, and real estate advertisements (Grishman, 2001). A sublanguage is characterized by a specialized vocabulary, semantic relationships, and in many cases specialized syntax.

Text Style

The input text provided by users might be structured, semi-structured, grammatical, ungrammatical, telegraphic and noisy style. Text Styles are very important factors in building NLP system. Systems developed for semi-structured text are not appropriate for free text and vice-versa.

Figure 5. Human Factors in NLP systems

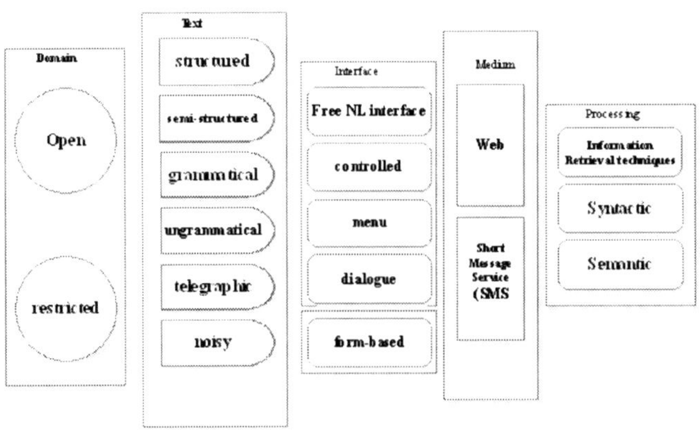

Developing information systems that depend on natural, spontaneous and unprocessed text requires techniques and approaches different from those used for edited text.

The Medium

Most of these systems are web-based, despite the fact that the mobile medium is becoming more popular among people, specifically in the developing countries. Recent data shows that the world mobile usage reaches 4 billion, while the world Internet usage is only 1.6 billion (2006). The cell-phone industry is one of the fastest-growing communication industries in the Arab world. According to ITU, the number of mobile users in the Arab world is 214 million with 62% teledensity. On the other hand, the number of Internet users is only 54 million with 15.84% teledensity. Building web-based NLP systems in developing countries will certainly lead to a smaller customer base of such systems. Unlike the Internet, which requires expensive equipment, national infrastructure, and primarily knowing how to read and write, the cell phone provides an accessible, relatively inexpensive solution to being part of what is happening around you without requiring a high level of literacy.

Interface Styles

As we have shown above, different types of interface styles can be used: free NL interface, controlled, menu, dialogue and form-based. Selecting the appropriate interface style is influenced mainly by the target group and medium.

Generally speaking, a NL interface is an interesting option because, if it could be built, it would offer many advantages. Firstly, it does not require any learning and training, because its structure and vocabulary are already familiar to the user. Secondly, natural language enables users to encode complex meanings. Thirdly, this type of interface is text-based, making it suitable for all types of devices and medium. In contrast, form-based or graphical user interfaces need more sophisticated and specific resources.

However, incorporating a NL interface requires translating ambiguous user's inputs into clear intermediate representations. Two main problems are associated with building such systems: the first one is handling linguistic knowledge and the second one is handling domain knowledge.

HOW TO EMPLOY HUMAN FACTORS NLP SYSTEMS?

The key success factors in building task-related NL systems are understanding how people encode their thoughts, and finding the right representation to model the concerned domain knowledge.

Corpus-Based Approach

Corpus analysis is used to identify the characteristics of the target texts in terms of textual, semantic, lexical and syntactic attributes.

A corpus-based approach will certainly lead to a better understanding of the sublanguage used in a given application such as SMS-based classified ads, and the way people encode their thoughts in this context (device, task, domain). In turn, this will help in selecting the right approach for the development of NL systems. As an example, systems developed for semi-structured text are not appropriate for free text and vice-versa. Developing information systems that depend on natural, spontaneous and unprocessed text requires techniques and approaches different from those used for edited text. Most of the current systems that process users queries and generate responses use shallow text processing techniques based on pattern extraction or information retrieval techniques (Benamara 2004).

Figure 6. The role of corpus in revealing human factors

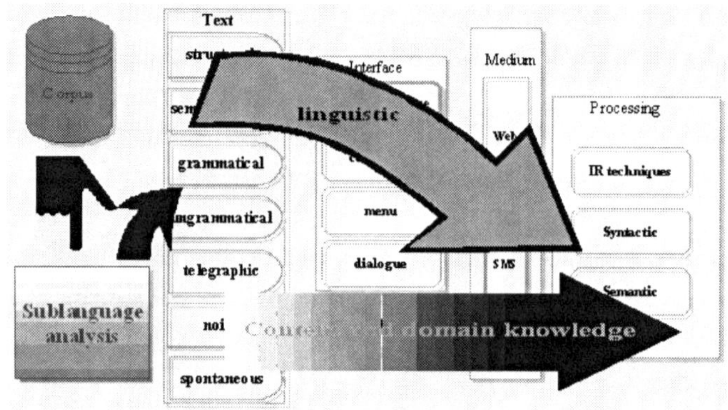

Sublanguage Features

The study of a sublanguage corpus is necessary for producing a sublanguage grammar. As presented in Figure 7, the analysis of the linguistic aspects and features of a sublanguage is needed to specify the sublanguage grammar (with the incorporation of the domain knowledge). Then, general linguistic knowledge and sublanguage grammar can be used to determine the best NL technique to use. Similarly, the sublanguage grammar and the domain knowledge are both indispensable in selecting the best content representation.

Most researchers studied sublanguages from the regularities within syntactic features (Bonzi, 1990). (Tagliacozzo, 1976) used the type-token ratio (TTR) in her analysis of different corpuses. She found that highly technical writing has a lower TTR than technical writing. (Kittredge, 1982) studied four technical sublanguages (weather bulletins, recipes, stock market reports, and aviation maintenance manuals) and found syntactic homogeneity within each sublanguage. (Bonzi, 1990) compared the sublanguages of hard sciences and social sciences in their use of various syntactic patterns.(Bruthiaux, 1994), studying classified ads, used the type-token ratio (TTR) in his analysis of different levels of text. He found that the jobs domain has a higher ratio than the cars and apartments domains. (Friedman, Kra et al. 2002) studied a clinical sublanguage and a second sublanguage, biomolecular literature. Both of them was studied by establishing semantic categories for the entities and relations in the domain, specifying semantic and syntactic co-occurrence patterns, and specifying target forms for each of the patterns. (Wermter and Hahn, 2004) studied a German-language corpus that contains different medical texts using TTR and sentence length.

Figure 7. NL development using sublanguage study

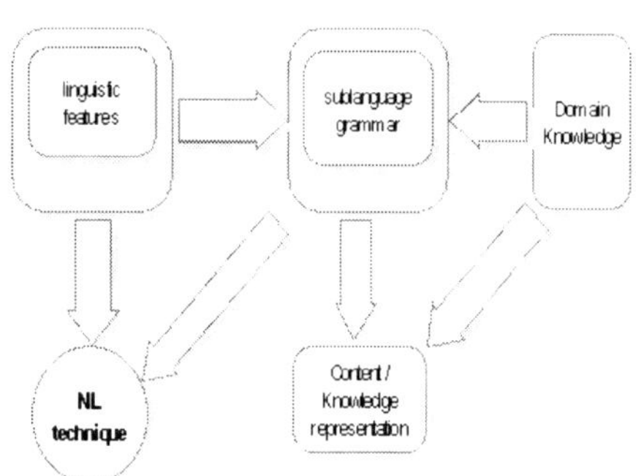

CASE STUDY: CATS SYSTEM

Overview

The classified ads transactional system (CATS) was built by the first author (Daoud, 2009) to

Figure 8. A "sell" post

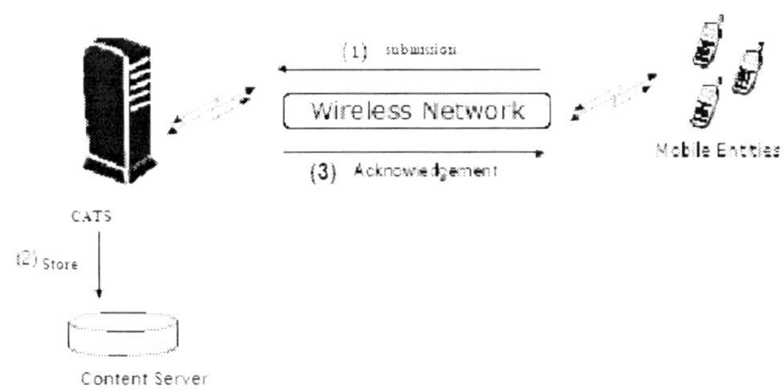

demonstrate the importance of employing human factors in building a NLP system. The system is currently deployed in Jordan with one of the main mobile operators.

CATS is a SMS-based classified ads selling and buying platform. It allows users to send classified ads of the articles/goods they would like to sell and to search for the goods/articles they desire using a full natural language interface. The system extracts content from both "sell" and "looking for" posts and transforms the natural language text into a corresponding content representation. For a "sell" post, the content representation is mapped into database records and stored into RDMS. For a "looking for" type of posts, the content representation is used to build a SQL query to retrieve information from the data that has previously been processed and stored in the RDMS. The current version of CATS is in Arabic and is restricted to the Cars and Real Estate sub-domains. However, the system is designed with a view of easy porting into other domains and other languages.

Human Factors Implemented in CATS

We have shown that different human factors influence the development of a successful and usable NL system. We will discuss them within the scope of CATS to demonstrate their importance:

The Classified Ads Domain

CATS handles classified ads only, hence it is a restricted domain system. The text processed by CATS has a limited vocabulary and syntax. Classified ads are an ideal domain for this experiment. In addition to the compatibility of e-commerce definition, the domain is well known by users, hence there is no need whatsoever for training them.

Figure 9. A "looking for" post

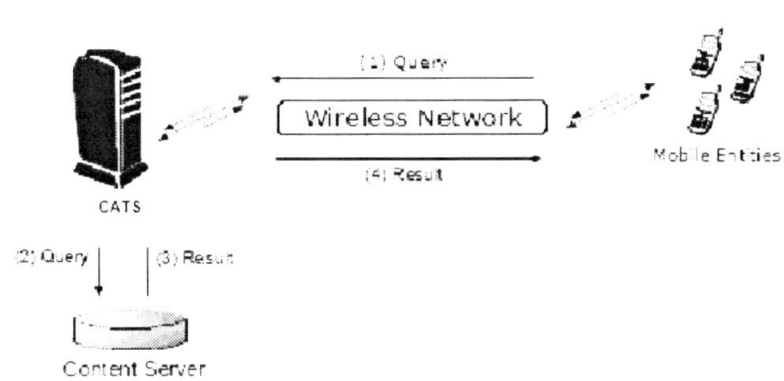

From the users' point of view, this domain is of high interest. Currently, most of the classified ads are published weekly through printed materials. You have to read thousands of classified ads to get what you are looking for. Looking at the web for ads, we find that the most popular publisher of one of the classified ads circulation has a website (www.ewaseet.com) that is updated weekly, without any searching capabilities. Hence, it is difficult to find what one is looking for. Furthermore, we discovered that most of the listed classified ads in the printed materials are posted by brokers and not by ordinary people. Hence, we thought of providing buyers and sellers with a more credible and transparent channel of communication, able of connecting them efficiently.

We picked as a beginning the Cars and Real Estate domains as both of them are very popular in Jordan and constitute the major shares of the classified ads industry.

The Medium Factor

Recent (September 2008) statistical information indicates that a significant 86 per cent of Jordanians currently subscribe to a mobile phone service. This is a massive increase from the 8 per cent and 28 per cent penetration rate at the end of 2000 and 2004 respectively. On the other hand, 24 percent of the Jordanians use Internet, according to recently available data.

More to the point, SMS is becoming one of the most popular channels for exchanging information. The most important factor that explains this enormous success is its simple, immediate, and confidential way to communicate. Moreover, it has played a major role in narrowing down the digital gap caused by the low level of Internet penetration in some countries. As an example, SMS enables communication between millions of mobile subscribers anywhere, anytime, and hence offers unmatched service coverage, beyond even that of the Internet, as mobile phone penetration is much higher than Internet penetration.

The Interface Style Factor

We are interested in developing a NL system that should target the largest possible group in Jordan where the penetration of mobile users is much higher than Internet. The system is restricted to specific classified ads domain. In this system users request information and the system provides them with responses to their requests. To broaden the system's user base, SMS would be the best medium to reach the largest possible number of users. Thus, it is not possible to interact with the system using this medium by a menu driven interface, or by form filling. Therefore, the only practical interaction style is spontaneous natural language text: the user should be allowed to post a complete sentence, without any restrictions on the used language. More to the point, natural language dialogue interfaces are inadequate and time consuming, since users need to send several messages to accomplish the task. On the contrary, a human language allows a wide range of expressions. It helps the user specify many parameters as needed in a single request, in natural language, thus avoiding the tiresome aspect of dialogue-based systems.

Thus, the above reasons justify our decision to select SMS as the medium of communication for this fully NL-based system. In other words, a full NL-based interface is the only interaction method that is practical and possible in SMS. The second reason is the popularity of the SMS medium, as people from different backgrounds can interact with CATS. The final reason is the mobility advantage of SMS: people can interact with the system anywhere and anytime. That encourages people to use this service.

The Text Style Factor

The shortage of data is one of the main obstacles in developing natural language systems. It is not easy to collect corpuses for restricted domains, especially if they must come from a very private medium of communication such as SMS. A very interesting aspect of CATS (Daoud, 2005) is the study of its natural "sublanguages". That provides an unprecedented opportunity to analyze the SMS-based restricted domain sublanguages. As a matter of fact, the dataset collected from CATS is unique and very interesting. Firstly, it contains real, spontaneous, and unedited text. Secondly, it is written by thousands of authors from a diversity of backgrounds. Thirdly, it covers different domains and topics.

The experiments conducted by (Sekine, 1997) clearly concluded that parsing a domain-specific text requires a grammar that suits that domain. Domain-specific grammars can produce better results than domain-independent grammars. From a practical perspective, corpus is the sources of these domain-dependent grammars.

Typology of SMS-Based Task-Oriented Sublanguages

To measure the lexical complexity of a SMS-based classified ads sublanguage, we use the type-token ratio (TTR). This ratio increases with the lexical complexity and richness of the text and decreases if more words repeat themselves and the lexical complexity is lower. We calculated the TTR for different corpuses for the sake of comparison.

We measure the language complexity by the length of the sentence in words, and we do comparisons between different domain and between SMS-based and Web-based sentences. Finally, finding the words frequency in a corpus identifies the nature of text (telegraphic or normal), in particular the less the percentage of function words in a corpus, the more fragmentary is its style.

The analysis of the sublanguage also includes the manual study of lexico-semantic patterns found in the posts. Our objective is extracting classes of objects that specify the domain knowledge described by the sublanguage.

General Corpus Statistics

The SMS-based corpus consists of posts from Cars and Real Estate domains collected during a limited period of CATS operation. We also collected SMS-based Job announcements sent to another mobile short number connected to a publisher of a printed circulation interested in the Jobs domain. The open domain sentences are the posts received by CATS or by the Job announcement short number and are not related to any mentioned domains or any related ones.

As shown in Figure 10, the length of sentences in the Cars domain is less than that of the Real Estate domain, compared to 7.3 words for TREC questions. We also find that the average sentence length is highest in the Job domain. These findings suggest that the language complexity, as measured by sentence length, is higher in the Jobs domain than in the Real Estate and Cars domains. In other words, the user needs a lesser amount of words to encode his thoughts in the Cars domain than in the Real Estate domain or in the Jobs domain.

We also find that the average length of SMS-based open domain sentences is only 6.42 words, making it the smallest among all other domains. When we compare SMS-based posts with Web-

Figure 10. Examined SMS-based corpus

Domain	Number of sentences	Sentence average length (words)	types	Tokens	TTR
Cars	771	9	1181	5875	.201
Real Estate	641	12.5	1441	6182	.233
Jobs	174	14.61	921	2452	.375
Open	231	6.42	1099	1538	.714

based posts, we find that the first are generally smaller than the second. This might be explained by the fact that the SMS medium imposes a length and cost constraint on the advertisement. On the other hand, the Web posters have no strict constraint, are encouraged to making their posts arbitrarily long, and therefore tend to include more irrelevant information.

We also calculated the type-token ratio (TTR), the ratio of the number of word types to the number of word tokens in the text, as a measure of range of vocabulary used. That is, the greater the range of possible referents, the greater the number of word types which will appear in a text. For a given number of tokens, a text tending toward greater explicitness will contain a higher number of types (Bruthiaux, 1994).

As shown in Figure 11 the least TTR value was for Cars at 0.201, then for Real Estate at 0.233, then for Jobs at 0.375. That suggests that the vocabulary is more limited in the Cars domain than in the Real estate domain, and in turn than in the Jobs announcement.

We also find that the TTR of the SMS-open domain is very high (0.714), suggesting a lexically complex and rich text. This indicates that posters were not focusing narrowly on particular topics of discussion so that the same words were not repeated often. These TTR results confirm the findings of (Bruthiaux 1994) for the same domains.

The TTR values of Web-based posts were even lower compared to SMS based ones, suggesting a higher lexical complexity and diversity in the SMS-based text.

The TTR of general Arabic corpus of nearly the same text length (number of tokens) is 0.539 as calculated in (Goweder and De Roeck, 2001). Comparing this result with TTRs found for the Cars, Real estate, and Jobs posts confirms the narrow scope and limited vocabulary characteristic of SMS-based restricted domains. On the other hand, the SMS-based open domain TTR is higher than that of (Goweder and De Roeck, 2001), suggesting a more topical diversity.

Word Frequencies

The top 50 most frequently used words by percentage in SMS-based Cars, Real Estate, Jobs and open domains are 53.77%, 45.76%, 43.27 and 22.37% respectively. These findings suggest that as we move from Cars to Real Estate, to Jobs and finally to the Open domain, the percentage of function words (such as prepositions) increases. This finding can be correlated with the TTR of each sub-domain, indicating a less telegraphic text as we move from the Cars domain to the open domain.

Lexical Characteristics

Although the vocabulary used is narrow and limited, users use different words to express the same concept. For example, to express the concept "more", users use around 30 words (including spelling variations).

We observe that some words in the Cars and Real Estate domains can have different meanings than in the open domain. Therefore, specialized dictionaries are required to process the text. For

Figure 11. Comparison of SMS based TTRs with the findings in Goweder and De Roeck (2001)

Domain	Types	Tokens	TTR	TTR of general Arabic
Cars	1181	5875	.201	0.539
Real Estate	1441	6182	.233	0.539
Jobs	921	2452	.375	0.579
open	1099	1538	.714	0.618

example, in the Cars domain *'duck'* denotes a Mercedes model, and a *'piece'* in the Real Estate domain means a land.

Multi-word concepts and terms are also very frequent to the extent that they appear in the top-most frequent words list.

In the Cars domain, named entities are references to Car Makes and Models. In the Real Estate domain, they are references to Locations. The study of the corpus of classified ads shows that Named Entities consist of one or more words. As Arabic is not like English in distinguishing named entities by capitalizing the first character, and sentences are very short, recognition of named entities is impossible without using lexical lookup.

The dataset under study is full of numerical values. In the Car domain, they represent price, year, motor size and sometime models for some car makes. In the Real Estate domain, they represent the price, area, number of bedrooms, etc. The posters encode numerical values differently. Some of them use non-arabic numerals such as "three thousands". Others use Arabic numerals such as "3000". Finally, some posters combine the two approaches and write expressions such as "3 thousands". Usually, numerical values are preceded by hint words and/or followed by unit words. But, it becomes problematic when users fail to write both hints words and unit words, as demonstrated by the post:

"For sale Mercedes 200 1999"

There are many variations of spelling of the Arabic text in the studied corpus. For example people write the Alef letter "ا", or with Hamza (ء) over it "أ" or under it "إ.". Also, we find confusions between the Ha' "ه" and Ta' "'ة", and between Ya' "ي" and Alef-Maqsoura "ى".

Another problem is the wrong insertions of spaces. In Arabic, spaces are normally used to separate words. After some Arabic letters, people tend to wrongly insert a space, or to (also wrongly) omit it (e.g., "أبو بكر" or "أبوبكر" (Abu-Baker)).

The inconsistency of the Arabic spelling of transliterated proper nouns is a major challenge. This appears frequently in the classified ads text where many of the proper names (car make and model as an example) are transliterated from other languages. This phenomenon is noticeable within unedited and spontaneous classified ads, reflecting the cultural and educational background of the text writer.

Syntactic Characteristics

The studied posts can have different syntactic structures caused by different word orders and grouping patterns of their constituents.

In some posts, we find that some constituents are not present because they do not interest the poster or are irrelevant for him, in cases such as "looking for a car above 2001". In this post, the user omits all other criteria that can restrict his query and mentions only one.

Other causes of omissions arise when information is deemed to be implicitly known, such as "looking for a Clio" in which "car" is omitted, or "for sale 500 square meter" in which "land" is omitted.

In some posts, we don't find any indication of the type ("sell" or "looking for"): "a Toyota Corolla above 99 and with less than 7000 dinar" because the poster thinks it can be known from the context of the post.

Semantic Characteristics

We have shown that the syntactic structure for different posts which express the same information can vary enormously.

Some posters encode the knowledge but at different levels of detail. For example: "looking for a CIVIC" or "A Japanese Honda Civic car for sale".

The use of generalization in the query is also presented in the studied corpus. For example, the use of a generalization concept for searching is quite frequent such as"looking for a French car", "looking for a villa in West Amman" or "looking for economical car". Usually these words ("French", "West Amman" and "economical") do not appear in the "sell" post since they are implicitly known.

The Main Outcome of Sublanguage Analysis

The data that we studied contains many alternative surface structures for the same utterance. We believe this phenomenon reflects the diversity of the posters. It was evident from looking at the posts that there was no unique underlying syntactic structure in the sublanguage used. Some posts consist of fragmented phrases (telegraphic) rather than fully-formed sentences. Other posts are more cohesive and some are full sentences. Obviously, syntax-based parsing based methods would not prove very useful in dealing with the given data. As an example, a traditional parser looking for object and subject will fail in analyzing the following post:

"Opel Astra station color red (power sunroof Center Electrical windows and mirrors check for sale"

Similarly, techniques used for semi-structured text relying on position, layout and format of text are bound to fail on the given data.

Therefore we can view a classified ads post as a sequence of properties restricting the main domain object (i.e. car, apartment). This statement is true for both Real Estate and Cars and for both "sell" and "looking for" posts. This information model is more efficient than relying on syntactic structures for the description of the SMS.

This approach of describing sentences semantically achieves better results than using pure syntactic description. They were introduced as an engineering methodology, which allows semantic knowledge to be easily included in the system (Androutsopoulos, Ritchie et al., 1995).

The study suggests also the need for a lexical lookup that is able to handle spelling variations. It should also be able to store a concepts hierarchy.

In the information structure with this sublanguage, it is also necessary to have a content representation that is able to model the post and normalize the knowledge in a post regardless of its original surface structure.

Hence, what is required is an additional level of abstraction that represents the underlying meaning of a post.

Formulating correct responses for users' queries is another motivation for defining a unique knowledge representation for both types of posts. Suppose we have the following "sell" post:

ادلخ يف لقتسم لزنم عيبلل "For sale an independent house in Khalda"

And that a poster sends the following query:

نامع برغ يف اليف بولطم "Wanted *a villa in the West of Amman"*

Relying only on bag of words for finding answers is insufficient, and of course will lead to totally unacceptable results, since none of the tokens in the "looking for" post match any of those in the "sell" post. This example shows clearly the need to transform both posts into a language-independent structure that captures the meaning. This will enable the system to correctly find matches, because posts with similar meaning will be recognized, regardless of how they are structured grammatically and which particular terms are used.

Semantic Classification

Each concept has a semantic category which is the direct outcome of the study of the corpus. For example, in the Cars domain, we have semantic categories for: vehicle type, car manufacturer, model, color, motor size unit, motor size hint word, price hint word, currency, features etc.

In the same manner, we have different categories for the Real Estate domain such as: property type, area, area unit, locations, floor, bedroom number, feature, etc.

The semantic classification includes semantic taxonomy. We use feature-based categorization of concepts, in which concepts are assigned to categories according to commonalities in specific features. For example, Renault, Peugeot and Citroen share the same feature in that they are all French cars. In a similar manner, Clio and Megan are car models that share the same manufacturer. Similarly, in the Real Estate domain, locations names are grouped together according to the larger area they belong to.

ASSESSMENT AND DISCUSSION

This service is currently available in Jordan, where thousand of people have already used it to sell or buys cars or properties. The number of posts received depends on many factors such as the season or the marketing campaign by the mobile operators. Usually after some marketing we get on average 1000 posts per day, otherwise we get 20 ~30 posts per day.

Evaluation

Because the CATS system is targeting end users, we performed an end-to-end evaluation of the system by surveying users directly. We have explained the system to a sample of around 200 users from different backgrounds, asking them to test the system by posting "sell" and "looking for" SMS messages.

Generally, the feedback was positive: 95% of the participants said that results were accurate. The rest said that the results should be more precise.

We have noticed that 70% of the messages are of the "looking for" type.

Daoud (Daoud, 2009) compared CATS with other similar systems and reported a 91% score for the f-measure. CATS, despite the free, spontaneous and noisy nature of its input, has surpassed other systems in quality.

CONCLUSION

We have shown in this chapter the importance of human factors for building NLP systems. This was demonstrated by the study of CATS which is built around the concept of human factors important to a developing country like Jordan. The corpus produced by CATS is unique and can be exploited in building spontaneous NLP systems. Additionally, we can explore different methodologies to build similar system.

Additionally, CATS shows the importance of Socially-Aware Language Understanding. Where the system should be able to handle the full range of dynamic situations in which language is used, to recognize, interpret, and respond appropriately in all the 'contexts' in which language is encountered, not just formal, well-written text.

Furthermore, CATS system could be ported to other domains using the same methodology. In other words, CATS system is designed with a view of easy porting into other domains and other languages. Specifically similar system that can handle spontaneous text can be very beneficial for people with disabilities. The methodology described above shows clearly the importance of corpus based approach in developing NLP system that can assist handicapped people.

A corpus-based method will certainly lead to a better understanding of the sublanguage used in a given application for people with speech difficulties or handicapped, and the way people encode their thoughts in this context (device, task, and domain). The present system should increase the communication rate of physically disabled individuals via NLP techniques. In turn, this will help in selecting the right approach for the development of NL systems. In CATS system, SMS has been selected as the medium of communication for this fully NL-based system. The developed system can handle spontaneous SMS text and show that human factors are key success elements. Consequently, disabled individuals could eventually interact freely with the system anywhere and anytime.

Additionally, the developed CATS shows the importance of Socially-Aware Language Understanding. The system should be able to handle the full range of dynamic situations in which language is used, to recognize, interpret, and respond appropriately in all the 'contexts' in which language is encountered, not just formal but in well-written text.

NLP systems promise to provide communicative abilities to populations who presently cannot communicate or communicate with great difficulty. Persons with special needs require protections as technologies advance their communicative capacities. However, these populations are also vulnerable to abuse. Persons with physical or mental differences are often the objects of ridicule or abuse. Consequently, they require special care to avoid bullying, teasing or any other type of emotional or possibly physical abuse (pornography, sex solicitation, etc.)

Therefore, our future work will concentrate on handling additional levels of interpretation beyond standard semantics, and can be thought of as self-adapting personal language processing, which will incorporate all the sets of features which convey meaning based on linguistic and paralinguistic cues that humans use in their social-communicative interactions, whether in speech or in every day written communications (e.g., email, text messaging, or instant messaging). Such cues include emphasis marking, and stylistically mediated pragmatic effects such as formality and partiality at the lexical level. These phenomena tend to be less formalized and less obviously rule-governed than syntax

ACKNOWLEDGMENT

Diese Arbeit ist Herrn Professor Dr. Willi Törnig zu seinem 81. Geburtstag gewidmet

REFERENCES

Allen, J. (1987). *Natural language understanding*. San Francisco: Benjamin/Cummings.

Allen, J. (1995). *Natural language understanding* (2nd ed.). San Francisco: Benjamin/Cummings.

Androutsopoulos, I., Ritchie, G. D., & Thanisch, P. (1995). Natural language interfaces to databases: An introduction. *Journal of Natural Language Engineering*, *1*(1). doi:10.1017/S135132490000005X

Appelt, D. E., & Israel, D. J. (1999, August). Introduction to information extraction technology: a tutorial. *Paper presented at the 16th International Joint Conference on Artificial Intelligence (IJCAI'99)*, Stockholm, Sweden.

Baletsa, G., Foulds, R., & Crochetiere, W. (1976). Design parameters of an intelligent communication device. *Paper presented at the 29th Annual Conference on Engineering in Medicine and Biology*.

Barrows, M. B. (2000). Limited parsing of notational text visit notes: Ad-hoc vs. NLP approaches. *Paper presented at the AMIA Symposium.*

Benamara, F. (2004). Cooperative question answering in restricted domains: The WEBCOOP experiment. *Paper presented at the ACL'04 Workshop on Question Answering in Restricted Domains*, Barcelona.

Bonzi, S. (1990). Syntactic patterns in scientific sublanguages: A study of four disciplines. *Journal of the American Society for Information Science American Society for Information Science*, *41*(2), 121–131. doi:10.1002/(SICI)1097-4571(199003)41:2<121::AID-ASI5>3.0.CO;2-S

Bourgeois, M. S. (1993). Effects of memory aids on the dyadic conversations of individuals with dementia. *Journal of Applied Behavior Analysis*, *26*, 77–87. doi:10.1901/jaba.1993.26-77

Bruthiaux, P. (1994). Functional variation in the language of classified ads perspectives. *Working Papers of the Department of English*, City Polytechnic of Hong Kong, 6(2), 21-40.

Camarata, S. M. (1996). *On the importance of integrating naturalistic language, social intervention, and speech intelligibility training. Positive behavioral support: Including people with difficult behavior in the community*. Baltimore, MD: Paul H. Brooks.

Chai, J., Horvath, V., Nicolov, N., Stys, M., Kambhatla, N., & Zadrozny, W. (2002). Natural language assistant - A dialog system for online product recommendation. *AI Magazine*, *23*(2), 63–75.

Charlop-Christy, M. H., LeBlanc, L. A., & Carpenter, M. H. (1999). Naturalistic teaching strategies to teach speech to children with autism: Historical perspective, development & current practice. *California School Psychologist*, *4*, 30–46.

Chelin, J., Kosseim, L., & Radhakrishnan, T. (2006). Using Natural Language Processing to assist the visually handicapped in writing compositions. *In Springer-Verlag (Ed.), Lecture Notes in Computer Science (LNCS)*, Book: Advances in AI (pp. 300-311). Berlin & Heidelberg: Springer-Verlag.

Copestake, A. (1997). Augmented and alternative NLP techniques for augmentative and alternative communication. *Paper presented at the ACL workshop on Natural Language Processing for Communication Aids*, Madrid, Spain.

Copestake, A. (2003). *Natural Language Processing*. Retrieved from http://www.cl.cam.ac.uk/teaching /2002/NatLangProc/nlp1-4.pdf

Copestake, A., & Flickinger, D. (1998). *Enriched Language Models for Flexible Generation in AAC Systems*. Stanford, CA: Stanford University.

Cowan, N. (2001). The magical number 4 in short-term memory: A reconsideration of mental storage capacity. *The Behavioral and Brain Sciences*, *24*, 87–185. doi:10.1017/S0140525X01003922

Daoud, D. (2005, November). Building SMS-based System using Information Extraction Technology. *Paper presented at the ACIDCA-ICMI'2005, Tozeur*, Tunisia.

Daoud, D. (2009). *Building e-commerce systems handling task-oriented spontaneous NL text*. Mauritius: VDM Publishing House Ltd.

Demasco, P., McCoy, K., Gong, Y., & Pennington, C. (1989). Towards more intelligent AAC interfaces: The use of natural language processing. In J.J. Presperin (Ed.), *Proceedings of the Twelfth Annual RESNA Conference* (pp. 27-30). Washington, D.C: RESNA Press.

Demasco, P. W., & F., M. K. (1992). Generating text from compressed input: An intelligent interface for people with severe motor impairments. *Communications of the ACM*, *35*(5), 68–78. doi:10.1145/129875.129881

Demasco, P. W., McCoy, K. F., Jones, M., Pennington, K., & Snyder, S. (1991). *Compansion: A Technique that Applies Natural Language Processing to Augmentative Communication*. Delaware, USA: Applied Science and Engineering Laboratories, University of Delaware.

Dijkstra, K., Bourgeois, M. S., Allen, R. S., & Burgio, L. D. (2004). Conversational coherence: Discourse analysis of older adults with and without dementia. *Journal of Neurolinguistics*, *17*, 263–283. doi:10.1016/S0911-6044(03)00048-4

Engels, R., & Bernt, B. (2000). *Information extraction: State-of-the-Art-Report*. On-To-Knowledge Consortium.

Ernst, T. (1991). *The Phrase Structure of English Negation*. Newark, DE: University of Delaware.

Fillmore, C. J. (1968). *The case for case*. New York: Holt, Rinehart, and Winston.

Fillmore, C. J. (1977). *The case for case reopened. Syntax and Semantics VIII: Grammatical Relations* (pp. 59–81). New York: Academic Press.

Foulds, R., Baletsa, G., & Crochetiere, W. (1975). The Effectiveness of Language Redundancy on Non-Vocal Communication. *In Proceedings of the Conference on Devices and Systems for the Disabled*, 82-86.

Friedman, C., Kra, P., & Rzhetskya, A. (2002). Two biomedical sublanguages: a description based on the theories of Zellig Harris. *Journal of Biomedical Informatics*, *35*(4), 222–235. doi:10.1016/S1532-0464(03)00012-1

Friedman. (2000). A broad coverage Natural Language processing system. *Paper presented at the AMIA Symp 2000.*

Gazdar, G., & Mellish, C. (1989). *Natural language processing in lisp, an introduction to computational linguistics*. New York: Addison-Wesley Publishing Company.

Goweder, A., & De Roeck, A. (2001, July). Assessment of a significant Arabic corpus. *Paper presented at the Arabic NLP Workshop at ACL/EACL 2001*, Toulouse, France.

Griffith, H. W. (1985). *Guide to Symptoms, Illness, and Surgery*. Los Angeles, CA: Body Press.

Grishman, R. (1986). *Computational linguistics: An introduction*. New York: Cambridge University Press.

Guarino, N., Masolo, C., & Vetere, G. (1999). OntoSeek: Content-based access to the web intelligent systems and their applications. *IEEE Intelligent Systems*, *14*(3), 70–80. doi:10.1109/5254.769887

Hammo, B., Abu-Salem, H., Lytinen, S., & Evens, M. (2002, July). QARAB: A question answering system to support the Arabic language. *Paper presented at the Computational Approaches to Semitic Languages workshop*, University of Pennsylvania, USA.

Hyuna, S., Bakkena, S., Friedman, C., & Johnson, S. (2004, September). Natural language processing challenges in HIV/AIDS clinic notes. *Paper presented at the World Congress on Medical Informatics (Medinfo 2004)*, San Francisco, California, USA.

Internet usage statistics. (2006). *Internet usage statistics.* Retrieved from http://www.internetworldstats.com/stats.htm

Joshi, A., Webber, B., & Sag, I. (1981). *Elements of discourse understanding*. Cambridge, UK: Cambridge University Press.

Katz, J. J., & Fodor, J. A. (1963). The structure of semantic theory. *Language*, *39*, 170–210. doi:10.2307/411200

Khoo, C. S.-G. (1997). The use of relation matching in information retrieval. *Electronic Journal*, 7(2).

Kittredge, R. (1982). Variation and homogeneity of sublanguages. In R. Kittredge & J. Lehrberger (Eds.), *Sublanguage: studies of language in restricted semantic domains*. Berlin and New York: de Gruyter.

Kittredge, R. I. (1982). Sublanguages. *American Journal of Computational Linguistics*, *8*(2), 79–84.

Koegel, L. K., Koegel, R. L., Harrower, J. K., & Carter, C. M. (1999). Pivotal response intervention I: Overview of approach. *Journal of the Association for Persons with Severe Disabilities*, *24*, 174–185. doi:10.2511/rpsd.24.3.174

Koegel, R. L., Camarata, S. M., Koegel, L. K., Ben-Tall, A., & Smith, A. E. (1998). Increasing speech intelligibility in children with autism. *Journal of Autism and Developmental Disorders*, *28*, 241–250. doi:10.1023/A:1026073522897

Koegel, R. L., Koegel, L. K., & Surratt, A. (1992). Language intervention and disruptive behavior in preschool children with autism. *Journal of Autism and Developmental Disorders*, *22*, 141–153. doi:10.1007/BF01058147

Koegel, R. L., O'Dell, M. C., & Koegel, L. K. (1987). A Natural language teaching paradigm for nonverbal autistic children. *Journal of Autism and Developmental Disorders*, *17*, 187–200. doi:10.1007/BF01495055

Kruschwitz, U., & Al-Bakour, H. (2005). Users want more sophisticated search assistants: Results of a task-based evaluation. *American Society for Information Science and Technology*, *56*(13), 1377–1393. doi:10.1002/asi.20230

Laski, K. E., Charlop, M. H., & Schreibman, L. (1988). Training parents to use the Natural Language Paradigm to increase their autistic children's speech. *Journal of Applied Behavior Analysis*, *21*, 391–400. doi:10.1901/jaba.1988.21-391

LeBlanc, L. A., Geiger, K. G., Sautter, R. A., & Sidener, T. M. (2006). Using the natural language paradigm (NLP) to increase vocalizations of older adults with cognitive impairments. *Research in Developmental Disabilities*, *28*, 437–444. doi:10.1016/j.ridd.2006.06.004

Lehrberger, J. (1982). Automatic Translation and the Concept of Sublanguage. In Kittredge, R., & Lehrberger, J. (Eds.), *Sublanguage: studies of language in restricted semantic domains*. Berlin, New York: Walter de Gruyter.

Levinson, S. (1983). Pragmatics. Cambridge: Cambridge University Press. Li, X., & Roth, D. (2001). Exploring evidence for shallow parsing. *Paper presented at the 2001 workshop on Computational Natural Language Learning*, Toulouse, France.

Long, B. (1994, May). *Natural language as an interface style*. Abstract retrieved from http://www.dgp.toronto.edu/ people/byron/papers/nli.html

Lytinen, S. L., & Gershman, A. (1986). ATRANS: automatic processing of money transfer messages. *Paper presented at the Fifth National Conference on Artificial Intelligence (AAAI-86)*, Philadelphia, PA.

McCoy, K. F. (1997). Simple NLP techniques for expanding telegraphic sentences. In *Proceedings of Natural Language Processing for Communication Aids, an ACL/EACL '97 Workshop*, Madrid, Spain, July 12.

McCoy, K. F., & Suri, L. Z.Natural Language Processing Principles for Improving Deaf Writing. *Rehabilitation R & D Progress Reports*. *Journal of Rehabilitation Research and Development*, *29*, 331.

Miller, G. A. (1956). The magical number seven, plus or minus two: Some limits on our capacity for processing information. *Psychological Review*, *63*, 81–97. doi:10.1037/h0043158

Mitamura, T. (1999, September). Controlled Language for Multilingual Machine Translation. *Paper presented at the Machine Translation Summit VII, Singapore*.

Newell, A. F., Arnott, J. L., Beattie, W., & Brophy, B. (1992). Effect of "{PAL{" word prediction system on the quality and quantity of text generation. *Augmentative and Alternative Communication*, 304-311. 35.

Pool, J. (2005). *Can controlled languages scale to the Web? DISCUSSION DRAFT ONLY*. Retrieved October 1 2006, from http://utilika.org/pubs/etc/ambigcl/

Riloff, E., Schafer, C., & Yarowsky, D. (2002, Aug. 2002). Inducing information extraction systems for new languages via cross-language projection. *Paper presented at the COLING-02*, Taipei.

Schank, R. (1975). *Conceptual Information Processing*. Amsterdam: North Holland.

Schank, R. (1991). *Where's the AI?*Evanston, IL: Northwestern University, Institute for the Learning Sciences.

Schwitter, R. (2004). Representing knowledge in controlled natural language: A case study. *Paper presented at the Knowledge-Based Intelligent Information and Engineering Systems, 8th International Conference, KES2004*, Wellington, New Zealand.

Sekine, S. (1997). The Domain Dependence of Parsing. *Paper presented at the Applied Natural Language Processing(ANLP'97)*, Washington D.C., USA.

Small, S., & Rieger, C. (1982). *Parsing and comprehending with word experts (a theory and its realization). Strategies for natural language processing*. New Jersey, NJ: Lawrence Erlbaum Associates.

Smeaton, A. F. (1992). Progress in the application of natural language processing to information retrieval tasks. *The Computer Journal*, *35*, 268–278. doi:10.1093/comjnl/35.3.268

Sowa, J. F. (1999). Relating templates to language and logic. In Pazienza, M. T. (Ed.), *Information extraction: Towards scalable, adaptable systems. Lecture Notes in Computer Science* (*Vol. 1714*). Springer-Verlag.

Stokes, T. F., & Baer, D. M. (1977). An implicit technology of generalization. *Journal of Applied Behavior Analysis*, *10*, 349–367. doi:10.1901/jaba.1977.10-349

Suri, L. Z., & McCoy, K. F. (1993a). *Correcting discourse-level errors in a CALL system for second language learners* (pp. 94–02). Newark, DE: Department of Computer and Information Sciences, University of Delaware.

Suri, L. Z., & McCoy, K. F. (1993b). *A methodology for developing an error taxonomy for a computer assisted language learning tool for second language learners*. Newark, DE: Department of Computer and Information Sciences, University of Delaware.

Swiffin, A. L., Arnott, J. L., & Newell, A. F. (1987). The use of syntax in a predictive communication aid for the physically impaired. *Paper presented at the Tenth Annual Conference on Rehabilitation Technology Washington, DC.*

Tagliacozzo, R. (1976). Levels of technicality in scientific communication. *Information Processing & Management*, *12*(2), 95–110. doi:10.1016/0306-4573(76)90040-6

Templeton, M., & Burger, J.D. (1983). Problems in natural-language interface to DSMS with examples from EUFID. *ANLP*, 3-16.

Tennant, H. R., Ross, K. M., Saenz, R. M., Thompson, C. W., & Miller, J. R. (1983). Menu-based natural language understanding. *Paper presented at the 21st annual meeting on Association for Computational Linguistics*, Cambridge, Massachusetts.

Thompson, C. W., Ross, K. M., Tennant, H. R., & Saenz, R. M. (1983). Building usable menu-based natural language Interfaces to databases. *Paper presented at the VLDB-1983.*

Van Dyke, J., McCoy, K., & Demasco, P. (1992). Using syntactic knowledge for word prediction. *Paper presented at the ISAAC-92. Abstract appears in Augmentative and Alternative Communication.*

Van Dyke, J. A. (1991). Word prediction for disabled users: Applying natural language processing to enhance communication. *Unpublished Thesis.* University of Delaware, Delaware, USA.

Vertan, C. (2004,December). Querying multilingual semantic web in natural language. *Paper presented at the Semantic Web Applications and Perspectives (SWAP)*, first Italian Semantic Web Workshop, Ancona, Italy.

Vertan, C., & Hahn, W. (2003). Menu choice translation: a flexible menu-based controlled natural language system. *Paper presented at the Controlled language translation Workshop, EAMT-CLAW-03, Dublin City University.*

Wermter, J., & Hahn, U. (2004). An Annotated German-Language Medical Text Corpus as Language Resource. *Paper presented at the 4th International Conference on Language Resources and Evaluation (LREC'04), Lisbon, Portugal.*

Wilks, Y. (1975). A preferential, pattern-seeking, semantics for natural language inference. *Artificial Intelligence*, *6*, 53–74. doi:10.1016/0004-3702(75)90016-8

Wright, A., Beattie, W., Booth, L., Ricketts, W., & Arnott, J. (1992). An integrated predictive word-processing and spelling correction system. *Paper presented at the 15th Annual RESNA Conference,* Washington, DC.

ADDITIONAL READING

Dempster, M. (2008). Using Natural Language Generation to Encourage Effective Communication in Non-Speaking People. *11th International Conference on Computers Helping People with Special Needs (icchp 08). Proceedings Young Researchers.* Gaizauskas, R. & Hepple, M. (2008). *POOL projects offered by a subset of lecturers in the Natural Language Processing.* Department of Computer Science, University of Sheffield, UK. Obtained from http://nlp.shef.ac.uk/projects/nlpproj_3rd_0708.html

Jurafsky, M., & Martin, J. H. (2000). *Speech and Language Processing: An Introduction to Natural Language Processing, Computational Linguistics, and Speech Recognition.* Upper Saddle River, NJ: Prentice Hall.

McCoy, K., Demasco, P., Jones, M., Pennington, C., & Rowe, C. (1990). Applying natural language processing techniques to augmentative communication system. In H. Karlgren (Ed.), *Proceedings of the Thirteenth International Conference on Computational Linguistics* (pp. 413-415). Morristown, NJ: Association of Computational Linguistics.

McCoy, K. F., & Demasco, P. (1995). Some Applications of Natural Language Processing to the Field of Augmentative and Alternative Communication. *In Proceedings of the IJCAI '95 Workshop on Developing AI Applications for Disabled People* (pp. 97-112). Montreal, Canada.

Newell, A. F., Langer, S., & Hickey, M. (1998). The role of natural language processing in alternative and augmentative communication. *Natural Language Engineering*, *4*, 1–16. doi:10.1017/S135132499800182X

Sager, N. (1981). *Natural Language Information Processing: A Computer Grammar of English and Its Applications*. London: Addison-Wesley.

Todman, J., & Alm, N. (1997). Pragmatics and AAC Approaches to Conversational Goals. *In Proceedings of ACL/EACL Workshop on Natural Language Processing for Communication Aids*. Madrid, Spain, 1-8.

Van Dyke, J. A. (1991). *A Syntactic Predictor to Enhance Communication for Disabled Users. Technical Report*. USA: Department of Computer and Information Sciences, University of Delaware.

Vanderheyden, P., & Pennington, C. (1998). An augmentative communication interface based on conversational schemata. In Mittal, V., Yanco, H., Aronis, J., & Simpson, R. (Eds.), *Assistive Technology and Artificial Intelligence: Applications in robotics, user interfaces and natural language processing* (pp. 109–125). New York, NY: Springer Verlag.

KEY TERMS AND DEFINITIONS

Natural Language: A system of arbitrary symbols through which humans communicate.

Natural Language Processing (NLP): A field of computer science and linguistics concerned with the interactions between computers and human (natural) languages.

Natural Language Interface (NL interface): Accepts users' inputs in natural language allowing interaction with some system, which then results in sufficient responses to input NL text or query statements.

Augmentative and Alternative Communication: (AAC): Communication for those with impairments or restrictions on the production or comprehension of spoken or written language. In other words, it is any method that supplements or replaces speech and writing when these are temporarily or permanently impaired and inadequate to meet all or some of a person's communication needs. Use of AAC involves selecting messages or codes from a set of possibilities. The user can use these elements alone or in combination in order to communicate a variety of messages. AAC may be unaided, or aided, involving high and low technology.

Human Factors: Involves the study of all aspects of the way humans relate to the world around them, with the aim of improving operational performance, safety, through life costs and/or adoption through improvement in the experience of the end user.

Corpus: A collection of linguistic data, either compiled as written texts or as a transcription of recorded speech. The main purpose of a corpus is to verify a hypothesis about language - for example, to determine how the usage of a particular sound, word, or syntactic construction varies.

Sublanguage: A sublanguage is the language of a restricted domain.

Short Messaging Service (SMS): Language usage over computer mediated discourses, like chats, emails and SMS texts, significantly differs from the standard form of the language. An urge towards shorter message length facilitating faster typing and the need for semantic clarity, shape the structure of this non-standard form known as the texting language.

Compilation of References

(1998). *Humanitarian Demining Development Technologies Catalog*. USA: Department of Defence.

Adelstein, F., & Joyce, R. A. (2007). File Marshal: Automatic extraction of peer-to-peer data. *Digital Investigation*, *4*(1), 43–48. doi:10.1016/j.diin.2007.06.016

Adlam, T., Gibbs, C., & Orpwood, R. (2001). The Gloucester smart house bath monitor for people with dementia. *Physica Medica*, *17*(3), 189.

Adlam, T., & Orpwood, R. (2003). Technology, autonomy and cognitive disability. In Proceedings of *UbiComp 2003: Ubiquitous Computing 5th International Conference*. Seattle, WA, USA

Akdeniz, Y. (2008). *Internet child pornography and the law: national and international responses*. Hampshire, England: Ashgate Publishing Limited.

Alcantara, M. (2009). El Bulling, Acoso Escolar. Innovación y experiencias Educativas. *CSI-CSIF Revista Digital, 55*.

Alexander, D. (1993). *Natural Disasters*. London: University College London Press ltd.

Allegrante, J. P. (2003). *Improving functional recovery after hip fracture*. Retrieved March 8, 2010 from http://clinicaltrials.gov/ct/show/ NCT00000436?order=1

Allen, J. (1987). *Natural language understanding*. San Francisco: Benjamin/Cummings.

Allen, J. (1995). *Natural language understanding* (2nd ed.). San Francisco: Benjamin/Cummings.

Allin, S. J., Bharucha, A., Zimmerman, J., Wilson, D., Roberson, M., Stevens, S., et al. (2003). Toward the automatic assessment of behavioral disturbances of dementia. *UbiComp 2003: Ubiquitous Computing 5th International Conference*.Seattle, WA, USA

Allwood, J. (1986). Mentalitet och språk – några reflektioner. In Sällström, P., (ed.) (1986) *Mentaliteter". Funderingar kring begreppet mentalitet och möjligheten av en forskning om mentalitet*, Åbo Akademi: Publications of the Research Institute of the Åbo Akademi Foundation.

Alonso, G. (2003). Human rights and undocumented migration along the Mexican-U.S. border. *UCLA Law Review. University of California, Los Angeles. School of Law*, *51*(October).

Alvi, S., DeKeseredy, W. S., & Ellis, D. (2000). *Contemporary social problems in North American society*. Toronto: Addison Wesley Longman.

Anderson, C. (2009). *Free: The Future of a Radical Price*. New York: Hyperion.

Anderson, R. N. (1999). Method for constructing complete annual u.s. life tables. national center for health statistics. *Vital and Health Stat*, Series 2(129).

Androutsopoulos, I., Ritchie, G. D., & Thanisch, P. (1995). Natural language interfaces to databases: An introduction. *Journal of Natural Language Engineering*, *1*(1). doi:10.1017/S135132490000005X

Appelt, D. E., & Israel, D. J. (1999, August). Introduction to information extraction technology: a tutorial. *Paper presented at the 16th International Joint Conference on Artificial Intelligence (IJCAI'99)*, Stockholm, Sweden.

Arellano, N. E. (2008). Winnipeg police combat online child porn with Image matching software. *IT Business online news,* Retrieved July, 13, 2009, from http://www.itbusiness.ca/it/client/en/ Home/News.asp?id=49700&bSearch=True

Asociación Mexicana de Internet. (AMIPCI). (2005). *Hábitos de los usuarios de internet en México.* Retrieved February 28, 2010, from http://www.amipci.org. mx/temp/ResumenEjecutivoEstudioAMIPCI2005-0315079001138390784OB.pdf

Asociación Mexicana de Internet (AMIPCI). (2009). *Hábitos de los usuarios de internet en México.* Retrieved February 28, 2010, from http://amipci.org.mx/estudios/temp/ RESUMENEJECUTIVOEstudioAMIPCI2009UsuariosdeinternetFINAL-0334725001245691260OB.pdf

Aviles, J. M. (2006). Bulling. Intimidación y maltrato entre alumnado. Retrieved May 1, 2010, From: www.educacionenvalores.org/ Bullying-Intimidacion-y-maltrato.html

Baletsa, G., Foulds, R., & Crochetiere, W. (1976). Design parameters of an intelligent communication device. *Paper presented at the 29th Annual Conference on Engineering in Medicine and Biology.*

Bancroft, L. (2002). *Why does he do that? Inside the minds of angry and controlling men*. New York: Berkley.

Banger, S. K. (2000). *Remote Sensing and Geographical Information System for natural disaster management*. Accessed March 27, 2010 from http://www.gisdevelopment.net/application/natural_hazards/overview/nho0011.htm

BankInfoSecurity. (2009) *Reports Shows Internal IT Attacks Rising*. (Rev. ed.). Retrieved February 1, 2010, from: http://www.bankinfosecurity.com/articles.php?art_id=56&search_keyword=internal+it+attacks+rising&search_method=exact

Bardram, J. (2004). The personal medical unit -- a ubiquitous computing infrastructure for personal pervasive healthcare. *In Proceedings of Ubiquitous Computing 6th International Conference*, Nottingham, UK

Bardram, J., & Bossen, C. (2003). Moving to get ahead: local mobility and collaborativework. In Kuutti, K., Karsten, E. H., Fitzpatrick, G., Dourish, P., & Schmidt K. (Eds), Proceedings of the *Eighth European Conference on Computer Supported Cooperative Work*, p. 355–374, Helsinki, Finland. Amsterdam: Kluwer Academic Publishers.

Bardram, J., Kolbeck, T. A. K., & Nielsen, C. (2003). Supporting local mobility in healthcare by application roaming among heterogeneous devices. In Proceedings of the *Fifth International Conference on Human Computer Interaction with Mobile Devices and Services*. Springer Verlag. 161-176.

Barrows, M. B. (2000). Limited parsing of notational text visit notes: Ad-hoc vs. NLP approaches. *Paper presented at the AMIA Symposium.*

Basili, V. R., Caldiera, G., & Dieter Rombach, H. (2004). The goal question metrics approach. Retrieved April 15, 2007, from http://wwwagse.informatik.uni-kl.de/pubs/repository/basili94b/encyclo.gqm.pdf

BBC Technology News. (2003, January 7). Teenager wins DVD court battle. http://news.bbc.co.uk/1/hi/technology/2635293.stm. London, UK.

Beck, U. (1995). Att uppfinna det politiska. Bidrag till en teori om reflexive modernisering, Uddevalla: Daidalos.

Belkasoft. (n.d.). *Forensic Investigation of Instant Messenger Histories*. Retrieved May 1, 2010, from http://www.forensicfocus.com/ forensic-investigation-of-instant- messenger-histories

Bellovin, S. M. (2006). On the brittleness of software and the infeasibility of security metrics. *IEEE Security & Privacy,* 1540-7993, p. 96.

Bellsey, B. (2009). *Where you are not alone*. Retrieved May 1, 2010, from: http://www.bullying.org

Benamara, F. (2004). Cooperative question answering in restricted domains: The WEBCOOP experiment. *Paper presented at the ACL'04 Workshop on Question Answering in Restricted Domains*, Barcelona.

Bergen, R. K. (1996). *Wife rape*. Thousand Oaks, CA: Sage.

Bergen, R. K., & Bogle, K. A. (2000). Exploring the connection between pornography and sexual violence. *Violence and Victims*, *15*, 227–234.

Berger, R. J. Searles, P., & Cottle, C.E. (1991). *Feminism and pornography*. New York: Praeger.

Berhe, A. A. (2007). The Contribution of Landmines to Land Degradation. *Land Degradation & Development*, *18*, 1–15. doi:10.1002/ldr.754

Betancourt, J. R., Green, A. R., Carrillo, J. E., & Annaneh-Firempong, O. (2003). Defining cultural competence: a practical framework for addressing racial/ethnic disparities in health and health care. *Public Health Reports*, *118*(4), 293–302.

Betowski, B. (2007). *1 in 3 boys heavy porn users, study shows*. Retrieved February 23 from http://www.eurekalert.org/pub_ releases/2007-02/uoa-oit022307.php.

Bevan, N. (1995). Measuring usabilty as quality of use. *Software Quality Journal*, *4*(2), 115–130. doi:10.1007/BF00402715

Bitney, J., & Title, B. (1997). *No bullying program: Preventing bullying/victim violence at school (programs directors manual)*. Central City, MN: Hazelden and Johnson Institute.

Blagden, P. M. (1993). Summary of UN Demining. *Proceedings of the International Symposium on Anti-Personnel Mines*, Montreux, April 1993, CICR/ICRC, pp 117-123.

Boatright, J. R. (2007). *Ethics and the Conduct of Business* (5th ed.). Upper Saddle River, NJ: Pearson Prentice Hall.

Boeringer, S. B. (1994). Pornography and sexual aggression: Associations of violent and nonviolent depictions with rape and rape proclivity. *Deviant Behavior*, *15*, 289–304. doi:10.1080/01639625.1994.9967974

Boise, L., Morgan, D. L., Kaye, J., & Camicioli, R. (1999). Delays in the diagnosis of dementia: perspectives of family caregivers. *American Journal of Alzheimer's Disease*, *14*(1), 20–26. doi:10.1177/153331759901400101

Bonzi, S. (1990). Syntactic patterns in scientific sublanguages: A study of four disciplines. *Journal of the American Society for Information Science American Society for Information Science*, *41*(2), 121–131. doi:10.1002/(SICI)1097-4571(199003)41:2<121::AID-ASI5>3.0.CO;2-S

Bourgeois, M. S. (1993). Effects of memory aids on the dyadic conversations of individuals with dementia. *Journal of Applied Behavior Analysis*, *26*, 77–87. doi:10.1901/jaba.1993.26-77

Bowker, L. H. (1983). *Beating wife-beating*. Lexington, MA: Lexington Books.

Boyer, N. R., Maher, P. A., & Kirkman, S. (2006). Transformative learning in online settings: The use of self-direction, metacognition, and collaborative learning. *Journal of Transformative Education*, *4*(4), 335–361. doi:10.1177/1541344606295318

Boyle, T. (2007). Small towns have higher crime rates. *Toronto Star*, June 29, A1.

Branigan, T. (2009, July 2). China's Green Dam internet fitering system will go ahead, official says. *Guardian*, pp. http://www.guardian.co.uk/ world/2009/jul/02/china-green-dam.

Brannigan, A., & Goldenberg, S. (1987). The study of aggressive pornography: The vicissitudes of relevance. *Critical Studies in Mass Communication*, *4*, 289–304. doi:10.1080/15295038709360135

Brante, T. (1980). *Vetenskapens struktur och förändring*. Lund: Doxa.

Braz, C., Seffah, A., & M'Raihi, D. (2007). Designing a trade-off between usability and security: A metrics based-Model. In Baranauskas, M., Palanque, P., Abascal, J., Junqueira Barbosa, S.(Eds.) *11th IFIP TC 13 Conference on Human-Computer Interaction, INTERACT 2007* (pp. 114–126). Rio de Janeiro. Lecture Notes in Computer Science, Springer.

Brennan, B. M. & and B.A. Woodruff, B. A.(2003). Death and Injury from Landmines and Unexploded Ordnance in Afghanistan. *J.Am.Med.Assoc*, *290*(5), 2003, pp.650–653. doi:10.1001/jama.290.5.650

Brownridge, D. A., & Halli, S. S. (2001). *Explaining violence against women in Canada*. Lanham, MD: Lexington Books.

Bruschini, C., De Bruyn, K., Sahli, H., & Cornelis, J. (1999). *EU-Report-EUDEM: The EU in HUmanitarian DEMining*. Brussels, Belgium: European Commision.

Bruschini, C., De Bruyn, K., Sahli, H., & Cornelis, J. (1999). EUDEM: The EU in Humanitarian DEMining. *EU report*, Brussels, 1999.

Bruthiaux, P. (1994). Functional variation in the language of classified ads perspectives. *Working Papers of the Department of English*, City Polytechnic of Hong Kong, 6(2), 21-40.

Bryant, C. G. A. (1975). Kuhn, Paradigms and Sociology. *The British Journal of Sociology*, *26*(3), 354–359. doi:10.2307/589851

Buchanan, J., & Young, L. (2000). Examining the relationship between material conditions, long term problematic drug use and social exclusion: A new strategy for social inclusion. In Bradshaw, J., & Sainsbury, R. (Eds.), *Experiencing Poverty* (pp. 120–143). London, UK: Ashgate Press.

Burgess, A. W., Hartman, C. R., McCausland, M. P., & Powers, P. (1984, May). Response patterns in children and adolescents exploited through sex rings and pornography. *The American Journal of Psychiatry*, *141*(5).

Burke, P. (1986). Strengths and weaknesses of the history of mentalities, in *History of European ideas*.

Burke, P. (1992/1990). *Annales-skolan. En introduktion*, Göteborg Bokförlaget Daidalos AB.

(1996). InBuschmann, F., Maunier, R., Rohnert, H., Sommerlad, P., & Stal, M. (Eds.). Pattern-oriented software architecture: *Vol. 1. A system of patterns*. Hoboken, NJ: Wiley.

Bustamante, J. (1997). *Cruzar la línea: La migración de México a los Estados Unidos*. Mexico City, Mexico: Fondo de Cultura Económica.

Bustamante, J. A. (2001). Proposition 187 and operation gatekeeper: Cases for the sociology of international immigration and human rights. *Migraciones Internacionales*, July - December, 1(1).

Buxton, W. (2007). *Sketching user experience: Getting the design right and the right design*. Morgan Kaufmann.

Cabezas, H. (2007). Detección de Conductas Agresivas "Bullyings" en escolares de Sexto a Octavo Año en Una Muestra Costarricense. *Revista Educación*, *31*(1), 123–133.

Cain, B., & Meidinger, T. (1996). The Improved Landmine Detection System. *EUREL*, *1996*, 188–192.

Calvillo Gámez, E. H., Cairns, P., & Cox, A. L. (2010). Assessing the Core Elements of the Gaming Experience. In Bernhaupt, R. (Ed.), *Evaluating Experience in Games*. London, UK: Springer. doi:10.1007/978-1-84882-963-3_4

Calvillo Gámez, E. H., & Cairns, P. (2008). Pulling the strings. A theory of Puppetry for the Gaming Experience. In S. Günzel, M. Liebe, & D. Mersch (Ed.), *Conference Proceedigns of the Philosophy of Computer Games* (pp. 308-323). Potsdam, Germany: Potsdam University Press.

Camarata, S. M. (1996). *On the importance of integrating naturalistic language, social intervention, and speech intelligibility training. Positive behavioral support: Including people with difficult behavior in the community*. Baltimore, MD: Paul H. Brooks.

Cameron, L., & Low, G. (1999). *Researching and applying metaphor*. Cambridge University Press.

Canadian Legal Information Institute. (2008). Criminal code of Canada – Bill C-46, *Statues and Regulations of Canada,* R.S.C. 1985, C-46. Retrieved [Aug., 4, 2009] from http://www.canlii.org/en/ca/laws/ stat/rsc-1985-c-c-46/latest/rsc- 1985-c-c-46.html.

Carroll, J. S. (2008). Pornography and emerging adulthood. *Journal of Adolescent Research*, *23*, 6–30. doi:10.1177/0743558407306348

Castells, M. (2001). *The Internet Galaxy: Reflections on the Internet, Business, and Society*. New York: Oxford University Press.

Castells, M. (2003). *La galaxia internet. reflexiones sobre internet, empresa y sociedad*. Barcelona, Spain: Random House Mondadori.

CDHDF. (2009). *Comisión de Derechos Humanos del Distrito Federal*. Retrieved March 8, 2010 from http://www.cdhdf.org.mx/

CEDOH. (2009). *Nociones fundamentales sobre la discriminación*. Retrieved on 30/04/2010, From http://www.cedoh.org/proyectos/ discriminacion/assets/Nociones.pdf

Chai, J., Horvath, V., Nicolov, N., Stys, M., Kambhatla, N., & Zadrozny, W. (2002). Natural language assistant - A dialog system for online product recommendation. *AI Magazine*, *23*(2), 63–75.

Charlop-Christy, M. H., LeBlanc, L. A., & Carpenter, M. H. (1999). Naturalistic teaching strategies to teach speech to children with autism: Historical perspective, development & current practice. *California School Psychologist*, *4*, 30–46.

Check, J. (1995). Teenage training: The effects of pornography on adolescent males. In Lederer, L., & Delgado, R. (Eds.), *The Price We Pay: The Case Against Racist Speech, Hate Propaganda and Pornography* (pp. 89–91). New York: Hill and Wang.

Chelin, J., Kosselm, L., & Radhakrishnan, T. (2006). Using Natural Language Processing to assist the visually handicapped in writing compositions. *In Springer-Verlag (Ed.), Lecture Notes in Computer Science (LNCS)*, Book: Advances in AI (pp. 300-311). Berlin & Heidelberg: Springer-Verlag.

Chiasson, S., Biddle, R., & van Oorschot, P. C. (2007). A second look at the usability of click-based graphical passwords. *In Symposium on Usable Privacy and Security* (1-12). Pittsburgh, Pennsylvania, ACM Press.

Chiasson, S., van Oorschot, P. C., & Biddle, R. (2006). A usability study and critique of two password managers. In *15th USENIX Security Symposium* (pp. 1-16), Vancouver, B.C., Canada, IEEE, Press.

Chiasson, S., van Oorschot, P. C., & Biddle, R. (2007). Graphical password authentication using cued click points. In Biskup, J., & López, J. (Eds.). *12th European Symposium On Research In Computer Security* (pp. 359-374). Dresden, Germany. Lecture Notes in Computer Science. New York: Springer.

Child Rights Information Network. (1990). *United Nations Convention on the Rights of the Child*. Retrieved August 4, 2009, from http://www.crin.org/docs/resources /treaties/uncrc.asp

Christensen, A. (1994). *Hemrätt i hyreshuset. En rättsvetenskaplig studie av bostadshyresgästens besittningsskydd*. Stockholm: Juristförlaget.

Claus, M. (2000). Internet changes porn scene at DePauw U. *High Beam Research*, December 12, 1. Retrieved August 26, 2009, from http://www.highbeam.com/ doc/ 1P1-37925753.html.

CONAPO. (2004). *There are in Mexico 7.9 million elders. Consejo Nacional de Poblacion. Technical Report*. Mexico: Mexico City.

Conway, J. (2004). Theory of games to improve artificial societies. *International Journal of Intelligent Games & Simulation*, *2*(1).

Copestake, A., & Flickinger, D. (1998). *Enriched Language Models for Flexible Generation in AAC Systems*. Stanford, CA: Stanford University.

Copestake, A. (1997). Augmented and alternative NLP techniques for augmentative and alternative communication. *Paper presented at the ACL workshop on Natural Language Processing for Communication Aids*, Madrid, Spain.

Copestake, A. (2003). *Natural Language Processing*. Retrieved from http://www.cl.cam.ac.uk/teaching /2002/ NatLangProc/nlp1-4.pdf

Cowan, N. (2001). The magical number 4 in short-term memory: A reconsideration of mental storage capacity. *The Behavioral and Brain Sciences*, *24*, 87–185. doi:10.1017/ S0140525X01003922

Cranor, L. F., & Garfinkel, S. (Eds.). (2005). *Security and usability: Designing secure systems that people can use*. Sebastopol, CA: O'Reilly Media.

Cranor, L. F. (2003). Designing a privacy preference specification interface: A case study. In *Workshop on Human-Computer Interaction and Security Systems, ACM CHI 2003*. Fort Lauderdale, ACM Press.

Creighton, S. J. (2003). *Child pornography: Images of the abuse of children*. NSPCC Research Department, NSPCC Information Briefings. Retrieved September 13, 2009, from http://wallis.kezenfogva.iif.hu/eu_ konyvtar/projektek/Daphne/www/documents/ projets/2003-017/ChildPornography.pdf

Current Population Survey. (2000). *Current population survey.* Retrieved February 28, 2010, From http://www.census.gov/apsd/ techdoc/cps/cpsmar00.pdf

Cyber-Ark. (2008). *The Global Recession and its Effect on Work Ethics*. Retrieved February 1 2010, from: http://cyber-ark.com

Cyberbulling (2010). *What is cyberbulling?* Obtained May 1, 2010, from: http://www.cyberbullying.info/whatis/whatis.php

Dabove, M. I. (2006). Discriminación y ancianidad. Reflexiones Filosóficas en torno al sistema jurídico Argentino. *Magazine Telemática de Filosofía del Derecho*, 153-164.

Dalton, B., & Taylor, K. (2000). Distributed Robotics Over the Internet. *IEEE Robotics & Automation Magazine*, *7*(2), 22–27. doi:10.1109/100.848264

Daly, K., & Chesney-Lind, M. (1988). Feminism and criminology. *Justice Quarterly*, *5*, 497–538. doi:10.1080/07418828800089871

Daoud, D. (2009). *Building e-commerce systems handling task-oriented spontaneous NL text*. Mauritius: VDM Publishing House Ltd.

Daoud, D. (2005, November). Building SMS-based System using Information Extraction Technology. *Paper presented at the ACIDCA-ICMI'2005, Tozeur*, Tunisia.

Data Report, CRYPSYS. (2008). Unique software provides child pornography protection: CRYPSYS Data Security is launching NetClean software to prevent the spread of child pornography in the workplace. *My News Desk*, Retrieved August 12, 2009, from www.mynewsdesk.com/se/pressroom/ netclean_technologies/pressrelease/download/resource_ attached_pdf_document/215913

Deazley, R. (2006). *Rethinking Copyright: History, Theory, Language*. Cheltenham: Edward Elgar Publishing Limited.

DeCSS Central. (n.d.). *DeCSS Central*. Retrieved from http://www.lemuria.org/DeCSS/

DeCSS Central. (n.d.). *DVD CCA Lawsuit(s)*. Retrieved from http://www.lemuria.org/DeCSS/cca.html

DeKeseredy, W. S. (2010). *Contemporary critical criminology*. London: Routledge.

DeKeseredy, W. S., Ellis, D., & Alvi, S. (2005). *Deviance and crime: Theory, research and policy*. Cincinnati, OH: LexisNexis.

DeKeseredy, W. S., & Schwartz, M. D. (1996). *Contemporary criminology*. Belmont, CA: Wadsworth.

DeKeseredy, W. S., & Schwartz, M. D. (1998a). *Woman abuse on campus: Results from the Canadian national survey*. Thousand Oaks, CA: Sage.

DeKeseredy, W. S., & Schwartz, M. D. (2009). *Dangerous exits: Escaping abusive relationships in rural America*. New Brunswick, NJ: Rutgers University Press.

DeKeseredy, W. S., Schwartz, M. D., Fagen, D., & Hall, M. (2006). Separation/divorce sexual assault: The contribution of male peer support. *Feminist Criminology*, *1*, 228–250. doi:10.1177/1557085106288862

DeKeseredy, W. S. (2009b). Patterns of violence in the family. In Baker, M. (Ed.), *Families: Changing trends in Canada* (6th ed., pp. 179–205). Toronto: McGraw-Hill Ryerson.

DeKeseredy, W. S., & Flack, W. F. Jr. (2007). Sexual assault in colleges and universities. In Barak, G. (Ed.), *Battleground criminal justice* (pp. 693–696). Westport, CT: Greenwood.

DeKeseredy, W. S., & Schwartz, M. D. (1998b). Male peer support and woman abuse in postsecondary school courtship: Suggestions for new directions in sociological research. In Bergen, R. K. (Ed.), *Issues in intimate violence* (pp. 83–96). Thousand Oaks, CA: Sage.

DeKeseredy, W. S. (2009a). Male violence against women in North America as hate crime. In B. Perry (Ed.), *Hate crimes, volume 3: The victims of hate crime* (pp. 151-172). Santa Barbara, CA: Praeger.

DeKeseredy, W.S. (in press). *Violence against women in Canada*. Toronto, Canada: University of Toronto Press.

Demasco, P. W., & F., M. K. (1992). Generating text from compressed input: An intelligent interface for people with severe motor impairments. *Communications of the ACM*, *35*(5), 68–78. doi:10.1145/129875.129881

Demasco, P. W., McCoy, K. F., Jones, M., Pennington, K., & Snyder, S. (1991). *Compansion: A Technique that Applies Natural Language Processing to Augmentative Communication*. Delaware, USA: Applied Science and Engineering Laboratories, University of Delaware.

Demasco, P., McCoy, K., Gong, Y., & Pennington, C. (1989). Towards more intelligent AAC interfaces: The use of natural language processing. In J.J. Presperin (Ed.), *Proceedings of the Twelfth Annual RESNA Conference* (pp. 27-30). Washington, D.C: RESNA Press.

Department of Defense, Humanitarian Demining Research and Development (R&D) Program (2002). Other Completed Mine/Vegetation Clearance Equipment. *Development Technologies Catalog 2001-2002*.

Dhamija, R. (2007). Security usability studies: Risk, roles and ethics. In *Workshop on Security User Studies, ACM CHI 2007*. San Jose, CA: ACM Press.

Dijkstra, K., Bourgeois, M. S., Allen, R. S., & Burgio, L. D. (2004). Conversational coherence: Discourse analysis of older adults with and without dementia. *Journal of Neurolinguistics*, *17*, 263–283. doi:10.1016/S0911-6044(03)00048-4

Dines, G. (2006). The white man's burden: Gonzo pornography and the construction of black masculinity. *Yale Journal of Law and Feminism*, *18*, 296–297.

Dines, G., & Jensen, R. (2008a). Pornography. In Renzetti, C. M., & Edleson, J. L. (Eds.), *Encyclopedia of interpersonal violence* (pp. 519–520). Thousand Oaks, CA: Sage.

Dines, G., & Jensen, R. (2008b). Internet, pornography. In Renzetti, C. M., & Edleson, J. L. (Eds.), *Encyclopedia of interpersonal violence* (pp. 365–366). Thousand Oaks, CA: Sage.

Directive 2001/29/EC of the European Parliament and of the Council of 22 May 2001 *on the harmonization of certain aspects of copyright and related rights in the information society*, (Infosoc).

Dix, A. (2007). Designing for Appropriation. *Proceedings of the BCS HCI 2007 Conference, People and Computers XXI. 2.* London, UK: BCS-eWik.

Domínguez, J. I., & Fernández De Castro, R. (2009). *United States and Mexico: Between Partnership and Conflict*. Abingdon, Oxfordshire: Routledge.

Donnermeyer, J. F., Jobes, P., & Barclay, E. (2006). Rural crime, poverty, and community. In DeKeseredy, W. S., & Perry, B. (Eds.), *Advancing critical criminology: Theory and application* (pp. 199–218). Lanham, MD: Lexington Books.

Donnerstein, E. (1984). Pornography: Its effect on violence against women. In Malamuth, N., & Donnerstein, E. (Eds.), *Pornography and sexual aggression* (pp. 53–81). Orlando, FL: Academic Press.

Doring, N. (2009). The Internet's impact on sexuality: A critical review of 15 years of research. *Computers in Human Behavior*, *25*, 1089–1101. doi:10.1016/j.chb.2009.04.003

Dourish, P. (2003). The appropriation of interactive technologies: Some lessons from placeless documents. *Compuer Supported Cooperative Work (CSCW)*, 465-490.

Drucker, P. (1992). *The age of discontinuity*. New York: Harper & Row.

Ds 1999:36 The Swedish Penal Code.

DSM-IV. (1995). *Manual diagnóstico y estadístico de los transtornos mentales (DSM-IV).* Madrid, Spain: Masson, S.A.

Durkheim, E. (1933). *Emile Durkheim on the division of labour in society: Being a translation of his De la division du travail social*, with an estimate of his work by George Simpson. New York.

DVD copy control Association. (n.d.). *Frequently Asked Questions (FAQ).* Retrieved september 3, 2009, from www.dvdcca.org: http://www.dvdcca.org/faq.html

DW – World. DE (2007). combating child porn conference focuses on new technology. *Deutsche Welle's online service.* Retrieved August 12, 2009, from http://www.dwworld.de/dw/ article/0,2790025,00.html GAO report (2002). *Combating child pornography: US Government Accountability Office (GAO) Report to the ranking minority member*, Committee on Government Reform, House of Representatives. Washington, DC: United States General Accounting Office.

Dworkin, A. (1994). Pornography happens to women. Retrieved August 15, 2009 from http://www.nostatusquo.com/ ACLU/ dworkin/PornHappens.html.

Eberhard, H. (2008). On the distinction between the concept of God and conceptions of God. *International Journal for Philosophy of Religion, 64*, 63–73. doi:10.1007/s11153-008-9163-z

Eckberg, D. L., & Hill, L. Jr. (1979). The Paradigm Concept and Sociology: A critical review. *American Sociological Review, 44*(6), 925–937. doi:10.2307/2094717

Economist Intelligence Unit. (2008). *Index of Democracy 2008.* London: The Economist.

E-México. (2009). *Portal e-México.* Retrieved February 28, 2010, from http://www.e-mexico.gob.mx/.

E-migra. (2006). *E-migra.* Retrieved February 28, 2010, from http://www.e-migra.org

Enciclomedia (2009). *Programa Enciclomedia.* Retrieved February 28, 2010, from http://www.sep.gob.mx/wb/sep1/programa_enciclomedia

Engels, R., & Bernt, B. (2000). *Information extraction: State-of-the-Art-Report.* On-To-Knowledge Consortium.

Erickson, P., Wilson, R., & Shannon, I. (1995). Years of healthy life. us department of health and human services, CDC. National Center for Health Statistics, Hyattsville Maryland, *Statistical Notes,* 7.

Ernst, T. (1991). *The Phrase Structure of English Negation.* Newark, DE: University of Delaware.

Espirit, H. P. C. N. (1997). Industrial R&D Requirements for Humanitarian Demining. (Available through <http://www.cordis.lu/espirit/ src/report43.html/).

Ewerman, A., & Hydén, H. (1997). *IT och social förändring.* Byggforskningsrådet.

Exif Version 2.2 (2002). *Exchangeable image file format for digital still cameras: Exif version 2.2.* Retrieved October 1, 2009, from http://www.digicamsoft. com/ exif22/exif22/html/ exif22_1.htm

Families (2002). *families, work and ist: a study of the interactions between family trends and new work methods in the information society.* Project supported under DG-Information Society's IST Programme; Project Coordinators: Work Research Centre, Dublin, Ireland. Results of Family Survey, Deliverable No. 3, February 2002, http://www.families-project.com/doc/families-d3-public.pdf

Fante, C. (2005). *Fenómeno Bullying. Como prevenir a violência nas escolas e educar para a paz.* Florianopolis, Brasil: Verus.

Ferguson, I. (1996). *A preliminary investigation into offensive and illegal content on the Internet: Deviant criminal pornography.* Ottawa: Department of Justice Canada.

Fillmore, C. J. (1968). *The case for case.* New York: Holt, Rinehart, and Winston.

Fillmore, C. J. (1977). *The case for case reopened. Syntax and Semantics VIII: Grammatical Relations* (pp. 59–81). New York: Academic Press.

Fine, B. (2002). Economic Imperialism and the New Development Economics as Kuhnian Paradigm Shift? *World Development*, *30*(12), 2057–2070. doi:10.1016/S0305-750X(02)00122-5

Finkenzeller, K. (1998). *RFID handbook*. New York: John Wiley and Son, Ltd.

Fisher, W. A., & Barak, A. (2001). Internet pornography: A social psychological perspective on Internet sexuality. *Journal of Sex Research*, *38*, 312–323. doi:10.1080/00224490109552102

Fishkin, K. P., Kautz, H., Patterson, D., Perkowitz, M., & Philipose, M. (2003). guide: towards understanding daily life via auto-identification and statistical analysis. *UbiComp 2003: Ubiquitous Computing 5th International Conference*.Seattle, WA, USA.

Fong, T. (2001). Collaborative Control: A Robot-Centric Model for Vehicle Teleoperation. *Doctoral dissertation. Techical report CMU-RI-TR-01-34*. Robotics Institute, CMU.

Foster, F. H., & Shook, R. L. (1993). *Patents, copyrights & trademarks*. New York: John Wiley & Sons.

Foulds, R., Baletsa, G., & Crochetiere, W. (1975). The Effectiveness of Language Redundancy on Non-Vocal Communication. *In Proceedings of the Conference on Devices and Systems for the Disabled*, 82-86.

Foundation, G. N. U. (n.d.). *GNU Operating System*. Retrieved from http://www.gnu.org

French, B. (2006). The business of land-mine clearing. *The Economics of Peace and Security Journal*, *1*(2), 54–57.

Friedman, C., Kra, P., & Rzhetskya, A. (2002). Two biomedical sublanguages: a description based on the theories of Zellig Harris. *Journal of Biomedical Informatics*, *35*(4), 222–235. doi:10.1016/S1532-0464(03)00012-1

Friedman. (2000). A broad coverage Natural Language processing system. *Paper presented at the AMIA Symp 2000*.

Fulmer, T. T., Feldman, P. H., Kim, T. S., Carty, B., Beers, M., Molina, M., & Putnam, M. (1999). An intervention study to enhance medication compliance in community-dwelling elderly individuals. *Journal of Gerontological Nursing*, *25*(8), 6–14.

Funk, R. E. (2006). *Reaching men: Strategies for preventing sexist attitudes, behaviors, and violence*. Indianapolis, IN: Jist Life.

García-Ruiz, M., Edwards, A., Aquino-Santos, R., Vargas Martin, M., & Mendoza-González, R. (2007). Using sonification to teach network intrusion detection: A preliminary usability study. In Montgomerie, C. & Seale, J. (Eds.), *World Conference on Educational Multimedia, Hypermedia and Telecommunications, EDMEDIA*'07. (pp. 849-857). Chesapeake, VA: AACE.

Garwin, R. L., & Jo, L. Husbands, J. L. (1997). *Progress in Humanitarian Demining: Technical and Policy Challenges*. Amaldi Conference, Paris, 20-22 November 1997. Accessed from: http://www.fas.org/rlg/de-mining.htm

Gazdar, G., & Mellish, C. (1989). *Natural language processing in lisp, an introduction to computational linguistics*. New York: Addison-Wesley Publishing Company.

Gelles, R. J., & Straus, M. A. (1988). *Intimate violence: The causes and consequences of abuse in the American family*. New York: Simon and Schuster.

Geneva Centre for Humanitarian Demining (GICHD). (2006). *Mechanical Demining Equipment Catalogue*. Geneva, Switzerland: Geneva Centre for Humanitarian Demining.

Geneva Centre for Humanitarian Demining (GICHD). (2007). *An Overview of the Guide to Mine Action and Explosive Remnants of War*. Geneva, Switzerland: Geneva Centre for Humanitarian Demining.

Geneva Centre for Humanitarian Demining (GICHD). (2010). *Land Release: A guide for mine and ERW affected countries*. Geneva, Switzerland: Geneva Centre for Humanitarian Demining.

Gerth, H. H., & Wright Mills, H. (1991). *(translated and edited) From Max Weber: Essays in Sociology*. London: Routledge.

Gillespie, I. (2008). Nowadays, it's brutal, accessible; pornography. *London Free Press*, June 11, A3.

Giordano, P. C. (1995). The wider circle of friends in adolescence. *American Journal of Sociology*, *101*, 661–697. doi:10.1086/230756

Gondolf, E. W. (1999). MCMI-III results for batterer program participation in four cities: Less "pathological" than expected. *Journal of Family Violence*, *14*, 1–17. doi:10.1023/A:1022843324943

González Velázquez, E. (2009). El trayecto y problemática que viven las personas que emigran de nuestro país a los Estados Unidos. In Rocha Silva, M. A., & Valencia Silva, M. M. (Eds.), *El impacto económico y social de la migración en Colima* (pp. 67–79). Mexico City, Mexico: Gasca.

Goodenough, O. R., & Decker, G. (2006), Why do People Steal Intellectual Property, in *The Gruter Institute Working Papers on Law, Economics and Evolutionary Biology,* Vol 4, Issue 1, Berkeley Electronic Press.

Goodman, J., Brewster, S., & Gray, P. (2004). Older people, mobile devices and navigation. HCI and the Older Population Workshop. *Paper presented at BCS HCI Workshop 2004*. pp. 13–14. UK: Leeds.

Google. (2009). *Orkut help forum*. Retrieved May 1 2010 from http://www.google.com/support/forum/p/ orkut/thread?tid=1792d1c9901741fe&hl=en

Goose, S. (2004). Overview of Antipersonnel Mine Stockpile Destruction. *ICBL Treaty Working Group*, June 2004.

Gossett, J. L., & Byrne, S. (2002). "Click here": A content analysis of Internet rape sites. *Gender & Society*, *16*, 689–709. doi:10.1177/089124302236992

Government Gazette. (2004). *Republic of South Africa Deeds Films and Publications Amendment Act*. Retrieved September 12, 2009, from http://www.info.gov.za/view/DownloadFileAction?id=67955

Goweder, A., & De Roeck, A. (2001, July). Assessment of a significant Arabic corpus. *Paper presented at the Arabic NLP Workshop at ACL/EACL 2001*, Toulouse, France.

Granovetter, M. S. (1973). The strength of weak ties. *American Journal of Sociology*, *78*(6), 1360–1380. doi:10.1086/225469

Greenberg, A. (2007, August 22). *America's Hackable Backbone*. Retrieved August 29, 2009, from Forbes: http://www.forbes.com/2007/08/22/ scada-hackers-infrastructure-tech-security-cx_ag_0822hack.html

Grieco, E., & Ray, B. (2004). Mexican immigrants in the US labor force. In *Migration Information Source*. Washington, USA: Migration Policy Institute.

Griffith, H. W. (1985). *Guide to Symptoms, Illness, and Surgery*. Los Angeles, CA: Body Press.

Grishman, R. (1986). *Computational linguistics: An introduction*. New York: Cambridge University Press.

Gronau, A. (1985). Women and images: Feminist analysis of pornography. In Vance, C., & Burstyn, V. (Eds.), *Women against censorship* (pp. 127–155). Toronto: Douglas and McIntyre.

Gros, B., & Bruschini, C. (1996). Sensor technologies for the detection of AP mines: a survey of current research and system developments. *In Proceedings of the 6th International Symposium on Measurement and Control in Robotics (ISMCR'96)*. Brussels, Belgium, May 1997, 564-569

Grossman, L. (2009, June 17). Iran Protests: Twitter, the medium of the movement. Washington, DC: *Time Magazine*.

Gualda Caballero, E., & Ruiz Garcia, M. (2004, July-December). Migración femenina de Europa del Este y mercado de trabajo agrícola en la provincia de Huelva, España. *Migraciones Internacionales*, *2*(4), 36–65.

Guarino, N., Masolo, C., & Vetere, G. (1999). OntoSeek: Content-based access to the web intelligent systems and their applications. *IEEE Intelligent Systems*, *14*(3), 70–80. doi:10.1109/5254.769887

Guermazi, R., Hammami, M., & Ben Hamadou, A. (2007). Combining classifiers for web violent content detection and filtering. In Shi, Y., Albada, G.D. van, Dongarra, J. & Sloot, P.M.A. (Eds.), *7th International Conference on Computational Science, ICCS 2007* (pp. 773-780). Beijing, China. Lecture Notes in computer Science.New York: Springer.

Gutiérrez Martín, A. (2003). *Alfabetización digital: algo más que ratones y teclas*. Barcelona, Spain: Gedisa.

Habermas, J. (1999). *Between Facts and Norms: Contributions to a Discourse Theory of Law and Democracy*. Cambridge, MA: MIT Press.

Habib, M. K. (2002b). Mine Clearance Techniques and Technologies for Effective Humanitarian Demining. *International Journal of Mine Action*, *6*(1), 2002.

Habib, M. K. (2007a, June). Humanitarian Demining: Reality and the Challenge of Technology - The State of the Arts. *International Journal of Advanced Robotic Systems*, *4*(2), 151–172.

Habib, M. K. (2007b). Controlled Biological and Biomimetic Systems for Landmine Detection. [Washington, DC: Elsevier Publisher.]. *Journal of Biosensors and Bioelectronics*, *23*, 1–18. doi:10.1016/j.bios.2007.05.005

Habib, M. K. (2001). Mine Detection and Sensing Technologies: New Development Potentials in the Context of Humanitarian Demining. *Proceedings of the IEEE International Conference of Industrial Electronics, Control and Instrumentation (IECON'2001)*, USA, 2001, pp. 1612-1621.

Habib, M. K. (2001a). New Potential and Development of Mine Detection and Sensing Technologies in the Context of Humanitarian Demining. *Proceedings of the 5th International Conference on Mechatronics Technology (ICMT'2001)*, 222-230.

Habib, M. K. (2001b). Service Robots in Humanitarian Landmine Clearance. *Proceedings of the International Symposium on Robots (ISR'2001)*, Seoul, Korea, 1333-1338.

Habib, M. K. (2002a). Mechanical Mine Clearance Technologies and Humanitarian Demining: Applicability and Effectiveness. *Proceedings of the 5th International Symposium on Technology and Mine Problem*, California, USA, April 2002.

Habib, M. K. (2004). Mechanization Technology of Mine Clearing Operations and Humanitarian Demining. *The Eleventh World Congress in Mechanism and Machine Science.* Tianjin-China, pp. 973-978.

Habib, M. K. (2008a). Humanitarian Demining: Difficulties, Needs and the Prospect of Technology. *IEEE International Conference on Mechatronics and Automation 'ICMA2008'*, Takamatsu-Kagawa, Japan, Paper Reference WC2-2.

Habib, M. K. (2008b). Humanitarian Demining: The Problem, Difficulties, Technologies and the Role of Robotics. In Maki K. Habib, (Ed.), *Humanitarian Demining, Innovative Solutions and the Challenges of Technology*, pp.1-56.ARS-pro literature Verlag Publishers.

Hammami, M., Chahir, Y., & Chen, L. (2003). WebGuard: Web based adult content detection and filtering system. *IEEE/WIC International Conference on Web Intelligence*, 2, 574-578, 13-17.

Hammo, B., Abu-Salem, H., Lytinen, S., & Evens, M. (2002, July). QARAB: A question answering system to support the Arabic language. *Paper presented at the Computational Approaches to Semitic Languages workshop*, University of Pennsylvania, USA

Hansen, D. E. (2008). Knowledge transfer in online learning environments. *Journal of Marketing Education*, *30*(2), 93–105. doi:10.1177/0273475308317702

Harmon, P. A., & Check, J. V. P. (1989). *The role of pornography in woman abuse*. Toronto: LaMarsh Research Program on Violence and Conflict Resolution, York University.

Hassenzahl, M., & Tractinsky, N. (2006). User experience -- a research agenda. *Behaviour & Information Technology*, *25*(2), 91–97. doi:10.1080/01449290500330331

Hayes, T., Pavel, M., Schallau, P. K., & Adami, A. M. (2003). Unobtrusive monitoring of health status in an aging population. *Ubiquitous Computing 5th International Conference, Seattle*, WA, USA.

Healy, A. & W., Webber, W. (1993). Sensors for the Detection of Land-based Munitions. *Naval Postgraduate School*, Monterey, California, N PS-M E-95-003, September 1993.

Healy, M. (2004). Child pornography: An international perspective. *Computer Crime Research Centre*. Retrieved August 4, 2009, from http://www.crime-research.org/articles/536/4

Heidegger, M. (1927). *Sein und Zeit*. Tübingen: Max Niemeyer.

Heilig, G. K. (1996) World population prospects: Analyzing the 1996 UN population projections. (International Institute for Applied Systems Analysis) *Report WP-96-146* of the IIASA LUC-Project. Retrieved September 1, 2009 from http://www.iiasa.ac.at/Research/ LUC/Papers/gkh1/chap1.htm.

Helm, P. (1996). Integrated Risk Management for Natural and Technological Disasters. *Tephra.*, *15*(1), 4–13.

Hernández, M., & Solano, I. (2007). CiberBullying, Un Problema de Acoso Escolar. *RIED*, 17-36.

Herrero, C., Cesar, P. & Vourimaa, P. (2003). Delivering MHP Applications into a Real DVB-T Network, OtaDigi. *In Proceedings of Telecommunications in Satellite, Cable and Broadcast Service (TELSIKS 2003)*, Serbia – Montenegro, p. 231-234.

Hewish, M., & Ness, L. (1995). Mine-detection Technologies. *International Defense Review.*, *28*(10), 40–46.

Hewish, M., & Ness, L. (1995). Mine-detection Technologies. *International Defense Review*, October 1995, pp. 40-46.

Hitch, P. J., Fielding, R. G., & Llewelyn, S. P. (1994, November). Effectiveness of self-help and support groups for cancer patients: A Review. *Psychology & Health*, *9*(6), 437–448. doi:10.1080/08870449408407470

Hoffman, D. L., & Novak, T. P. (1996). Marketing in Hypermedia Computer- Mediated Environments: Conceptual Foundations. *Journal of Marketing*, *60*(3), 211–231. doi:10.2307/1251841

Horten, M. (2008). *The 'Telecoms Package' and the copyright amendments – a European legal framework to stop downloading, and monitor the Internet*. University of Westminster. http://www.laquadrature.net/wiki/images/0/0e/Telecom_package.paper.monica.horten.28.june.2008.v5.pdf (last visited 1 March 2010).

Horton, R. (1982). Tradition and modernity revisited, in Hollis and Lukes (eds.). *Rationality and relativism*, Mit Press.

Howard, P., Rainie, L., & Jones, S. (2001). Days and nights on the Internet: The impact of a diffusing technology. *The American Behavioral Scientist*, *45*(3), 383–404.

Hugenholtz, B. (2000). Why the Copyright Directive is Unimportant, and Possibly Invalid, *European Intellectual Property Review, at 501–502*.

Humanitarian Mine Action Equipment Catalogue (1999), German Federal Foreign Office.

Humanitarian Mine Action Equipment Catalogue (1999). *German Federal Foreign Office.*

Huntington, S. P. (2000). *Reconsidering Immigration Is Mexico a Special Case?* Retrieved May 1, 2010, from http://www.cis.org/articles/2000/ back1100.html

Hydén, H. (2002). *Normvetenskap*, Lund studies in Sociology of Law.

Hyuna, S., Bakkena, S., Friedman, C., & Johnson, S. (2004, September). Natural language processing challenges in HIV/AIDS clinic notes. *Paper presented at the World Congress on Medical Informatics (Medinfo 2004)*, San Francisco, California, USA.

Ibrahim, A. (2009). Detecting and preventing the electronic transmission of illegal images. *Unpublished Master's thesis*. Faculty of Engineering and Applied Science, University of Ontario Institute of Technology, Oshawa, Canada.

IFRC (International Federation of Red Cross and Red Crescent Societies). (1995). World Disasters Report 1995, *from*http://www.ifrc.org/pubs/wdr/95.

INASP. (2003). *Income generation: experiences from public library systems in eastern, central and southern Africa*. Oxford, UK: Oxford University Press.

Independent Reviewer of Terrorism Legislation. (2007). *The Definition of Terrorism*. London: The Home Department.

Instituto Nacional de Geografía. Estadística e Informática (INEGI). (2009a). Retrieved February 28, 2010, from http://www.inegi.org.mx.

Instituto Nacional de Georgrafía. Estadística e Informática. (INEGI). (2000). Retrieved February 28, 2010, from http://www.inegi.gob.mx/est/contenidos /espanol/rutinas/ept.asp?t=medu09&c=3277

Instituto Nacional de Georgrafía. Estadística e Informática (INEGI). (2009b). Retrieved February 28, 2010, from http://mapserver.inegi.gob.mx/geografia /espanol/estados/col/ubic_geo.cfm? c=1203&e=06&CFID=591044&CFTOKEN=29002896

International Campaign to Ban Landmines (ICBL), (2008). *Landmine Monitor Report 2008: Toward a Mine-Free World*. Landmine Monitor, 2008.

International Campaign to Ban Landmines (ICBL), (2009). *Landmine Monitor Report 2009: Toward a Mine-Free World -- Special 10 Years Review*. Landmine Monitor, 2009.

International Committee of Red Cross. (1996a). Antipersonnel Mines: An Overview. Geneva, September 1996. (See also: http://www.icrc.org/).

International Committee of Red Cross. (1996b). Antipersonnel Mines- Friends or Foe? ICRC Publication, Ref. 0654, Geneva, 1996.

International Committee of Red Cross. (1998). The Silence Menace: Landmines in Bosnia and Herzegovina. ICRC Publication, Ref. 2160, Geneva, 1998.

International Committee of Red Cross. (2009). Anti-Personnel Mines: Overview of the problem, Geneva, 2009. (Available through <http://www.icrc.org/web/eng/siteeng 0.nsf/htmlall/mines-fac-cartagena-021109?opendocument).

International Mine Action Standard (IMAS). (2003). *Glossary of Mine Action Terms, Definitions and Abbreviations*. IMAS 04.10, Second Edition. 01 January 2003

Internet usage statistics. (2006). *Internet usage statistics*. Retrieved from http://www.internetworldstats.com/stats.htm

Internet Watch Foundation. (2009). *Protection of children Act 1978 (England & Wales)*. Retrieved September 12, 2009, from http://www.iwf.org.uk/police/page.22.36.htm

Internet World Stats. (2009, June). Retrieved October 1, 2009, from http://www.internetworldstats.com / stats.htm

Intill, S. (2003). *Ubiquitous computing technology for just-in-time motivation of behavior change*. S.M. Thesis, Massachusetts Institute of Technology.

Investigation Software Press. (2008). BlueBear Law Enforcement Services Helps Reduce Your Child Exploitation Investigation Time by Over 85%, *PoliceOne resource for Police and Law Enforcement*. Retrieved September 5, 2009, from http://www.policeone.com/police-products/investigation/Investigative-Software/press-releases/1708908-BlueBear-Law-Enforcement-Services-Helps-Reduce-Your- Child-Exploitation-Investigation-Time-by-Over-85/

I-SAFE. (2010). *Cyber Bullying: Statistics and Tips*. Retrieved May 1 2010, from http://www.isafe.org/channels/sub.php? ch=op&sub_id=media_cyber_bullying

ISO International Organization for Standardization. (1998). *International ISO standard* 9241-11, Ergonomic requirements for office work with visual display terminals (VDTs) - Part 11: Guidance on usability. 1st edition, Switzerland.

Jackson, B. A. (2009). Organizational Decisionmaking by Terrorist Groups. In P. K. Davis, & K. Cragin, *Social Science for Counterterrorism: Putting the Pieces Together.* Santa Monica: RAND Corporation.

Jenkins, P. (2003). *Beyond tolerance: Child pornography on the internet*. New York: New York University Press.

Jensen, R. (1995). Pornographic lives. *Violence Against Women*, *1*, 32–54. doi:10.1177/1077801295001001003

Jensen, R. (1996). Knowing pornography. *Violence Against Women*, *2*, 82–102. doi:10.1177/1077801296002001005

Jensen, R. (2007). *Getting off: Pornography and the end of masculinity*. Cambridge, MA: South End Press.

Jensen, R., & Dines, G. (1998). The content of mass-marketed pornography. In Dines, G., Jensen, R., & Russo, A. (Eds.), *Pornography: The production and consumption of inequality* (pp. 65–100). New York: Routledge.

Jimison, H. B., Pavel, M., & Pavel, J. (2003). Adaptive interfaces for home health. *Ubiquitous Computing 5th International Conference*, Seattle, WA, USA.

Johnson, R. L. (2004, Feb.). Saha, S., Arbelaez, J. J., Beach M. C., & Cooper, L. M. (2004). Racial and ethnic differences in patient perceptions of bias and cultural competence in health care. *Journal of General Internal Medicine*, *19*, 101–110. doi:10.1111/j.1525-1497.2004.30262.x

Johnston, J., Eloff, J., & Labuschagne, L. (2003). Security and human computer interfaces. *Computers & Security*, *22*(8), 675–684. doi:10.1016/S0167-4048(03)00006-3

Jones, K. J., & Belani, R. (2005). *Web browser forensics*, Part 1 and Part 2. Retrieved October 1, 2009, from http://www.securityfocus.com/ infocus/1827 and http://www.securityfocus. com/infocus/1832

Jordan, Z. (2006). A view at cyberporn and its influence on aggression against women. Unpublished manuscript. Ames, Iowa: Iowa State University.

Joshi, A., Webber, B., & Sag, I. (1981). *Elements of discourse understanding*. Cambridge, UK: Cambridge University Press.

Julian, B. R. (1990). Are earthquakes chaotic? *Nature*, *345*, 481–482. doi:10.1038/345481a0

Karau, S. J., & Williams, K. D. (1993). Social loafing: A meta-analytic review and theoretical integration. *Journal of Personality and Social Psychology*, *65*, 681–706. doi:10.1037/0022-3514.65.4.681

Katz, J. (2006). *The macho paradox: Why some men hurt women and how all men can help*. Naperville, IL: Sourcebooks.

Katz, J. J., & Fodor, J. A. (1963). The structure of semantic theory. *Language*, *39*, 170–210. doi:10.2307/411200

Kautz, H., Fox, D., Etzioni, O., Borriello, G., & Arnstein, L. (2002). An overview of the assisted cognition project. In AAAI-2002 *Workshop on Automation as a Caregiver: The Role of Intelligent Technology in Elder CareMelno Park, CA:* AAAI Press.

Kelly, C. C. (2007). *The Manhattan Project: The Birth of the Atomic Bomb in the Words of Its Creators, Eyewitnesses and Historians*. New York: Black Dog and Leventhal.

Kendall, L. (2003). Cyberporn. In Kimmel, M. S., & Aronson, A. (Eds.), *Men and masculinities: A social, cultural, and historical encyclopedia* (*Vol. 1*, p. 193). Santa Barbara, CA: ABC-CLIO.

Kevin, J. (2002) FBI Arrests 40 in Child Porn Sting, *US Today*. Retrieved September 3, 2009, from http://www.usatoday.com/tech/news /2002/03/18/net-porn.htm

Khoo, C. S.-G. (1997). The use of relation matching in information retrieval. *Electronic Journal*, 7(2).

Kidd, C. (1999). The aware home: a living laboratory for ubiquitous computing research. *In Proceedings of 2nd International workshop on Cooperative Buildings (CoBuild99). Lecture Notes in Computer Science* 1670, Springer-Verlag, Berlin. p. 191-198.

Kienzle, D. M., & Elder, M. E. (2002). Security patterns for web application development. *Final Technical Report*. Retrieved March 10, 2009, from http://www.scrypt.net/~celer/ securitypatterns/final%20report.pdf

King, C. (1997). Mine Clearance in the Real World. *Proceedings of the International Workshop on Sustainable Humanitarian Demining,* Zagreb (SusDem'97), pp. S2.1-8.

Kirk, J. (2007). Commerce bank thwarts a major database hack. *Computerworld*, *41*(42), 16–16.

Kirkpatrick, G. (2005). 'Online 'chat' facilities as pedagogic tools: A case study'. *Active Learning in Higher Education*, *6*, 145. doi:10.1177/1469787405054239

Kittredge, R. I. (1982). Sublanguages. *American Journal of Computational Linguistics*, *8*(2), 79–84.

Kittredge, R. (1982). Variation and homogeneity of sublanguages. In R. Kittredge & J. Lehrberger (Eds.), *Sublanguage: studies of language in restricted semantic domains*. Berlin and New York: de Gruyter.

Klein, M. (1997). Coordination Science: Challenges And Directions. In Conen, W. & Neumann, G. (eds.) *Coordination Technology for Collaborative Applications*. Lecture Notes in Computer Science, LNCS, Springer Publishing.

Koch, W. (2008). Software tracks child porn traffickers online. *USA Today,* Retrieved August 12, 2009, from http://www.usatoday.com/news/nation/ 2008-04-15-childporn-side_N.htm

Koegel, L. K., Koegel, R. L., Harrower, J. K., & Carter, C. M. (1999). Pivotal response intervention I: Overview of approach. *Journal of the Association for Persons with Severe Disabilities*, *24*, 174–185. doi:10.2511/rpsd.24.3.174

Koegel, R. L., Camarata, S. M., Koegel, L. K., Ben-Tall, A., & Smith, A. E. (1998). Increasing speech intelligibility in children with autism. *Journal of Autism and Developmental Disorders*, *28*, 241–250. doi:10.1023/A:1026073522897

Koegel, R. L., Koegel, L. K., & Surratt, A. (1992). Language intervention and disruptive behavior in preschool children with autism. *Journal of Autism and Developmental Disorders*, *22*, 141–153. doi:10.1007/BF01058147

Koegel, R. L., O'Dell, M. C., & Koegel, L. K. (1987). A Natural language teaching paradigm for nonverbal autistic children. *Journal of Autism and Developmental Disorders*, *17*, 187–200. doi:10.1007/BF01495055

Koops, B., Lips, M., Prins, C., & Schellekens, M. (2006). *Starting Points for ICT Regulation: Deconstructing Prevalent Policy One-Liners (Information Technology and Law)* (*Vol. 9*). Asser Press.

Krafka, C., Linz, D., Donnerstein, E., & Penrod, S. (1997). Women's reactions to sexually aggressive mass media depictions. *Violence Against Women*, *3*, 149–181. doi:10.1177/1077801297003002004

Kravets, D. (2008). Pirate Bay Says It Can't Be Sunk, Servers Scattered Worldwide, *Wired Blog Network*, 1 February 2008, available at http://blog.wired.com/27bstroke6/2008/02/the-pirate-bay.html.

Kruschwitz, U., & Al-Bakour, H. (2005). Users want more sophisticated search assistants: Results of a task-based evaluation. *American Society for Information Science and Technology*, *56*(13), 1377–1393. doi:10.1002/asi.20230

Kuhn, T. S. (1970). *The Structure of Scientific Revolutions*. Chicago: University of Chicago Press.

Kullenberg, C. (2009). The Social Impact of IT: Surveillance and Resistance in Present-Day Conflicts How can activists and engineers work together? *FIfF-Kommunikation.* 37–40.

Laerhoven, V. K., & Lo Benny, P. L. J. W. P., Ng, S. T., King R., Kwan, S., Gellersen, H. W., Sloman, M., Wells, O., Needham, P., Peters, N., Darzi, A., Toumazou, C., & Yang, G. Z (2004). Medical healthcare monitoring with wearable and implantable sensor. *Ubiquitous Computing 6th International Conference*, Nottingham, UK, Proceedings.

Lakoff, G., & Johnson, M. (1980, 2003). *Metaphors we live by*, Chicago and London: The University of Chicago Press.

Larsson, S. & Svensson, M. (forthcoming). Compliance or Obscurity? Online anonymity as a consequence of fighting illegal file sharing.

Larsson, S. (2005). *Musikupphovsrätten i ett samhälle under förändring – Internet och fildelning ur ett rättssociologiskt perspektiv (Intellectual Property Rights in a Transitional Society – Internet and File Sharing from a Sociology of Law Perspective),* thesis for Master of Laws, University of Lund.

Larsson, S. (2008). *Between Daring and Deliberating, 3G as a sustainability issue in Swedish spatial planning*, Blekinge Institute of Technology, Licentiate Dissertation Series No. 2008:02, School of Technoculture, Humanities and Planning.

Larsson, S. (2009). The darling conceptions of your time, or: why Galileo Galilei sings so sadly in the chorus. In Eide, S. R. (ed.) *FREE BEER 1.0*, an anthology from the presenters at Free Society Conference 2008.

Larsson, S. (In press). The Path Dependence of European Copyright, *SCRIPT:ed.*

Larsson, S., & Hydén, H. (2008, July). *Different darling conceptions tied to different societal systems. The case of illegal file sharing in a brave new world.* Paper presented at Research Committee Sociology of Law annual meeting, Law and Justice in the Risk Society, Milan and Como, Italy.

Laski, K. E., Charlop, M. H., & Schreibman, L. (1988). Training parents to use the Natural Language Paradigm to increase their autistic children's speech. *Journal of Applied Behavior Analysis*, *21*, 391–400. doi:10.1901/jaba.1988.21-391

LeBlanc, L. A., Geiger, K. G., Sautter, R. A., & Sidener, T. M. (2006). Using the natural language paradigm (NLP) to increase vocalizations of older adults with cognitive impairments. *Research in Developmental Disabilities*, *28*, 437–444. doi:10.1016/j.ridd.2006.06.004

Lee, P. Y., Hui, S. C., & Fong, A. C. M. (2003). A structural and content-based analysis for web filtering. *Internet Research: Electronic Networking Applications and Policy*, *13*(1), 27–37. doi:10.1108/10662240310458350

Lee, H., & Nam, T. (2007). P2P Honeypot to prevent illegal or harmful contents from spreading in P2P network. *The 9th International Conference on Advanced Communication Technology*, 1, Feb 2007, 497-501

Legislative Review, I. C. M. E. C. (2006). *Legislation across the world on child pornography.* International Centre for Missing & Exploited Children. Retrieved September 5, 2009, from http://www.internationalresource-centre.org/ missingkids/servlet/PageServlet?Language Country=en_X2&PageId=3428

Lehman, P. (2006). Introduction: "A dirty little secret" – Why teach and study pornography? In Lehman, P. (Ed.), *Pornography: Film and culture* (pp. 1–24). New Brunswick, NJ: Rutgers University Press.

Lehrberger, J. (1982). Automatic Translation and the Concept of Sublanguage. In Kittredge, R., & Lehrberger, J. (Eds.), *Sublanguage: studies of language in restricted semantic domains*. Berlin, New York: Walter de Gruyter.

Leifer, L. J., Toye, G., & Van der Loos, M. (1996). Integrating the socio-technical framework of human service through the Internet World Wide Web. *Robotics and Autonomous Systems*, *18*(1-2), 117–126. doi:10.1016/0921-8890(96)00019-X

Lessig, L. (2008). *Remix: making art and commerce thrive in the hybrid economy*. New York: Penguin Press.

Levinson, S. (1983). Pragmatics. Cambridge: Cambridge University Press. Li, X., & Roth, D. (2001). Exploring evidence for shallow parsing. *Paper presented at the 2001 workshop on Computational Natural Language Learning*, Toulouse, France.

Lindlof, T. R. (1992). Computing tales: parents' discourse about technology and family. *Social Science Computer Review*, *10*(3), 291–309. doi:10.1177/089443939201000301

Linux Outlaws. (2009). *Spore on Linux Howto*. Retrieved from http://linuxoutlaws.com/spore

Litman, J. (2006). *Digital Copyright: Protecting Intellectual Property on the Internet, the Digital Millennium Copyright ACT, Copyright Lobbyists Conquer the Book Description, Prometheus books. Maine, H. (1861, reprint 1959). Ancient Law*. London: The World Classics.

Long, B. (1994, May). *Natural language as an interface style.* Abstract retrieved from http://www.dgp.toronto.edu/ people/byron/papers/nli.html

Longino, H. (1980). What is pornography? In Lederer, L. (Ed.), *Take back the night: Women on pornography* (pp. 40–54). New York: William Morrow.

Luo, V. C. (2007). Tracing USB Device artefacts on Windows XP operating system for forensic purpose. *In Proceedings of the 5th Australian Digital Forensics Conference*, Dec 2007, 210-218

Lytinen, S. L., & Gershman, A. (1986). ATRANS: automatic processing of money transfer messages. *Paper presented at the Fifth National Conference on Artificial Intelligence (AAAI-86)*, Philadelphia, PA.

Maalla N. (2009). Promotion and protection of all human rights, civil, political, economic, social and cultural rights, including the right to development: Report of the special rapporteur on the sale of children, child prostitution and child pornography. *United Nations Human Rights Council, Twelve Session, Agenda item 3.*

Machel, G. *(1996). Impact of armed conflict in children: Land-mines: A deadly inheritance.* Retrieved March 12, 2010, from http://www.unicef.org/graca/mines.htm

Macy, M., & Willer, R. (2002). From factors to actors. [England.]. *Annual Review of Sociology*,28.

Madoc-Jones, I., & Buchanan, J. (2004, December). Indigenous people, language and criminal justice: The experience of first language Welsh speakers in Wales. *Criminal Justice Studies*, *17*(4), 353–367.

Madoc-Jones, I., Buchanan, J., & Senior, P. (2003). Delivering the diploma in probation studies in North Wales and dyfed powys: On-line learning explored. *British Journal of Community Justice*, *2*(2).

Maidment, M. R. (2006). Transgressing boundaries: Feminist perspectives in criminology. In W.S. DeKeseredy and B. Perry (Eds.), *Advancing critical criminology: Theory and application* (pp. 43-62). Lanham, MD: Lexington.

Makinen, G. (2002). *The Economic Effects of 9/11: A Retrospective Assessment.* Washington, DC: Congressional Research Service.

Maroney, T. (2006). A Proposed Taxonomy of an Emerging Field. In *Law and Human Behavior* (*Vol. 30*). Law and Emotion.

Marsh, R. T. (1997). *Critical Foundations: Protecting America's Infrastructures.* Washington, DC: The White House.

Martínez Pizarro, J. (2001 July - December). Reflexiones sobre la gobernabilidad de la migración internacional en América Latina. In *Migraciones Internacionales, 1*(1) (pp. 82–110). Mexico: Colegio de la Frontera Norte.

Martins, H. (1972). The Kuhnian "Revolution" and its Implications for Sociology. In Nossiter, (Eds.), *Imagination and Precision in the Social Sciences. London: Faber & Faber, Morgan, G. (1999). Organisationsmetaforer.* Lund: Studentlitteratur. [*Images of Organization*, 1997, Sage Publications Inc.]

Matheson, S. G., & Lager, P. (2007). A model for developing international student exchanges. *Social Work Education*, *26*(3), 280–291. doi:10.1080/02615470601049867

McCarthy, J., & Wright, P. (2004). *Technology as Experience*. Cambridge, USA: MIT Press.

McCormack, P. (1997). Undernutrition in the elderly population living at home in the community: a review of the literature. *Journal of Advanced Nursing*, *26*(5), 856–863. doi:10.1046/j.1365-2648.1997.00415.x

McCoy, K. F., & Suri, L. Z.Natural Language Processing Principles for Improving Deaf Writing.*Rehabilitation R & D Progress Reports. Journal of Rehabilitation Research and Development*, *29*, 331.

McCoy, K. F. (1997). Simple NLP techniques for expanding telegraphic sentences. In *Proceedings of Natural Language Processing for Communication Aids, an ACL/EACL '97 Workshop*, Madrid, Spain, July 12.

McCrory, A., & Judge, J. (2000). Ubiquitous? pervasive? sorry, they don't compute. *Computerworld*, *34*(12), 44.

McFee, J. E. (1996). Multisensor mine detector for peacekeeping: Improved landmine detector concept. *SPIE Technical Conference,* 2765.

Meek-Prieto, M. (2008). *Just age playing around? How second life aids and abets child pornography. North Carolina Journal of Law & Technology, 9* (88th ed.). NC: JOLT Online.

Memory Alpha. (2009). *Memory Alpha wiki (Star Trek World)*. Retrieved May 1 2010, from: http://memory-alpha.org/ wiki/Portal:Main

Mendoza González, R., Vargas Martin, M., Muñoz Arteaga, J., Álvarez Rodríguez, F. J., & Ochoa Ortíz Zezzatti, C. A. (2009). Web service-security specification based on usability criteria and pattern approach. *Journal of Computers*, *4*(8), 705–712.

Mendoza González, R., Muñoz Arteaga, J., Vargas Martin, M., Álvarez Rodríguez, F. J., & González Calleros, J. (2009). a pattern methodology to specify usable security in websites. In *3rd International Workshop on Secure Systems Methodologies using Patterns, SPattern'09*. Linz, Austria. IEEE Press.

Mendoza González, R., Vargas Martin, M., Muñoz Arteaga, J., Álvarez Rodríguez, F. J., & Garcia-Ruiz, M. A. (2009). A measurement model for secure and usable e-commerce websites. In *22nd IEEE Canadian Conference on Electrical and Computer Engineering, CCECE/ CCGEI 2009* (pp. 77-82). St. John's, Newfoundland and Labrador, Canada. IEEE Press.

Migración internacional, Derechos Humanos y desarrollo en América. (2006). *Migración internacional, derechos humanos y desarrollo en América*. Montevideo, Uruguay: Naciones Unidas/CEPAL.

Migrants Portal, E. -México program (2009). Retrieved February 28, 2010 from http://www.e-mexico.gob.mx/ wb 2/eMex/eMex_eMigrantes

Mihailidis, A., Fernie, G. R., & Barbenel, J. C. (2001). The use of artificial intelligence in the design of an intelligent cognitive orthosis for people with dementia. *Assistive Technology*, *13*, 23–39.

Miller, G. A. (1956). The magical number seven, plus or minus two: Some limits on our capacity for processing information. *Psychological Review*, *63*, 81–97. doi:10.1037/h0043158

Miller, M. E. (2010, 02 02). *"Twitteros" are Mexico's Latest Outlaws*. Retrieved 02 02, 2010, from CBS News: http://www.cbsnews.com/stories/2010/ 02/02/world/ main6166446.shtml

Mills, C. W. (1959). *The sociological imagination*. New York: Oxford University Press.

Ministerio (2009). Ministério público pede que Google explique crimes no Orkut (in Portuguese). *Folha Online*. March 10, 2006.

Ministry of Public Security of the State Council of the Chinese Government. (1997). *Computer Information Network and Internet Security, Protection and Management (translation of the US Embassy*. Beijing.

Mitamura, T. (1999, September). Controlled Language for Multilingual Machine Translation. *Paper presented at the Machine Translation Summit VII, Singapore*.

Morfitt, K., & Valli, C. (2005). After conversation – a forensic ICQ logfile extraction tool. *In Proceedings of the 3rd Australian Computer, Network & Information Forensics Conference*. School of Computer and, Information Science, Edith Cowan University, Perth, Western Australia, 2005, 54-61.

Morozov, E. (2009). Iran: downside to the "twitter revolution". *Disent* (fall), 10-15.

Muñoz, J., Mendoza, R., Vargas Martin, M., Vanderdonckt, J., & Álvarez, F. (in press). a methodology for designing information security feedback based on user interface patterns. *Advances in Engineering Software*.

Muñoz, J., Mendoza, R., Vargas Martin, M., Vanderdonckt, J., Álvarez, F., & González Calleros, J. (2008). A method to design information security feedback Using patterns and HCI-security criteria. In Lopez Jaquero, V., Montero Simarro, F., Molina Masso, J.P. & Vanderdonckt, J. (Eds.), *7th International Conference on Computer-Aided Design of User Interfaces, CADUI 2008* (pp. 283-294). Albacete, Spain. Lecture Notes in Computer Science, Springer.

Mykytyn, J. (2004). *The Unknown Crisis: Child Pornography on the Internet Anti Child Porn Organization*. Retrieved September 15, 2009, from http://www.antichildporn.org/ whiteppr.htm

Mynatt, E. D., Essa, I., & Rogers, W. (2000). Increasing the opportunities for aging in place. *In Proceedings of CUU Conference*, 1-7. New York:ACM Press.

Nacho'n, C. T. (2000). *Environmental Aspects of the International Crisis of Antipersonnel Landmines and the Implementation of the 1997 Mine Ban Treaty: Thematic Report. Landmine Monitor Report 2000.* NY: International Campaign to Ban Landmines.

Nacional de Derechos Humanos, C. *(CNDH).* (2009). Retrieved March 8, 2010 from http://www.cndh.org.mx/

Narayanan, A., & Shmatikov, V. (2009). De-anonymizing social networks. *In IEEE Symposium on Security and Privacy.*

National Commission on Terrorist Attacks Upon the United States. (2004). *The 9/11 Comission report.* Washington, DC: US Congress.

National Infrastructure Protection Plan: Partnering to enhance protection and resiliency. Washington, DC: Department of Homeland Security.

Newell, A. F. (2004). HCI and older people. HCI and the older population workshop. *In Proceedings of BCS HCI.* pp. 29–30. Leeds, UK.

Newell, A. F., Arnott, J. L., Beattie, W., & Brophy, B. (1992). Effect of "{PAL{" word prediction system on the quality and quantity of text generation. *Augmentative and Alternative Communication*, 304-311. 35.

Nicoud, J. D. (1997). Vehicles and Robots for Humanitarian Demining. *The Industrial Robot, 24*(2), 164–168. doi:10.1108/01439919710165725

Nicoud, J.-D. (1996a). A Demining Technology Project. *Proceedings of the International Conference on Detection of Abandoned Land Mines (MD'96)*, Edinburgh UK, October, 1996, 37-41.

Nielsen, J. (2001). Usability metrics. Retrieved September 2, 2007, from http://www.useit.com/alertbox/20010121.html

Ochoa, A. (2008). Social data mining to improve bioinspired intelligent systems. In Giannopoulou, E. G. (Ed.), *Data Mining in Medical and Biological Research*. Vienna, Austria: In-Tech.

Ochoa, A., et al. (2007). Baharastar – simulador de algoritmos culturales para la minería de datos social. *In Proceedings of COMCEV'2007.*

Ochoa, A., et al. (2009) Six degrees of separation in a graph to a social networking. *In Proceedings of ASNA'2009*, Zürich, Switzerland.

OECD. (2003). *Emerging Risks in the 21st century.* Paris: OECD.

Olweus, D. (1998). *Bullying at school: What we know and what we can do* [Spanish edition: Conductas de acoso y amenaza entre escolares]. Madrid, Spain: Ediciones Morata.

O'Malley, T. J. (1993). Seek and Destroy - Clearing Mined Land. *Armada International, 17*(1), 6–15.

O'Malley, T. J. (1993). Seek and Destroy - Clearing Mined Land. *Armada International, 17*(1), 6–15.

OpenNet Initiative. (2009). *China's Green Dam: The Implications of Government Control Encroaching on the Home PC.* OpenNet Initiative Bulletin.

Papera, S. (1996). *The connected family.* Marietta, GA: Longstreet Press Inc.

Park, S., Mackenzie, K., & Jayaraman, S. (2002). The wearable motherboard: a framework for personalized mobile information processing (PMIP). In *Proceedings of the 39th Annual Design Automation Conference* (New Orleans, Louisiana, USA, June 10 - 14, 2002). DAC '02. New York: ACM.170-174. DOI= http://doi.acm.org/10.1145/513918.513961

Parsonage, H. (n.d.). *The forensic recover of instant messages from MSN messenger and windows live messenger.* Retrieved October 1, 2009, from http://computerforensics.parsonage.co.uk/ downloads/MSNandLiveMessenger ArtefactsOfConversations.pdf

Patrick, A. O., & McBride, S. (2008). Showdown Looms over Pirated-Media Directory. Swedish prosecutors target organizers of pirate bay, a huge file-sharing guide, *The Wall Street Journal,* 11 January 2008.

Patterson, D. D. F., Kautz, H., & Philipose, M. (2003). Expressive, tractable and scalable techniques for modeling activities of daily living. *Ubiquitous Computing 5th International Conference*, Seattle, WA, USA.

Pécoud, A., & De Guchteneire, P. (2005, July – December). Migración sin fronteras: Una investigación sobre la libre circulación de personas. *Migraciones Internacionales*, *3*(2), 137–166.

Pérez, C., & Santín, D. (2006). *Data mining. Soluciones con Enterprise Manager*. Alfa Omega Grupo Editor, S.A. de C.V., Mexico

Physicians for Human Rights. (1993). *Landmines, A Deadly Legacy. Human Rights Watch*. New-York.

Ponce, J., Hernández, A., Ochoa, A., Padilla, F., Padilla, A., Alvarez, F., & Ponce de León, E. (2009). Data mining in web applications. In Ponce, J., & Karahoka, A. (Eds.), *Data Mining and Discovery Knowledge in Real Life Application*. Vienna, Austria: In-Tech.

Pool, J. (2005). *Can controlled languages scale to the Web? DISCUSSION DRAFT ONLY*. Retrieved October 1 2006, from http://utilika.org/pubs/etc/ambigcl/

Progress Report, ICMEC. (2006). *Child pornography not a crime in most countries*. International Centre for Missing & Exploited Children. Retrieved September 5, 2009, from http://www.icmec.org/en_X1/pdf/ SummerNewsletter2006formatted.pdf

Prop.2004/05:110 Upphovsrätten i informationssamhället – genomförande av direktiv 2001/29/EG, m.m.

Prop. 1999/2000:86 *Ett informationssamhälle för alla.*

Protégeles. (2009). *Protegeles línea de denuncia*. Retrieved May 1, 2010, from http://www.protegeles.com/

Racism in Brazilian. (2009). *Racism in Brazilian Orkut*. Retrieved May 1, 2010, from http://zonaeuropa.com/20050326_2.htm

Racismo na internet (2006). *Racismo na internet chega à Justiça* (in Portuguese). Estadão. Retrieved July 10, 2007, from http://www.estadao.com.br/tecnologia/ internet/noticias/2006/fev/01/97.htm.

Renner, K. (1949). *The Institutions of Private Law and their Social Functions*. London: Routledge.

Reynolds, R. (1998). An *introduction to cultural algorithms. Cultural algorithms repository*. Retrieved May 1, 2010, from http://www.cs.wayne.edu/~jcc/car.html

Riisgaard, H. T. (2004). Interacting without keyboard and mouse. *Ubiquitous Computing 6th International Conference*, Nottingham, UK, Proceedings.

Riloff, E., Schafer, C., & Yarowsky, D. (2002, Aug. 2002). Inducing information extraction systems for new languages via cross-language projection. *Paper presented at the COLING-02*, Taipei.

Robinson, P. H., & Dubber, D. M. (1999, March 12). *An Introduction to the model Penal Code*. Retrieved September 1, 2009, from PennLaw. University of Pennsylvania Law School: http://www.law.upenn.edu/fac/ phrobins/intromodpencode.pdf

Rocha Silva, M. A., Contreras-Castillo, J., & Acosta-Díaz, R. (2008). *Las nuevas tecnologías de la comunicación entre los migrantes*. Colima, Mexico: Archivo Histórico del Municipio de Colima y Secretaría de Cultura del Gobierno del Estado de Colima.

Rodríguez, R., Seoane, A., & Pereira, J. (2006). Niños contra niños: El bullying como transtorno emergente. *Anal Pediátrico*, *64*(2), 162–166. doi:10.1157/13084177

Rodríguez, M., Favela, J., Preciado, A., & Vizcaíno, A. (2004). An agent middleware for supporting ambient intelligence for healthcare. Paper Presented at *ECAI 2004 Second Workshop on Agents Applied in Health Care*. Valencia, Spain.

Rogers, E. M. (1962). *Diffusion of innovations*. New York: New York Press.

Ropelato, J. (2006). *Internet pornography statistics*. Retrieved October 1, 2009, from http://internet-filter- review.toptenreviews.com /internet-pornography-statistics.html

Rosoff, H., & von Winterfeldt, D. (2005). *A Risk and Economic Analysis of Dirty Bomb Attacks on the Ports of Los Angeles and Long Beach*. Los Angeles: Center for Risk and Economic Analysis of Terrorism Events. University of Southern California.

Russell, D. E. H. (1993). *Against pornography: The evidence of harm*. Berkeley, CA: Russell.

Russell, D. E. H. (1998). *Dangerous relationships: Pornography, misogyny, and rape*. Thousand Oaks, CA: Sage.

Sachpazidis, I. (2001). @Home telemedicine. *In Proceedings of Telemed 2001 Conference*. Telematik im Gesundheitswesen, 9th - 10th November 2001, Berlin, Germany.

Sachpazidis, I. (2002). @HOME: A modular telemedicine system Mobile Computing in Medicine. *In Proceedings of the 2 Workshop on mobile computing,* Heidelberg, Germany

Safari, R., Foad, D., & Majid, S. (2004). Reconfigurable fabric vest for fatal heart disease prevention. *Ubiquitous Computing 6th International Conference*, Nottingham, UK, Proceedings.

Sanday, P. R. (1990). *Fraternity gang rape*. New York: New York University Press.

Santos, B. de Sousa (1995). *Toward a New Common Sense: Law, Science and Politics in the Paradigmatic Transition*. New York: Routledge.

Sanz, D., Díaz, P., & Aedo, I. (2005). Security modeling in information web systems. In Díaz, P., Montero, S., & Aedo, I. (Eds.), *Web Engineering and Design Patterns*. Pearson Education. (In Spanish)

Schank, R. (1975). *Conceptual Information Processing*. Amsterdam: North Holland.

Schank, R. (1991). *Where's the AI?*Evanston, IL: Northwestern University, Institute for the Learning Sciences.

Schiffman, L. G., & Kanuk, L. L. (1997). *Comportamiento del consumidor*. Mexico City, Mexico: Prentice Hall.

Schwartz, M. D. (1987). Censorship of sexual violence: Is the problem sex or violence? *Humanity & Society*, *11*, 212–243.

Schwartz, M. D., & DeKeseredy, W. S. (1997). *Sexual assault on the college campus: The role of male peer support*. Thousand Oaks, CA: Sage.

Schwartz, M. D., & DeKeseredy, W. S. (1998). Pornography and the abuse of Canadian women in dating relationships. *Humanity & Society*, *22*, 137–154.

Schwitter, R. (2004). Representing knowledge in controlled natural language: A case study. *Paper presented at the Knowledge-Based Intelligent Information and Engineering Systems, 8th International Conference, KES2004*, Wellington, New Zealand.

Scott, R. T. (2009). *Child pornography gets new weapons, The Times-Picayune*. Retrieved February 4, 2010, from http://www.nola.com/crime/i ndex.ssf/2009/11/post_72.html

Segal, L. (1993). Does pornography cause violence? The search for evidence. In P.C. Gibson & R. Gibson (Eds.), *Dirty looks: Women, pornography, power* (pp. 5-21). London: British Film Institute.

Sekine, S. (1997). The Domain Dependence of Parsing. *Paper presented at the Applied Natural Language Processing(ANLP'97)*, Washington D.C., USA.

Sharma, D. C. (2006). Online Technologies kill distance in learning managing participation in online journalism courses. *Journal of Creative Communications*, *1*(1), 75–81. doi:10.1177/097325860500100105

Sieber, A. (1995). Localisation and identification of anti-personnel mines. Joint Research Centre, *European Commission, EUR 16329N*.

Silva, J. A. (2001). Do hipertexto ao algo mais: usos e abusos do conceito de hipermídia pelo jornalismo on-line. In: A. Lemos, & M. Palacios (orgs.) *Janelas do Ciberespaço - Comunicação e Cibercultura*. Sao Paulo, Brasil: Editora Sulina, pp. 128-139.

Simon, S. (2004). Hardcore porn hits the heartland: Rural superstores are "doing great." *Concord Monitor*. Retrieved on December 7, 2004 from http://www.cmonitor.com/apps/pbcs.dll/ article?AID=/20041207/REPOSITORY/412070332/1014.

Small, S., & Rieger, C. (1982). *Parsing and comprehending with word experts (a theory and its realization). Strategies for natural language processing*. New Jersey, NJ: Lawrence Erlbaum Associates.

Smeaton, A. F. (1992). Progress in the application of natural language processing to information retrieval tasks. *The Computer Journal*, *35*, 268–278. doi:10.1093/comjnl/35.3.268

Smith, K. (1996). *Environmental Hazards: Assessing Risk and Reducing Disaster* (2nd ed.). London/U.S.A./Canada: Routledge.

SOU 1956:25 *Upphovsmannarätt till litterära och konstnärliga verk.* Lagförslag av Auktorrättskommittén.

SOU2003:35 Upphovsrätten i informationssamhället - genomförande av direktiv 2001/29/EG, m.m.

Southworth, C., Tucker, S., Fraser, C., & Shulruff, T. (2008). High-tech violence against women. In Renzetti, I. C. M., & Edleson, J. L. (Eds.), *Encyclopedia of interpersonal violence* (pp. 329–330). Thousand Oaks, CA: Sage.

Sowa, J. F. (1999). Relating templates to language and logic. In Pazienza, M. T. (Ed.), *Information extraction: Towards scalable, adaptable systems. Lecture Notes in Computer Science* (*Vol. 1714*). Springer-Verlag.

Stack, S., Wasserman, I., & Kern, R. (2004). Adult social bonds and use of Internet pornography. *Social Science Quarterly*, *85*, 75–88. doi:10.1111/j.0038-4941.2004.08501006.x

Stanford, V. (2002). Using pervasive computing to deliver elder care. *IEEE Pervasive Computing: Mobile and Ubiquitous Systems*, *1*(1), 10–13.

Starr, C. (1969). Social benefits versus technological risks. *Science*, *165*(3899), 1232–1238. doi:10.1126/science.165.3899.1232

Stefan, C. & Andrew, Gn (2003). Creating distributed support systems to enhance the quality of life for people with cognitive disabilities. *Ubiquitous Computing 5th International Conference, Seattle, WA, USA.*

Stenchion, P. (1997). Development and disaster management. *The Australian Journal of Emergency Management*, *12*(3), 40–44.

Stokes, T. F., & Baer, D. M. (1977). An implicit technology of generalization. *Journal of Applied Behavior Analysis*, *10*, 349–367. doi:10.1901/jaba.1977.10-349

Stolzoff, N. C., Shih, E. C., & Venkatesh, A. (2000). *The home of the future: an ethnographic study of new information technologies in the home Project NOAH Center for Research on Information technology and Organizations.* Irvine: University of California.

Sumi, H., Carlos, G., Youssef, K., Choonhwa, L., Hicham, E. Z., & William, M. (2003). External representations in ubiquitous computing design. *Ubiquitous Computing 5th International Conference*, Seattle, WA, USA

Sundqvist, G. (2001). *Bredbandspolitik – En tekniksociologisk analys av kommunala bredband.* STS Research reports no 2, Avdelningen för teknik- och vetenskapsstudier, Göteborgs universitet.

Suri, L. Z., & McCoy, K. F. (1993a). *Correcting discourse-level errors in a CALL system for second language learners* (pp. 94–02). Newark, DE: Department of Computer and Information Sciences, University of Delaware.

Suri, L. Z., & McCoy, K. F. (1993b). *A methodology for developing an error taxonomy for a computer assisted language learning tool for second language learners.* Newark, DE: Department of Computer and Information Sciences, University of Delaware.

Svensson, M., & Larsson, S. (2009). *Social Norms and Intellectual Property. Online norms and the European legal development.* Research Report in Sociology of Law. Lund University.

Swiffin, A. L., Arnott, J. L., & Newell, A. F. (1987). The use of syntax in a predictive communication aid for the physically impaired. *Paper presented at the Tenth Annual Conference on Rehabilitation Technology Washington, DC.*

Tagliacozzo, R. (1976). Levels of technicality in scientific communication. *Information Processing & Management*, *12*(2), 95–110. doi:10.1016/0306-4573(76)90040-6

Taibi, T. (2006). Formalising design patterns composition. *IEE Proceedings. Software*, *153*(3). doi:10.1049/ip-sen:20050072

Tank, R., & Williams, P. A. H. (2008). The impact of U3 devices on forensic analysis. *In Proceedings of the 6th Australian Digital Forensics Conference*, 199-205

Taylor, M. (2009, August 13). China drops Green Dam web filtering system. *Guardian*, pp. http://www.guardian.co.uk/technology/2009/ aug/13/china-drops-web-censorship.

Telefonica (2003). *Libro blanco del hogar digital y las infraestructuras comunes de comunicaciones*. Retrieved march 8, 2010 from http://sociedadinformacion.fundacion. telefonica.com/DYC/SHI/TelefonicaSI/Publicaciones/ seccion=1242&activo=1&idioma=es_ES&id=2009100111090001.do

Templeton, M., & Burger, J.D. (1983). Problems in natural-language interface to DSMS with examples from EUFID. *ANLP*, 3-16.

Tennant, H. R., Ross, K. M., Saenz, R. M., Thompson, C. W., & Miller, J. R. (1983). Menu-based natural language understanding. *Paper presented at the 21st annual meeting on Association for Computational Linguistics*, Cambridge, Massachusetts.

The history of the CD - The beginning. (2005). Retrieved september 23, 2009, from Philips Research: http://www.research.philips.com/technologies/projects/cd/index.html

The National Men's Resource Center (Menstuff). (2004). Mine Fields. http://www.menstuff.org/issues /byissue/minehistory.html. Accessed on March 12th 2010.

The Statute of Anne. (1710, April 10). Retrieved juin 7, 2009, from The Avalon Project. Yale Law School: http://avalon.law.yale.edu/1 8th_century/anne_1710.asp

The Swedish Copyright Act, Lag (1960:729) om upphovsrätt till litterära och konstnärliga verk.

Thimbleby, H. (2008). Feature: Ignorance of interaction programming is killing people. *interactions, 15*(5), 52-57.

Thompson, N. (2006). *Anti-discriminatory practice* (4th ed.). Basingstoke, UK: Palgrave Macmillan.

Thompson, C. W., Ross, K. M., Tennant, H. R., & Saenz, R. M. (1983). Building usable menu-based natural language Interfaces to databases. *Paper presented at the VLDB-1983.*

Toronto Police. (2006). Child Exploitation Tracking System (CETS) fact sheet. *The Toronto Police,* Retrieved August 22, 2009, from http://www.torontopolice.on.ca/media/text/20060831-cetsfactsheet.pdf

Tsai Tse, M., Jiann-Tsuen, L., Jane, Y., & Hsu, M. (2004). Context-aware authorization for integrated healthcare service. *Ubiquitous Computing 6th International Conference*, Nottingham, UK

Tu maestro en línea (2010). *Tu maestro en línea.* Retrieved April 23, 2010 from http://168.255.111.18/contacto_maestro/index.jsp

UD Center for Universal Design. (2008). The Universal Design Principles. *North Carolina State University.* Retrieved May 15, 2009 from http://www.design.ncsu.edu/cud/ about_ud/udprin ciples.htm

UN-Economic and Social Council (UN-ECS) (2005). Third Annual Report of the Information and Communication Technologies Task Force. *E/2005/71.* New York United Nations Human Settlements Program (UN-HABITAT) (2009). *Global Report on Human Settlements: Planning Sustainable Cities*, Technical Report.

UNESCO. (2010). *UNESCO Program: Alphabetization as part of the liberty portal. Retrieved May 1, 2010, from* http://portal.unesco.org/education

UNHRC. (2009). *The United Nations Human Rights*. Retrieved on 30/04/2010, From http://www.un.org/en/rights/

United Nations. (1948, December 10). *The Universal Declaration of Human Rights*. Retrieved September 22, 2009, from Un.org: http://www.un.org/en/documents/udhr/

United Nations. (1998, July 17). *Rome Statute of the International Criminal Court*. Retrieved september 12, 2009, from UN.org: http://www.un.org/children/conflict/ key documents/english/romestatuteofthe7.html

United Nations. (2009). *The Universal Declaration of Human Rights*, Art. 2. Retrieved on 17/10/2009, From http://www.un.org/es/documents/udhr/

UNPA. (2009). *The United Nations Programme on Ageing*. Retrieved on 30/04/2010, From http://www.un.org/esa/socdev/ageing/

US Census Bureau. (2001). *US census bureau*. Retrieved February 28, 2010, from http://www.census.gov/prod/2002pubs /01statab/stat-ab01.html

US Department of Defense. (1999). Humanitarian Demining Development Technologies. *Catalogue*, USA, 1998.

US Department of Health. (2001). *Mental health: Culture, race, and ethnicity* (supplement). Retrieved October 5, 2009, from: http://www.ncbi.nlm.nih.gov/books/bv.fcgi?rid=hstat5.chapter.971

US Department of State. (1994). Hidden Killers: The Global Landmine Crisis. *Report to Congress*, Washington D. C., Publication 10225, December 1994 (See also: http://www.state.gov/)

Ustaoglu, Y. (2009). Simulating the behavior of a minority in Turkish society. *In Proceedings of ASNA'2009.*

Valdez, S. I., Hernández, A., & Botello, S. (2009). Approximating the search distribution to the selection distribution. *In Proceedings of GECCO 2009*, 461-468.

van Dongen, W. S. (2007, June). Forensic artefacts left by Windows Live Messenger 8.0. *Digital Investigation*, *4*(2), 73–87. doi:10.1016/j.diin.2007.06.019

van Dongen, W. S. (2007, November). Forensic artefacts left by Pidgin Messenger 2.0. *Digital Investigation*, *4*(3-4), 138–145. doi:10.1016/j.diin.2008.01.002

Van Dyke, J. A. (1991). Word prediction for disabled users: Applying natural language processing to enhance communication. *Unpublished Thesis*. University of Delaware, Delaware, USA.

Van Dyke, J., McCoy, K., & Demasco, P. (1992). Using syntactic knowledge for word prediction. *Paper presented at the ISAAC-92. Abstract appears in Augmentative and Alternative Communication.*

Van Westen, C. J. (1993). Remote Sensing and Geographic Information Systems for Geological Hazard Mitigation. *ITC-Journal*, *4*, 393–399.

Vazquéz, C. Rodolfo. (1997). Las mediciones de la emigración de México a Estados Unidos. In Bustamante, J.A., Delalunay, D. & Santibáñez, J. (Eds). *Taller de medición de la migración internacional*. México: Colef/ORSTOM.

Venkatesh, A. (1996). Computers and other interactive technologies for the home. *Communications of the ACM*, *39*(12), 47–54. doi:10.1145/240483.240491

Vertan, C. (2004,December). Querying multilingual semantic web in natural language. *Paper presented at the Semantic Web Applications and Perspectives (SWAP)*, first Italian Semantic Web Workshop, Ancona, Italy.

Vertan, C., & Hahn, W. (2003). Menu choice translation: a flexible menu-based controlled natural language system. *Paper presented at the Controlled language translation Workshop, EAMT-CLAW-03, Dublin City University.*

Vidal, J. (2007). *Fundamentals of multiagent systems with NetLogo examples*. Retrieved May 1, 2010, from: http://jmvidal.cse.sc.edu/papers/mas.pdf

Villegas Santana, F. J. (2008). Conferencia: Migración Mexicana hacia Estados Unidos. In Rocha Silva, M. A., & Valencia Silva, M. (Eds.), *El impacto económico y social de la migración en Colima* (pp. 51–66). Mexico City, Mexico: Gasca.

Vlad, C., & Felix, R. (2003). The chatty environment – providing everyday independence to the visually impaired. *Ubiquitous Computing 5th International Conference*, Seattle, WA, USA.

Wallace, R. (2003). Online learning in higher education: A review of research on interactions among teachers and students. *Education Communication and Information*, *3*(2), 241–280. doi:10.1080/14636310303143

Wallerstein, I. (1974). *The modern world-system*. New York: Academic P.

Walther, J. B., Slovacek, C. L., & Tidwell, L. C. (2001). Is a picture worth a thousand words: Photographic images in long-term and short-term computer-mediated communication. *Communication Research*, *28*, 105–134. doi:10.1177/009365001028001004

Wein, L. M., & Liu, Y. (2005). Analyzing a bioterror attack on the food supply: The case of botulinum toxin in milk. *Proceedings of the National Academy of Sciences of the United States of America*, *12*(28).

Weiser, M. (1991). The computer for the 21st century. *Scientific American*, *265*(3), 66–75. doi:10.1038/scientificamerican0991-94

Weiser, M. Ubiquitous computing. (1993). *IEEE Computer "Hot Topics"*. October.

Wermter, J., & Hahn, U. (2004). An Annotated German-Language Medical Text Corpus as Language Resource. *Paper presented at the 4th International Conference on Language Resources and Evaluation (LREC'04), Lisbon, Portugal.*

Whitehead A. D & Ryan M. P. (2001). Method and device for classifying internet objects and objects stored on computer-readable media, *Patent Document No: 7383282*, Application No: 09/978,182, Application Date: Oct. 17, 2001.

Whitney, L. (2009, August 18). *iTunes reps 1 in every 4 songs sold in U.S.* Retrieved September 18, 2009, from Cnet News: http://news.cnet.com/8301-13579_3-10311907-37.html

Wilks, Y. (1975). A preferential, pattern-seeking, semantics for natural language inference. *Artificial Intelligence*, *6*, 53–74. doi:10.1016/0004-3702(75)90016-8

William, J., & Goose, S. (1998). The international campaign to ban landmines. In Tomlin, B. W. (Ed.), *To Walk without Fear: The Global Movement to Ban Landmines*. New York: Oxford University Press.

Williams, K. R., & Guerra, N. G. (2007). Prevalence and predictors of internet bullying. *The Journal of Adolescent Health*, *41*, S14–S21. doi:10.1016/j.jadohealth.2007.08.018

WINE HQ. (n.d.). *Spores on Wine*. Retrieved from http://appdb.winehq.org/objectManager.php?sClass=application&iId=8185

Winograd, T., & Flores, F. (1986). *Understanding computers and cognition: a new foundation for design*. Ablex Publishing Corporation.

Winslow, C. P. (1997). *Sowing the Dragon's Teeth: Landmines and the Global Legacy of War*. Boston: Beacon Press.

WIP. (2003). *Findings from the first World Internet Report*. UCLA Center for Communications Policy.

Wolak, J., Mitchell, K., & Finkelhor, D. (2006). *Online victimization of youth: Five years later*. The Crimes Against Children Research Center, University of New Hampshire. Retrieved January 10, 2010, from http://www.missingkids.com/en_US/publications/NC167.pdf

Wolchok, S., Yao, R., & Halderman, J. (2009). *Analysis of the Green Dam Censorware System*. http://www.cse.umich.edu/~jhalderm/pub/gd/.

Wolf, H., & Foritzheim, K. (1997). Web Video: A Tool for WWW Based Teleoperation. *Proceedings of the IEEE International Symposium on Industrial Electronics (ISIE'97)*. July 1997, SS268-SS273

World Internet Institute report (2008). *Sverige i Världen 2008* [Sweden in the World 2008].

Wortley, R., & Smallbone, S. (2006). Child Pornography on the Internet. *U.S. Department of Justice Office of Community Oriented Policing Services*, Retrieved September 10, 2009, from http://www.cops.usdoj.gov/files/ric/ Publications/e04062000.pdf

Wright, W. (2008). *Spores*. Maxis Electronic Arts.

Wright, A., Beattie, W., Booth, L., Ricketts, W., & Arnott, J. (1992). An integrated predictive wordprocessing and spelling correction system. *Paper presented at the 15th Annual RESNA Conference,* Washington, DC.

You, K. S., & Lee, H. (2006). The physical, mental, and emotional health of older people who are living alone or with relatives. *Archives of Psychiatric Nursing*, *20*(4), 193–201. doi:10.1016/j.apnu.2005.12.008

Zerbisias, A. (2008). Packaging abuse of women as entertainment for adults: Cruel, degrading scenes "normalized" for generation brought up in dot-com world. *Toronto Star*, January 26, L3.

Zittrain, J., & Edelman, B. (2003). *Empirical Analysis of Internet Filtering in China. Harvard Law School, The Berkman Center for Internet and Society*. Cambridge, USA: Harvard Law School.

About the Contributors

Miguel Vargas Martin, BSc, MASc, PhD, PEng is an Associate Professor at the University of Ontario Institute of Technology (UOIT). Before joining UOIT, he was a post-doctoral researcher at Alcatel Canada and Carleton University (Ottawa). He is a licensed Professional Engineer in Ontario and holds a PhD in Computer Science, a Master's degree in Electrical Engineering, and a Bachelor of Computer Science. He has reported his work in over fifty journals, book chapters, conference papers, technical reports, and pending patents, and so far has supervised over 20 students at the graduate and undergraduate level. His current research interests include virtual learning environments, computer forensics, mitigation of denial-of-service attacks, security and human computer interaction, hidden communication channels, and web modeling and optimization.

Miguel A. Garcia-Ruiz graduated in Computer Systems engineering and obtained his MSc in Computer Science from the University of Colima, Mexico. He received his Ph.D. in Computer Science and Artificial Intelligence at the University of Sussex, UK. He took a virtual reality course at Salford University, UK, and a graphics techniques internship at the Madrid Polytechnic University, Spain. Miguel is an Associate Professor of Computer Science with the College of Telematics of the University of Colima. He has published more than sixty scientific papers in major journals, book chapters and two books, and directed a video documentary on virtual reality. His research interests include virtual reality and usability of multimodal human-computer interfaces. Currently, Miguel is a Visiting Professor at the University of Ontario Institute of Technology, Canada.

Arthur Edwards is a Senior Professor/Researcher at the College of Telematics of the University of Colima, Mexico, where his primary interest is Computer Assisted Language Learning, multimedia applications, collaborative learning environments, educational information systems, virtual reality applications and wireless and mobile learning systems.

* * *

Ricardo Acosta-Diaz is a Full time professor at the University of Colima and is the current director of the School of Telematics. He obtained his BA in Computer Systems at the technological School in Colima, Mexico and his Master in Computer sciences from the Research and Education Center in Ensenada Baja California in México (CICESE). He is studying his third PhD year in Telematics Engineering at Vigo University in Spain. His academic experiences include the teaching of undergraduate and graduate students in software development processes and techniques and also in educational technology.

Francisco Javier Álvarez, PhD., is Professor in Computer Science at Universidad Autónoma de Aguascalientes. He holds a Ph.D. in Engineering (UNAM, México, 2004), He has a number of collaborations with very important researches in prestigious universities around the world. Dr. Álvarez Rodriguez has reported his work in journals, book chapters, conference papers, and technical reports, and so far has supervised almost 30 students at the graduate and undergraduate level. His current research interests include software engineering (methodologies, metrics, among others), and distance education (educational internet technologies, learning environments, learning objects, among others.

Julian Buchanan is Professor of Criminal and Community Justice at Glyndwr University, Wales. He is a qualified probation officer who worked in Liverpool, England. He joined academia in 1996 and worked at the University of Liverpool and University of Central Lancashire before joining Glyndwr University in 2000. As a Probation Officer he spent his later years working in a multi racial area (Wavertree/Toxteth in Liverpool). His main interests centre upon the identification of injustice, discrimination and social exclusion for particular groups. He is concerned about the treatment of Black and Minority Ethnic people in the criminal justice system, as well as the treatment of people who use illicit drugs. He has researched and published widely. For more information see http://julianbuchanan.wordpress.com/

Eduardo H. Calvillo Gámez is a member of the New Information Technology Faculty at Universidad Politécnica de San Luis Potosí. He holds a Ph.D. in Human Computer Interaction from University College London, a degree of Masters of Science in Electrical Engineering from Tufts University and a degree of Bachelors of Science in Electronic Engineering from Universidad Autónoma de San Luis Potosi. Eduardo's research interests are User Experience, Video Games, Input Devices and Epistemology of Human Computer Interaction.

K. P. Chow is the Associate Professor of Computer Science at the University of Hong Kong. His areas of research interest are computer security and digital forensics. Dr. Chow has been a member of the Program Committee of the international computer forensic workshop SADFE and the conference chairman of the Sixth IFIP WG 11.9 International Conference on Digital Forensics. In the past few years, Dr. Chow has been invited to be a computer forensic expert to assist the Court and to give advice to counsels on understanding and interpreting digital evidence for both criminal and civil proceedings in Hong

Juan Contreras-Castillo is a Full time professor at the University of Colima and current Director at the Department of Lifelong and Distance Learning in the Autonomous University of the Mexican State. He obtained his BA in Computer systems from the School of Mechanical and Electronic Engineering in the University of Colima in México and his Master and PhD. In Electronics and Telecommunications in the Research and Education Center in Ensenada Baja California in México (CICESE). His academic experience includes teaching undergraduate and graduate courses and research in educational technology, distance learning and mobile computing to support the teaching and learning process.

Pedro Damián-Reyes is a professor of computer science at the School of Telematics in University of Colima; he leads the Multimedia Laboratory. His research interests include ubiquitous computing, context-aware computing and software development. He holds a BSc from the Instituto Tecnológico de

Colima, MSc from the University of Colima and Phd from the Centro de Investigación Científica y de Educación Superior de Ensenada (CICESE) in Baja California, México. damian@ucol.mx

Daoud M. Daoud received his BSc degree in electrical and computer engineering from Kuwait University in 1988, his MSc in Computing Science from Glasgow University- UK and his PhD in Computing Science from Joseph Fourier University – France. Daoud is currently serving as an assistant professor at Princess Sumaya University for Technology. He also served in Institute of Advanced Studies- United Nations University (1998-1999). He also worked as a principal investigator for the Arabic part of Universal Networking Language project (1996-1999). He also served as a director for Next Generation Services department at Paltel (1999-2001). His main research interests are Natural Language Processing, machine translation, Information Extraction and Information Retrieval. (email: daoud@batelco.jo)

Walter S. DeKeseredy is Professor of Criminology, Justice and Policy Studies at the University of Ontario Institute of Technology (UOIT). He has published 15 books and over 100 scientific journal articles and book chapters. In 2008, the Institute on Violence, Abuse and Trauma gave him the Linda Saltzman Memorial Intimate Partner Violence Researcher Award. He also jointly received the 2004 Distinguished Scholar Award from the American Society of Criminology's (ASC) Division on Women and Crime and the 2007 inaugural UOIT Research Excellence Award. In 1995, he received the Critical Criminologist of the Year Award from the ASC's Division on Critical Criminology (DCC) and in 2008 the DCC gave him the Lifetime Achievement Award.

Samir Abou El-Seoud received his BSc degree in Physics, Electronics and Mathematics from Cairo University in 1967, his Higher Diplom in Computing from Technical University of Darmstadt (TUD) -Germany in 1975 and his Doctor of Science from the same University (TUD) in 1979. Professor El-Seoud helds different academic positions at TUD Germany. Letest Full-Professor in 1987. Outside Germany Professor El-Seoud spent different years as a Full-Professor of Computer Science at SQU – Oman and Qatar University and acted as a Head of Computer Science for many years. At industrial institutions, Professor El-Seoud worked as Scientific Advisor and Consultant for the GTZ in Germany and was responsible for establishing a postgraduate program leading to M.Sc. degree in Computations at Colombo University / Sri-Lanka (2001 – 2003). He also worked as Application Consultant at Automatic Data Processing Inc., Division Network Services in Frankfurt/Germany (1979 – 1980). Professor El-Seoud joined PSUT in 2004. Currently, he is the Chairman of the Computer Science Dept. at PSUT. (email: selseoud@psut.edu.jo)

Nirmala Gopal, PhD has been an educator and academic for 20 years. She is currently a Senior Lecturer at the Department of Criminology, at the School of Sociology & Social Studies, University of KwaZulu Natal. She studied at teaching pedagogy at the University of Durban Westville. She is passionate about innovative teaching methodologies and believes that students should be exposed to new and exciting ways if learning. She has found Blackboard interesting for both herself and her students and is able to benchmark her students on an international level. This new way of teaching has brought the world to a single place. She enjoys co-operating with the various academics that are committed to making teaching and learning exciting.

Maki K. Habib obtained his PhD Eng in Intelligent and Autonomous Robots, University of Tsukuba-Japan. He was selected Research scientist at RIKEN-Japan, Senior researcher at RISO Laboratories-Japan, and Visiting researcher at EPFL-Switzerland. He was a Visiting expert under Asian Development Bank (ADB), Associate Professor at UTM-Malaysia, and a Senior Manager at MCRIA, Malaysia. Then, he was Senior research scientist with GMD-Japan, leading the Telecooperation group. He was Associate Professor with Monash University and leading the Mechatronics Engineering program. He was appointed as full Professor of Robotics and Mechatronics at Swinburne University. Then, he was Invited Professor at KAIST, Korea. He was a Visiting Professor at Saga University, Japan, and currently a full Professor at AUC, Egypt. He edits three books, published 10 book chapters and has more than 190 papers published in internationally recognized Journals and Conferences. His main area of research interest is focusing on human adaptive and friendly Mechatronics, autonomous navigation, service robots and humanitarian demining, intelligent control, telecooperation, distributed teleoperation and collaborative control, wireless sensor networks and ambient intelligence, biomemtic robots, industrial automation.

José Alberto Hernández Aguilar he received PhD Grade in 2008 at Universidad Autónoma del Estado de Morelos (UAEM). He received the MBA degree in 2003 at Universidad de las Américas (UDLA), A.C. Since 2002, he is part time professor at UDLA, México, D.F., for the IT Career, and since 2007 year, he is part time professor at the Sciences Faculty at UAEM. Areas of interest: Databases, Artificial Intelligence, Online Assessment Systems and Marketing Research.

Håkan Hydén is professor of sociology of law at the Department for Sociology of Law at Lund University and holder of the Torsten and Ragnar Söderberg Foundations professorship in memory of Samuel Pufendorf. His main research interest is "norms" (social and legal) and the development of a norm science, which he has advocated the last fifteen years. The last years professor Hydén's research has focused social and legal changes over time, although his research spans the fields of occupational safety and health, human rights–partly through an education on children's rights, focusing on schools in different parts of the developing world–and sustainable development, all areas in which he has authored, co-authored and edited a number of publications over three decades.

Amin Ibrahim (amin.ibrahim@uoit.ca) is a sessional instructor in the Department of Business and IT at the University of Ontario Institute of Technology (UOIT). He received his Master's degree from the Department of Electrical and Computer Engineering at UOIT (2009) and his Bachelor's degree in Computer Engineering from the University of Toronto (2004). His past research work includes combating child pornography at the network level and image steganography. He is currently enrolled as a PhD student in the Department of Computer Engineering at UOIT where he is conducting research to create a novel method of authentication system (authentication with 'what you don't know'). His research interests include image processing, pattern recognition and computer security.

Pierre K.Y. Lai is currently a Ph. D. student with the Department of Computer Science at the University of Hong Kong. Pierre was working on secure multi-party computation during her master degree. She has been a member of the technical program committee of the International Conference on Internet Monitoring and Protection since 2007. Her research interests include peer-to-peer networks investigation, live systems forensics and cryptography.

Frank Y.W. Law has been working in the Hong Kong Police Force since 1998, and has been involved in technology crime related policing since 2001. He is currently the technology crime trainer of the Hong Kong Police Force and is responsible for providing various types of training in technology crime investigation and computer forensics to both local and overseas law enforcement units. Frank is currently a part-time Ph.D. student with the Department of Computer Science at the University of Hong Kong. His research interests include live systems forensics, digital forensics and digital timestamp analysis.

Stefan Larsson is a researcher in sociology of law specialised in Internet in relation to social norms and law. He has an educational background with a Master of Laws and a Master of Social sciences, both from Lund University in Sweden, and is currently a PhD Candidate in Sociology of Law at Lund University in Sweden finishing his thesis on copyright law, social change and norms during 2010. He has also done research on the Swedish development of 3G mobile telephony and regulation for wind power development within the field of spatial planning and has a licentiate in technology in spatial planning from the Blekinge Institute of Technology. He is a member of a project "Cyber norms" located at Lund University, studying the creation of social norms in young web culture.

Ricardo Mendoza-González, PhD. is a Researcher/Professor at Instituto Tecnológico de Aguascalientes (Mexico). He holds a Ph.D. in Computer Science (Universidad Autónoma de Aguascalientes, México, 2009), a Master's degree in Computer Science (Universidad Autónoma de Aguascalientes, México, 2007), and a Bachelor of Computer Science (Instituto Tecnológico de Aguascalientes, México, 2004). PhD. Mendoza-González has reported his work in journals such as Advances in Engineering Software, and prestigious international conferences such as IEEE-CCECE, CADUI, ICIMP, and IASTED-CNIS. He collaborates with researchers of the University of Ontario Institute of Technology (UOIT), and Universidad Autónoma de Aguascalientes (México). His current research interests include several topics on: human-computer interaction, information security, usability, artificial intelligence, and software engineering.

Jaime Muñoz Arteaga, PhD., is Professor in Computer Science at Universidad Autónoma de Aguascalientes. Universidad Autónoma de Aguascalientes. He is a researcher in Human-Computer Interaction, and web technologies. He holds a Ph.D. in Computer Science, Human-Computer Interaction (University Toulouse 1 (UT1), Toulouse, France, 2000). He has a number of collaborations with very important researches in prestigious universities around the world. Dr. Muñoz Arteaga has reported his work in journals, book chapters, conference papers, and technical reports, and so far has supervised over 40 students at the graduate and undergraduate level. His current research interests include several topics on: human-computer Interaction, mobile technologies, software engineering, and artificial intelligence.

Rodrigo Nieto-Gomez is a researcher at the Center for Homeland Defense and Security at the Naval Postgraduate School in Monterey, CA. His fields of research include the geopolitical implications of Homeland Security/Defense and National Security with a regional focus for North America, border security, discourse analysis and the implications of new technologies for security and defense policies. He has collaborated with the Center for Civil- Military Relations to teach Homeland security issues outside the United States. He obtained his PhD (summa cum laude) in geopolitics at the Institut Francais de Geopolitique of the University of Paris, from where he also holds a master's degree in the same discipline. He is also a Mexican attorney-at-law, specializing in international public and private law inside the NAFTA region.

Carlos Alberto Ochoa Ortiz-Zezzatti (Bs'94-Eng. Master'00-PhD'04-Postdoctoral Academia Research'06 & Industrial Postdoctoral Research'09). He has 1 book, and 7 chapters in books related with AI. He has supervised 12 thesis of PhD, 13 thesis of master and 27 thesis of Bachelor. He participated in the organization of HAIS'07, HAIS'08, HAIS'09, HAIS'10, ENC'06, ENC'07, ENC'08, MICAI'08, MICAI'09 & MICAI'10. He is a permanent reviewer at Computers in Human Behavior Journal of Elsevier.

Patrik Olsson is an Assistant Professor of Criminology, Justice and Policy Studies at the University of Ontario Institute of Technology (UOIT). His research agenda concerns the rights of children and youth as well as socio-legal aspects of information technology and social change. A more predominant focus has been on the rights of children and adolescents, e.g. child labor, human trafficking, children in conflict with the law, children in prison, the right to education, the principles of nondiscrimination, and child participation. Dr. Olsson has since the 1990s conducted extensive research in relation to the socio-legal situation for exposed children in the MERCOSUR countries in South America, in Central America and Southeast Asia.

Felipe Padilla Díaz received the B.S. degree in computer science engineering from the Instituto Tecnológico de Queretaro in 1985, received the M.S. degree in enterprise management from the Universidad Autónoma de Aguascalientes in 1999, and the PhD. in engineering (Artificial Intelligence) from the UNAM. His research interests include Evolutionary Computation, Data Mining and learning classifier systems.

Julio Cesar Ponce Gallegos received the B.S. degree in computer science engineering from the Universidad Autónoma de Aguascalientes in 2003, and the M.S. degree in computer sciences from the Universidad Autónoma de Aguascalientes in 2007. He is currently working towards the PhD. Degree in Ant Colony Optimization from the UAA. He is currently an professor in the Universidad Autónoma de Aguascalientes. His research interests include Evolutionary Computation, and Data Mining.

María Alejandra Rocha Silva is a Full time professor at the University of Colima in the University of Colima, Mexico. She obtained her PhD Educational Technology from University of Salamanca in Spain. Her academic experience includes teaching undergraduate and graduate courses and research in educational technology, distance learning and mobile computing to support the teaching and learning process.

Cornel Samoila (csam@unitbv.ro) is a graduate of the Polytechnic Institute of Brasov-ROMANIA, specializing in Machines and Equipment for Hot Processing. After graduating in 1964 until today, he has been affiliated with Transylvania University of Brasov (former Polytechnic Institute). He has been an assistant, lecturer, assistant professor and full professor and worked primarily in industrial furnaces, project management, maintenance and logistics. His research areas include heat transfer, materials, Science and Engineering. He has also been the author of 12 books, 245 published papers, 89 research contracts and 25 patents. His academic recognitions include: Corresponding Member of the Romanian

Technical Science Academy; Member in the Romanian National Research Council, Honored Member in the Israeli Association of Engineers and Architects, Former President Vice-president of ASM-International-Romanian Chapter. Cornel Samoila has also served as Vice-Dean of the IT Faculty-University of Brasov, Vice-Rector of Transylvania University of Brasov, Dean of Science Materials and Engineering Faculty of Transylvania University of Brasov, City Counselor 1992-1996. Finally, he has been awarded the National Order "Cultural Merit" in Knight Degree for research activity

Pedro C. Santana is a lecturer in Human-Computer Interaction (HCI) and Software Engineering at the School of Telematic Engineering of the Universidad de Colima. He is currently pursuing a PhD in Telematics at the Universidad de Vigo and he received a Masters degree (with honors) in Computer Science at Universidad Autónoma de Baja California (UABC), Mexico in 2005. His research interests are on Interactive Digital Television, HCI, Software Engineering, Ubiquitous Computing, Autonomous Agents and Medical Informatics. psantana@ucol.mx

Alex Stefan (alex_stefan@bloomfield.edu) is currently the Director for International Programs at Institute for Technology and Professional Studies-Bloomfield College, NJ, USA , He has completed his studies at in electronic engineering at Technical Military Academy and it is specialized in wireless systems design focus on RADAR systems (with a doctorate in microwave circuits design), computer networking, security networking. After his Master in Business Administration and Master in Technology Management at NYU-Poly, he combined the technical management and CISCO security management with wireless systems. The last periods he was developing programs and researches on Clean-Tech, with focus on monitoring solar technologies. He is vice-president of European Education Resources, Bloomfield College's ONG and coordinator of Regional Cisco Academy.

Hayson K.S. Tse is a barrister in the Prosecutions Division, Department of Justice, Hong Kong Special Administrative Region, China, since 1995. He has been prosecuting technology crime since 2000. He is now the Deputy Section Head for the Computer Crime Team. He prosecutes trials and appeals, and provides advice to law enforcement agencies. He also attends part-time at the University of Hong Kong to read for a Ph.D. in Computer Science. His research interests include digital forensic, computation models of legal argument, and argument structuring system.

Kenneth W.H. Tse is a part-time MPhil student at the University of Hong Kong majoring in Computer Science. His research interest is digital forensics. Kenneth is currently working as a security consultant in eWalker Consulting Limited, in which he participated in security assessments, audit, and penetration test exercises for Government departments and other private sectors.

Doru Ursutiu (udoru@unitbv.ro) is the Manager of Center for Valorization and Transfer of Competence CVTC from at Transylvania University of Brasov (relbv.unitbv.ro) and manager of CVTC Creativity Laboratory. He completed his studies of Physics at the "Babes-Boliay" University in Cluj, Romania and is specialized in Open and Distance Learning (LOLA course - Hariot Watt University UK) and Noise in Electronic Systems. The key qualifications of Prof. Ursutiu are Physics, Electronics (with a doctorate in Noise and Fluctuations in Electronic Systems and Devices), Graphical Programming, Remote and Virtual Instrumentation. He is member of Executive Committee - Central European Chapter of Association for the Advancement of Computing in Education (AACE) and Vice-President of International Association of Online Engineering IAOE. In the same time he is member of Scientific Committee and reviewer for many international conferences and publications (ICL, IMCL, EURODL, IJIM, IJ-SoTL, etc.)

Stephen T. Wilson is a clinical social worker, and adjunct faculty at the University Of Washington School Of Social Work. I am also a doctoral student at Smith College School for Social Work. My focus areas include clinical social work and cross cultural treatment with adolescents and their families. I am also interested in how children understand race. At this point in my research career, I am interested in military families and their responses/adjustment to the current military involvement in the Gulf War and Afghanistan. I am a consultant for a number of agencies, where the focus is on working with persons of African descent. In addition I have provided training and supervision to multiple child caring agencies around appropriate interventions and interactions with children and adolescents.

Index

Breinigsville, PA USA
05 November 2010

248705BV00003B/3/P

9 781609 600945